REFUSE-DERIVED FUEL PROCESSING

REFUSE-DERIVED FUEL PROCESSING

Floyd Hasselriis

To: Robert W. Simkins
with best regards
Floyd Hasselriis

BUTTERWORTH PUBLISHERS
Boston · London
Sydney · Wellington · Durban · Toronto

An Ann Arbor Science Book

Ann Arbor Science is an imprint of Butterworth Publishers.

Library of Congress Cataloging in Publication Data

Hasselriis, Floyd.
Refuse-derived fuel processing.

"An Ann Arbor science book."
Includes index.
1. Refuse as fuel. I. Title.
TP360.H37 1984 662'8 83-12968
ISBN 0-250-40314-5

Butterworth Publishers
80 Montvale Avenue
Stoneham, MA 02180

10 9 8 7 6 5 4 3 2 1

Printed in the United States of America

ABOUT THE AUTHOR

Floyd Hasselriis began his career in thermal engineering after graduating from Columbia University and becoming a licensed engineer in the Merchant Marine, operating and maintaining steam turbine-driven cargo ships.

After World War II he taught mechanical engineering at the Cooper Union and graduate thermodynamics at Stevens Institute. He was awarded a Masters in ME by the University of Delaware in 1950. Ghostwriting a book on High Temperature Water led to his joining American Hydrotherm where, after being project engineer on several coal-burning central heating plants for the U.S. Air Force, he became Chief Engineer of the industrial division, specializing in industrial heating and temperature control systems.

Seeing a need for a company able to design, build and operate waste-to-energy plants, he joined Combustion Equipment Associates in 1971, starting with industrial fumes and wastes and leading to the design and development of the refuse processing systems that were built at East Bridgewater, MA and Bridgeport, CT. As manager of Corporate Technology, he led in the development of effective primary and secondary trommels, shredding operations, air classifiers, conveying and drying systems, as well as methods for storage, feeding and burning of both fluff and powdered RDF.

CEA was bankrupted by numerous delays, cost overruns and the burden of supporting the costs of operation before revenues began to build up, and was forced to shut down Bridgeport just when the successful production and burning of Eco-Fuel had been demonstrated over the course of a year.

Floyd Hasselriis, a registered professional engineer in four states, has been active in the ASME as a member of the Research Committee on Industrial and Municipal Wastes and the Solid Waste Processing Committee, and in the ASTM in Committee E-38 on Resource Recovery, where he has been active in the committees on Energy and Unit Operations, participating in the writing and review of numerous standards relating to RDF production. At present he is a consultant in the field of refuse processing and burning, as well as in mass-burning of municipal refuse and destruction of hazardous liquid wastes. He has published widely in these fields.

CONTENTS

Preface xiii

Acknowledgements xvii

1. What Is Municipal Solid Waste? 1

Sources of Municipal Solid Waste 3
Characterization of Municipal Solid Waste 5
Major Characteristics of Municipal Solid Waste 6
Properties of Special Interest in Processing 7
Moisture Content 8
Combustible Matter 9
Gross and Net (Higher and Lower) Heating Value 11
Methods of Determining the Heating Value of MSW and RDF 13
As-received, Dry, and Moisture- and Ash-Free Heating Values 14
Higher Heating Value 14
Lower Heating Value 16
Density 17
Particle Size of Refuse Components 17
Chemical Composition 19
Sampling and Testing of Refuse 19
Analysis of MSW and RDF Streams 27
Mean Values of Properties 30
Variability of Properties of MSW 30
Standard Deviations of Refuse Properties 30
Standard Error 33
Confidence Interval 33
Variations in Heating Value 34
Seasonal Variations 35
The Effect of Mixing on Variance of RDF 39

2. Transport, Storage and Handling 43

Properties of MSW and Its Products that Influence Handling 44
Collection of MSW 46
Estimating Solid Waste Quantity and Ton-Miles to Disposal 51
Variation in Daily, Weekly and Annual Refuse Disposal 51
Transfer Stations 52

Receiving, Handling, Storage and Feeding 56
Storage Pits for Incinerators 57
Tipping Floors 57
Inspecting and Picking the Refuse 60
Feeding the Plant 60
Front-loader Capacity and Horsepower 61
Innovative Systems for Feeding MSW 61
Storage and Feeding of Shredded Refuse 62
Screw Conveyer Capacities 66
Mechanical Conveyors 66
Capacity of Mechanical Conveyors 69
Transfer Points 71
Layout and Angle of Inclination of Conveyors 72
Housekeeping 73
Dust Collection 73
Maintenance 74
Design of Belt Conveyors 74
Rates of Spillage from Belt Conveyors 78
Design of Vibrating Conveyors 81
Plant Operation and Maintenance 85

3. Size Reduction 87

Types of Size Reduction Machinery 88
Cutters 88
Flail Mills 92
Hammermills 92
Densifying Machinery 93
Application of Size Reduction Machinery to MSW 93
Hammermills with Grates 93
Hammermills without Grates 94
Rotary Grinding Mills 95
General Nature of the Products of Size Reduction Machinery 96
History of Research in Size Reduction in the Field of Waste Processing 97
Particle Size Distribution of Raw and Shredded MSW 98
Particle Size Distributions of Raw and Shredded Refuse 108
Energy Required for Shredding or Grinding 109
Work Requirements 109
Bond Work Index and Rosin-Rammler Size Distribution 112
Power Required for Size Reduction 113
Specific Energy and Specific Capacity 115
Comparison of Bond Work Index and Rupture Energy Models 120
Summary 121
Factors Affecting Shredder Performance 122
Feed Size 123

Flowrate 123
Holdup 123
Moisture Content of the Feed 124
Shredder Grate Size 126
Air Velocity through Shredder 127
Hammer Wear versus Particle Size 127
Relative Velocity (Rotor Speed) 128
Analysis of Shredder Performance and Efficiency 128
Mechanical Efficiency 130
Performance Curves 131
Performance of Rotary Shear-Type Shredders 136
Performance of Shredders in Refuse-Fuel Production 137
Summary and Conclusions 138

4. Metals Recovery and Wet Processes 141

Unit Operations as Part of the Total Process 142
Analysis and Evaluation of the Performance of Unit Operations 143
Properties of the Feed 143
Sampling 144
Capacity 144
Energy 144
Operation and Maintenance 144
Measures of Efficiency and Effectiveness 144
Thermal Efficiency 144
Mechanical Efficiency 145
Energy Efficiency 145
Effectiveness 145
Quality 146
General Separation 146
Recovery and Purity 146
Efficiency 148
Ferrous Materials Recovery 149
Magnetics 149
Ferrous Metals Recovery Systems 151
Aluminum Recovery 158
Aluminum Recovery Systems 159
Wet Processes 161
Hydrasposal® Process 161
Wet Separation Equipment 163
Performance of Metal and Glass Recovery Equipment 167
Removal Efficiency Increases with Concentration 169
Other General Considerations in Materials Separation 169
The Contradiction between Capacity and Quality 169
Multiple Stages Increase Recovery at the Expense of Quality 170

5. Air Classification 171

How Air Classifiers Work 171
Nomenclature 172
Types of Air Classifiers 173
Air Classifiers Applied to Resource Recovery 174
  Vertical (Rader Pneumatics) 174
  Zig-Zag Vertical 174
  Inclined Vertical (CEA Carter-Day) 177
  Vibrolutriator® (Triple/S Dynamics) 177
  Horizontal Classifier (Bureau of Mines) 177
  Horizontal Classifier (The Boeing Company) 177
  Ferrous Concentrator (Triple/S Dynamics) 177
  Nonferrous Separator (MAC Equipment) 178
  Rotary Drum (Raytheon) 178
History of Air Classification in the Resource Recovery Industry 178
  U.S. Bureau of Mines, Salt Lake City 178
  National Center for Resource Recovery 181
  University of California, Berkeley 181
  Duke University, Durham, North Carolina 181
  Other Installations and Research Programs 182
  Conclusions 182
Basic Principles of Air Classifier Operation 182
  Basic Equations 183
  Dynamics of Particles under Gravity, without Drag 184
  Particle Acceleration Caused by Drag Forces 186
Application of Basic Principles to Air Classifiers 188
  Horizontal Air Classifier 188
  Vertical Air Classifier 190
  Determining Lifting Velocities 192
  Effect of Density on Lifting Velocity 194
Theoretical versus Actual Performance 199
  Single-Stage Vertical Classifier 199
  Multistage Vertical Classifier 202
  University of Eindhoven, The Netherlands 204
Actual Performance of Air Classifiers 210
  Americology Tests at Chemung County 211
  Field Tests of Various Types of Classifiers by Midwest Research Institute and Cal Recovery 213

6. Size Separation 225

Screening 225
Nomenclature 225
  Yield 225

Capacity of a Screen 226
Screen Efficiency 226
Specific Capacity 228
Specific Area 229
Types of Screens 229
Screens Used in Resource Recovery 229
History of Screening in the Resource Recovery Industry 230
Principles of Size Separation by Screening 233
Effect of Particle Size Distribution on Screening 239
Influence of Half Size and Oversize Material on Capacity 241
Empirical Estimates of Screen Capacity 242
Application of Basic Principles to Screen Design and Operation 243
Flat-Deck Screens 243
Disc Screens 243
Rotary Screens (Trommels) 247
Behavior of Particles in a Rotary Cylinder 250
How Probability Determines Screening Efficiency 253
Actual Performance of Screens in Resource Recovery 257
Pilot-Plant Tests in Minneapolis 257
Tests of Primary Trommel at Recovery I, New Orleans 264
Trommel Tests at East Bridgewater 267
Bridgeport Trommel Tests 267
Tests of Trommels by Cal Recovery 268
Tests of Disc Screens by SPM Group in Denver 270
Summary 272

7. Selecting Unit Operations and Optimizing the Process **275**

History of Process Selection in Resource Recovery 275
Pilot-Plant Test Data 277
Matrix Approach to Process Selection 280
Example: Mass Balance of a Primary Trommel 280
Example: Splits of an Air Classifier 283
Example: Flowsheet of a Complete Pilot Plant 283
Matrix Analysis of Materials Flow 285
Computer-Aided Analysis of Refuse Processing 288
Benefits of Computer Matrix Analysis 290
Optimizing Operations and the Overall Process 293
Optimizing Energy and Power Recovery 296
Summary 297

Appendix A Conversion Factors 299
Appendix B ASTM Standards for Resource Recovery 301
Appendix C Element Concentrations in RDF and RDF Ash 303
Appendix D Mathematics 305

Appendix E Calculation of Displacement of Particles in a Rotary Drum Classifier 315
Appendix F Analysis of Behavior in a Rotary Cylinder 317
Appendix G Sampling and Analysis of Refuse and Refuse Products 323

Index 335

PREFACE

This book is the product of a remarkable decade in which our society realized that it was suffering from pollution from our wastes, limited space and limited resources. This realization led to intensive investigation into these problems, development of schemes for their solutions, demonstration of their success or failure and analysis of the first round or two of experimentation. The terms resource recovery, conservation, energy recovery and waste-to-energy became the buzz words of the decade.

This effort moved so rapidly that there was not sufficient time to evaluate the results of early trials before proceeding, and often no time to make the small improvements that might mean success. Technical planning did not anticipate failure, and financial planning provided few contingencies for errors in early data, and especially for the vagaries of social and political life.

The '70s was a decade of government assistance: the enthusiasm of youth was recruited and established to lead the movement, and plenty of money was committed to research, planning and sponsorship. Projects were started by literature searches of the state-of-the-art and contracted out to consulting engineers and vendors who had to cope as well as possible in a new field. Others were started by forward-thinking engineers and entrepreneurs. Some succeeded and some failed.

Today, a mass of information, some hardly examined and vast experience have been generated, which can now be used to attack the problems that refuse to go away.

It should not be surprising that many projects failed technically or financially, although this came as a shock to many. Well established technologies require contingencies of about 20% to allow for unknowns, not to be spent until the plant has been tested. Experience has taught us that new technologies require more than 50% contingency, but in most cases such allowances were not provided for in original financial plans.

The resource recovery projects that survived and eventually prospered were those whose sponsors continued to give support in the face of early problems and threats of failure. Retrofit has always been needed in new technologies, but often not provided for in the budgets or pride of the designers or builders. Rome wasn't built in a day, but by patient building and rebuilding. One wonders how many arches were replaced until those that have lasted until today were built.

This book incorporates a great deal of retrospection: reviews of old data, of early and later trials, and perspectives of the very process of evolution and learning that took place. The historical notes are offered to point out the extent to which early investigators had either sound insights or narrow conceptions, as inevitably is the case. Often a small misconception or oversight caused years of frustration before the solution was discovered. Solutions to problems often first require dozens of remedies to be tested. The last one tried may be the one that works; it's seldom the first.

The tendency to fall back on experience and proven techniques can be as disastrous as pushing ahead with untried ideas. The tendency to repeat someone else's error can be checked by asking him why he did it and whether he would do it again.

The first draft of this book covered the broad gamut from waste to energy, including processing, combustion and emissions. However, for practical considerations it has been divided into two books. This volume covers the processing of municipal solid waste (MSW), while the second covers the various methods of recovery of energy by combustion, as well as the consequences, including emissions.

Chapter 1 attempts to answer the question, "What is MSW?" Drawing from the extensive test data generated in the last decade, which view the picture from many dimensions. Indeed, MSW is a commodity that is produced by all classes of people and has continuous variety and intriguing properties. When its properties are understood, the fickle MSW can be handled by tolerant processing operations without insufferable indigestion.

Chapter 2 reviews the process of handling the stream of refuse from curbside to processing plant, and its products within the plant, to the bitter or happy ending.

Chapter 3 looks at processing as a system of complex operations, each depending on what it receives and how it treats the whole stream and its various ingredients.

Chapter 4 investigates what happens when we reduce the size of refuse to make it more manageable; how shredding and other size-reducing operations affect the various materials in refuse; and what the products are like.

Chapter 5 examines what happens when materials of mixed sizes are faced with choices of falling through holes: some will always fall through, some will occasionally and some never. With this information, separation processes can be made to be selective to a certain extent, which depends on the amount of patience and perseverance we are willing to devote.

Chapter 6 becomes interested in how things fly, why, how, why not and under what conditions. By selecting the light in spirit, we collect the energetic materials and reject the wet clumsy ones. While making this selection, however, many items slip by. The problem is, is it worth trying to get them? In fact, which ones should we reject?

Running through the book like a hooded ghost is the probability distribution, which appears on plain graph paper as a figure leaning toward larger sizes, and on properly sophisticated paper as an erect and symmetrical figure. It appears that nature wants all materials that have been broken in spirit to have this shape, for our studies keep finding it wherever we turn.

To understand the strange materials we waste, we have to use statistics, which is a technique for finding out how wide the hooded ghost is, that is, how certain are we that, in the dark, we know where this phantom is? Perhaps we can tell by how loud it screams or how many darts hit the target. All this is necessary so that we don't waste too much effort in finding where the phantom is: "more or less there, I'm not quite sure, but if you hit THERE, you are almost certain to be on target."

"Optimizing" means that every time we try something and study the results we can make some improvement in the way it is done and the results we get. Rather than "trial and error," we prefer "better and better." We must know just what we mean by good before we can do better, and what is best so that we don't waste time improving perfection while imperfection lies unaided.

Effectiveness and efficiency are key words that tell us how near to perfection we are, based on a definition of perfect that we must set.

We cannot tell how well are are doing without testing. The American Society of Testing and Materials (ASTM) sets up standardized methods that are equally good (or good enough) in the opinion of a producer, buyer or user, as carried out by laboratory technicians who can reproduce the same data from testing the same material. The process of developing these standards is a small model of the process of developing the method for making the material being tested, and has been subjected to similar trial and error (or better and better) as resource recovery.

This remarkable decade of development has been a great social effort, involving thousands of dedicated people who worked hard and long because they believed themselves to be associated with a worthwhile, socially approved effort. Many joined this crowd of movers, many fell away. Some have lasted through the storm, many died by the wayside. However, they have left a legacy that will not be forgotten.

In compiling this book, I have drawn from the work of many others, often combining it with the results of my personal experience and investigation. Discovery of what it is all about is stimulated by being caught in a bind, either by choice or by accident. It starts by investigating a small aspect of "all," but the solutions are usually evasive until a broader understanding is achieved. Many of us, reflecting alone or in discussion with others, have slowly discovered the archeology and other mysteries of refuse. One of us leads the next, who leads us further, giving insights and understandings that reveal more and more of the truth.

ACKNOWLEDGEMENTS

It would be impossible to acknowledge all those who have contributed to this book and to my own understanding of the subject. The references include many of the most active in publication, so do not need additional mention here. Those who have not been mentioned deserve the most honor here because their work and its results were the incentive and the reward.

The encouragement of Herbert Hollander led me to participate in an Engineering Foundation Conference on research needs in waste-to-energy and, later, to become involved with the ASME Research Committee on Industrial and Municipal Waste. Then I began to delve deeper and deeper into the mysteries of refuse and the needs for further research. The activities of the ASME Solid Waste Processing Division have been stimulating to all of us.

The Engineering Foundation Conference on waste-to-energy brought together many practitioners in the field to discuss the problems in the area: here I met Aarne Vesilind, who ultimately asked me to write this book on unit operations.

Those who participated in the ASTM Committee E-38 on Resource Recovery, and especially those who saw the standards through to publication, played a special role of "keeping the industry together," by developing test standards for energy, materials and unit operations, as well as health and safety. I must mention the sincere dedication of Dr. John Love in driving Committee E-38-01 (Energy) to produce a complete set of standards for testing RDF and his guidance of the Laboratory Advisory Committee in completing robin tests and other investigations needed to understand the special characteristics of RDF.

The leadership of Dr. Eugene Domalski of the National Bureau of Standards (NBS) has contributed immensely both to the scientific knowledge of municipal waste through work done under the encouragement of the Research Committee, on calorimetry and fundamentals of combustion, as well as to the ASTM Committee on Energy.

The steady efforts of the EPA MERL group in Cincinnati, under the leadership of Carleton Wiles deserves special mention. A small mountain of data on the generation of energy from prepared fuels was produced by this dedicated group, and will stand the test of time. The contractors who did the work deserve much credit, including Midwest Research, Cal Recovery and the National Center for Resource Recovery (NCRR). The Department of Energy, Office of Municipal Waste, contributed substantially to promoting waste-to-energy, under the persistent guidance of Donald K. Walter.

The NCRR, under Dr. Harvey Alter, sponsored a wide range of research for ten years and made a major contribution to ASTM E-38. The enthusiasm and assistance of Jay Campbell and Marc Renard are especially acknowledged.

The dedicated groups of people who have carried their projects from concept to completion and beyond—this "stick-to-it-ness"—is what makes success in spite of travail. I particularly want to mention some remarkable people at CEA who devoted themselves to resource recovery and later were stricken by fatal heart attacks: John Converse, Vincent DiRoucco, Bobby Lahiri and Joe Lyons. Among these major contributors in developing waste-to-energy systems were Herbert Benningson, Robert Bedell, Kutty Menon, Dave Bergstrom, Wayne Wellan and John Peteanu. Mike Magoulas provided the devotion, drive and understanding needed to forge ahead from stage to stage at Brockton and Bridgeport. Bob Davis never stopped contributing fresh thinking and ways of doing things better. The early and devoted efforts of Tom Lamb, Ravi Nadkarni and others at Arthur D. Little in getting Eco-Fuel from concept to reality must be mentioned, as well as those of Norman Lyshkow, who worried the Eco-Fuel process from laboratory to production.

Dr. Ellis Blade deserves a special mention for his interest and help in unraveling the statistical mysteries of refuse and in checking my derivations of equations for the movement of particles in air streams, and particularly for his advice to look at the data, and not to smooth the information and obscure what it has to tell you.

CHAPTER 1

What Is Municipal Solid Waste?

Municipal solid waste (MSW) is a stream of matter generated and disposed of by our society, a never-ending proof of our customs, wasteful and otherwise. Disposal of this stream in an environmentally acceptable manner is a serious problem, but also can provide an opportunity to recover many benefits while reducing negative values.

On the average, each of us generates 3-5 lb/day (1.5-2.5 kg) of MSW. An elaborate system has been set up to collect it, transport it and, finally, "dispose" of it, hopefully where it can be left safely.

For many years it was left here and there (mostly there) without much concern. It often came back to haunt us, however, as our civilization moved out "there" and built homes and factories over empty or waste lands, and when the wastes seeped into our drinking water. If we burned it, the air was contaminated.

The development of sensitive sampling and testing devices made possible the measurement of extremely low levels of contaminants. Our knowledge of toxicology also advanced and we began to equate chemicals and trace elements with toxicity and hazards to life, although we ourselves are made of chemicals and trace elements and can have toxic reactions from both excesses and insufficiencies. Rightly or wrongly, our information ran ahead of our knowledge and understanding of harm to the environment and human life, often because harmful effects may take many years to appear.

The awareness of actual and potential hazards to life and environment drove us to consider the alternate methods available for disposal:

- Bury wastes in landfills that are carefully planned, built and operated.
- Burn wastes in incinerators with carefully designed furnaces, operating controls and methods for controlling atmospheric, land and water emissions.
- Process wastes to remove valuable materials for reuse, recover combustible materials to make energy and carefully landfill the residues.

These alternatives are all expensive and necessary. The choice depends on the local circumstances.

Recovery of energy and materials creates revenues that help pay for the costs of acceptable disposal methods, making this method more and more economical as landfill and fuel costs increase.

Short-term alternatives are easier to carry out because the "system" has already accepted them and is committed to them. Longer-term alternatives must be worked out, planned and sold to the many involved parties as the best course of action, as well as being preferable to no action. To do this, the planners and decision-makers must have facts and sound information that they and the public can accept. The first step in planning and selecting alternate disposal methods is to obtain knowledge of the nature of the refuse stream—where it comes from, what it is, how fast it comes, and what can and should be done with it from point of collection to final disposal.

MSW is a flowing stream. It is heterogeneous and always changing in nature. To design plants and equipment that will deal with it we must have or obtain enough information about it. These are typical questions to be answered:

- What types and quantities of MSW will be received?
- At what rates will these types arrive?
- What properties does MSW have as it is received?
- How do its properties vary: hourly, daily, weekly and seasonally?
- How do its properties change during processing?
- How can its properties be changed during processing?
- What are the properties that produce economic values?
- What unwieldy or hazardous objects must be removed?
- What contaminants should be removed?
- What tests can be performed to obtain answers to the above questions?
- What range of variations should we expect and what confidence do we have that our estimates are correct?

A great deal of testing has been carried out to answer these questions and to determine the properties of a complex mixture of materials that are continually changing in relative concentration. We find the following results:

- The composition of MSW in the United States varies significantly from place to place due to seasonal variations in yard waste, dirt and moisture content, and differing ratios of household, industrial and commercial wastes.
- There has been a steady increase over the years in the plastics content and, consequently, the heating value of refuse.
- Commercial and industrial wastes are substantially different from household wastes in moisture, ash and heating value, as well as in terms of shredding and processing properties.
- The nature of refuse received at a given location can vary substantially from truckload to truckload, between economic levels, from urban to suburban and from locality to locality. Major differences are found in food waste, cardboard and newspaper content, moisture, metals, textiles and oversize wastes.
- Whole truck tests reveal the variations with which the processing plant will have to deal.
- At any given moment, the various properties of any sample of MSW will vary at least ±25% from the mean or average value. After a maximum amount of grinding, drying and homogenization, the variation in heating value of the combustible fraction of MSW can be reduced to less than ±5%.

- The high degree of variability of MSW requires large numbers of samples to be collected and analyzed to estimate its properties with confidence. For a sample to be representative, it must have a large enough number of particles: this determines the required sample weight.

This chapter will examine the major properties of MSW, the ways it is measured and the variations that can be expected in this highly heterogeneous material. These measured properties will be the basis for evaluating the property changes that will take place as the MSW is processed into refuse-derived fuels (RDF) and other useful products.

SOURCES OF MUNICIPAL SOLID WASTE

The waste stream can contain everything and anything that can be discarded, including most solid materials, as well as liquid and gaseous materials in containers.

Estimates of the quantities of residential and commercial solid wastes generated in the U.S. in 1971 are shown in Table 1-1. The product source categories are divided into kinds of materials (components) and summarized on "as generated" and "as disposed" bases (the difference being the metals that are not disposed of by households). The total average for the entire country was estimated to be 3.3 pounds per person.

"Municipal solid waste" is a term often restricted to household waste, as distinguished from commercial and industrial waste. In the design of waste processing plants, consideration must be given to all possible sources of refuse and the consequences of the ingredients from these sources on the function of the plant.

The term "household waste" is commonly applied only to garbage discarded in residential neighborhoods and, hence, is limited in size and unit quantity to the size of the bags that people can easily handle. A typical measure is the garbage bag or can, which can range up to 60 gallons in capacity and be equipped with wheels, and which is filled gradually by the householder or apartment dweller. The amount the garbage collector or truck can handle is the ultimate limitation on the size of a unit of household refuse.

Another category of household refuse is the "oversize bulky waste" (OBW), which is placed alongside the garbage can. This is generally limited in size by the hopper of the refuse collection vehicle, which generally is not more than six feet wide and is about three feet in diameter. This volume may be consumed by corrugated boxes, branches of small trees, furniture, and the like, including tires, rugs, blankets, rocks and building blocks. Bundles of newspapers and magazines are likely items, as well as bags of leaves and even dirt and soil. Occasionally, it includes dead animals. Demolition debris is common, including steel lath, plaster, bricks, cinder blocks and miscellaneous plywood and lumber shapes.

Many "objectionable materials," such as cans containing gasoline, paint, thinner, and aerosol, appear in household wastes, often because one is prohibited from disposing of them elsewhere.

Table 1-1. Composition of MSW As-Generated and As-Disposed [1] [a]

	Product Category (millions of tons, as-generated wet weight) [b]							Totals			
								As-Generated Wet Weight [b]		As-Disposed Wet Weight [c]	
Material Category	*Newspapers, Books Magazines*	*Containers, Packaging*	*Major Household Appliances*	*Furniture, Furnishings*	*Clothing, Footwear*	*Food Products*	*Other Products*	*Millions of Tons*	*Percentage*	*Millions of Tons*	*Percentage*
Paper	9.8	19.1	tr.	tr	–	–	8.3	37.2	29.0	44.9	34.9
Glass	–	12.2	0.1	tr	–	–	1.0	13.3	10.4	13.5	10.5
Metals		5.9	2.1	0.1			4.0	12.1	9.6	12.6	9.8
Ferrous	–	(5.2)	(1.8)	(0.1)	–	–	(3.7)	(10.8)	(8.6)		
Aluminum	–	(0.7)	(0.1)	tr	–	–	(0.1)	(0.9)	(0.7)		
Other nonferrous	–	–	(0.2)	tr	–	–	(0.2)	(0.4)	(0.3)		
Plastics	–	2.7	0.1	0.1	tr.	–	1.5	4.4	3.4	4.9	3.8
Rubber and leather	–	tr.	tr.	tr.	0.7	–	2.6	3.3	2.6	3.4	2.6
Textiles	–	0.1	tr.	0.6	0.5	–	0.9	2.1	1.6	2.2	1.7
Wood	–	1.8	tr.	2.6	–	–	0.5	4.9	3.8	4.9	3.8
Total nonfood product waste	9.8	41.7	2.3	3.4	1.2	–	18.9	77.5	60.5	86.5	67.3
Food Waste	–	–	–	–	–	22.8	–	22.8	17.8	19.1	14.9
Total product waste	9.8	41.7	2.3	3.4	1.2	22.8	18.9	100.3	78.3	105.6	82.2
Yard waste								26.0	20.2	20.9	16.3
Miscellaneous Inorganics								1.9	1.5	2.0	1.6
Grand Total								128.2	100.0	128.5	100.0

[a] Net solid waste disposal defined as net residual material after accounting for recycled materials diverted from waste stream.

[b] "As-generated" weight basis refers to an assumed normal moisture content of material in its final use prior to discard: for example, paper at an "air-dry" 7% moisture; glass and metals at 0%. Total waste, including food and yard categories, estimated at 26% moisture.

[c] "As-disposed" basis assumes moisture transfer among materials in collection and storage but no net addition or loss of moisture for the aggregate of materials.

Such items as refrigerators, washing machines, hot water tanks, oil burners, dryers and mattresses are often also a part of the municipal stream and may be collected separately or emerge from the packer trucks along with the more benign materials.

The normal materials in the MSW stream are paper, cardboard, plastics of various shapes, food wastes, ferrous, nonferrous and mixed cans and heavy metals, glass ranging from sand to large jugs, yard waste ranging from dirt and leaves to twigs, shrubs, branches and tree trunks, and a myriad of odd items such as shoes, toys, small appliances, textiles, rugs and blankets.

Industrial and commercial wastes are usually not picked up by municipal collectors but are, nonetheless, to be expected in some quantity. If subject to reasonable size limitations they may not be too objectionable. Commercial haulers may be permitted to bring such wastes with a clear understanding as to what is allowable and with an absolute right of refusal enforced by loss of this permission on violation. These wastes may contain food, oils, plywood, construction materials, wheels, car parts (including gasoline cans and tanks), shock absorbers, springs and construction debris, which can be hazardous and damaging to machinery.

As all wastes must be delivered somewhere, one must consider all possible types to avoid being surprised or having to accept material that is not suitable for this type of disposal.

Looking at the total stream, including household, commercial and industrial wastes, it is clear that objectionable materials may be either removed at the source or rejected by the collector. Otherwise, they must be identified on the floor of the receiving facility, removed and placed in a temporary or permanent location, such as the landfill.

This book does not give much attention to the objectionable materials (OM) and oversize bulky wastes (OBW), except to point out that they *will* come in and must be found and removed by a suitable method. Generally, the refrigerators and washing machines, hot water heaters, bed springs, oil burners and boilers, doors, windows, etc., are placed in a container body for transport to a landfill. They may be beneficially picked before ultimate disposal.

CHARACTERIZATION OF MUNICIPAL SOLID WASTE

The results of many surveys have been published giving the constituents of MSW, based on sampling in many localities, and showing regional sociological and seasonal variations. The U.S. Office of Solid Waste Management collected data generated by methodical sampling, segregation and weighing, as shown in Table 1-2. Data were collected under these categories:

- Household
- Commercial
- Industrial
- Agricultural
- Institutional

Table 1-2. Various Descriptions of Municipal Solid Waste

Type of Waste	*Constituent*	*Proximate Analysis*	*Ultimate Analysis*	*Major Chemicals*	*Minor Trace Elements*
Household	Newspaper	Moisture			
	Other paper				
Commercial	Cardboard	Volatile	Carbon	SiO_2	Zn
	Plastic film		Hydrogen	Ai_2O_3	Mn
Industrial	Plastic, rigid	Fixed carbon	Oxygen	CaO	Pb
	Food		Nitrogen	Fe_2O_3	Cu
Institutional	Yard		Sulfur	NaO	Cr
	Wood		Chlorine	MgO	Sn
Construction		Ash	Ash	TiO	Ni
				K_2O	Cd
Demolition	Magnetics	Noncombus-			
	Nonferrous	tibles			
Agricultural	Glass				
	Sand				
	Miscellaneous				
Oversize	OBW				

- Demolition and construction
- Street and alley
- Tree and landscaping
- Park and beach
- Catch basin
- Sewage solids

Per capita rates of generation were determined from these studies, as well as data on the distribution among the various sources: household, commercial, industrial and agricultural.

In addition, the waste has been characterized by sorting samples into categories, such as paper, cardboard, plastics, textiles, food, wood, glass, ferrous metal, aluminum and other nonferrous metal, sand, grit and miscellaneous. These categories have not been selected consistently in all testing programs. Niessen and Chansky [2] have summarized these early studies.

MAJOR CHARACTERISTICS OF MUNICIPAL SOLID WASTE

The sources and constituents of MSW determine the properties that have to be dealt with in a processing plant by the various unit operations. (The ASTM *Thesaurus of Resource Recovery Terminology* is recommended for a more complete treatment of this subject [3].)

Some of the more important properties are worth discussing at this point, in preparation for a more detailed analysis of the way these properties influence the requirements and behavior of unit operations devices in regard to MSW and its processed streams.

Properties of Special Interest in Processing

The properties of MSW that influence the products (be they energy, fuel, paper, metals or glass) and the processing operations are of prime importance in regard to energy and materials recovery.

The *components* or *constituents* are of interest for their positive and negative characteristics, depending on how troublesome they are to process and their potential value.

The *combustible materials* must be identified, as well as the noncombustible moisture, glass, dirt and metals. The combustible fraction (cellulose, plastic and textiles) and the recoverable materials (corrugated cardboard, ferrous and nonferrous metals and glass) also have decidedly negative values, which must be considered in processing. Corrugated cardboard and textiles are tough to shred, and the textiles, if not well shredded, tangle and knot and wrap around rotating devices. The metals include springs and wires, which attach themselves to rags, creating "scarecrows," which tend to plug conveying systems and rotary valves. Combustibles include not only the desirable paper products, but also the hazardous solvents, which can cause deflagrations and, possibly, explosions. *Noncombustibles* include glass, sand and dirt, which are abrasive to equipment and cause slagging of boilers.

The ash of the combustible fraction contains elements that cause complicated chemical reactions in slag and fouling deposits in boilers, ash destined for landfill and particulate matter that must be collected by emission control equipment. Included in this ash are metals that vaporize, partly escaping emission controls and partly condensing on particulate matter. Some of these metals and trace elements in the ash can be harmful to the environment and to human health if emitted in sufficient quantities.

The *chemical* or *elemental composition* of the component fractions, as well as the mixture, of MSW is an important determinant of the heating value of the MSW and the energy products.

The moisture content of the components can be highly variable, ranging from dry paper to 75% moisture food and yard waste. To obtain the chemical energy of these materials, the water must first be driven off, requiring heat input and time, while reducing the net energy output.

Density affects the size of transport and handling equipment, as well as storage volume. Density changes with the pressure applied and relaxes after it is released.

Particle size and particle size distribution (PSD) are important properties of the individual constituents and of the total mixture.

The mechanical properties, such as tensile and compressive strain and yield points, and the energy required to stretch or compress materials, are important when considering compaction and tearing operations.

In proximate analysis, volatile matter is obtained by weight loss, by gradually heating the dried sample to 600 ±50°C in a platinum crucible with a close-fitting cover to 950 ±20°C, holding for 7 minutes, and cooling without disturbing the cover. The residue is defined as fixed carbon plus ash. The fixed carbon is calculated by subtracting moisture and ash from the sample weight.

Ultimate analysis determines carbon, hydrogen, sulfur and nitrogen, as well as oxygen by difference, as found in the gaseous products of complete combustion of a dried sample. Moisture and ash are determined from a separate sample.

The ash residue, assumed to contain only oxidized forms, can be analyzed for elemental composition, including the major elements, oxides of silicon, aluminum, calcium, iron, sodium, potassium, titanium and phosphorus, and the minor and trace elements, including zinc, lead, copper, chromium, cadmium, and so forth. Some of these properties are treated in detail in the following section.

Moisture Content

Moisture content is the most important property of MSW and RDF due to its influence on the processing, handling and burning of these materials. The normal moisture of the paper fraction ranges from 5 to 15%, and that of food and yard waste ranges up to 86%.

Moisture content can be expressed either on a dry basis (the weight of moisture contained in the material divided by the weight of dry material) or on a wet basis (the weight of moisture divided by the as-received weight of moisture plus dry matter). In this book the wet basis is used.

The weight fraction of moisture is determined by heating the solid material sufficiently to drive off the moisture, with the weight loss then considered to be the moisture content.

Not all the water contained in hygroscopic materials such as the paper, cardboard, food, yard waste and other materials found in MSW is readily driven off. Consider the following:

- Surface moisture is removed when the particle has reached temperatures in excess of 100°C
- Equilibrium moisture is retained within the paper in proportion to the vapor pressure of water in the ambient atmosphere; hence, low humidity and high temperatures are needed to extract all the moisture.
- Chemically bound water is driven off only by heating to the charring point (for cellulosic materials).
- Cellular moisture may not be released unless the cells are broken or exploded by temperatures well above 100°C.

By heating the particle of solid to 105–110°C for a sufficient time, the free, extraneous and surface moisture can be driven off, and most, if not all, of the equilibrium moisture.

The time required to dry cellulosic materials depends on the thickness of the material, the relative humidity of the atmosphere (relative vapor pressure), the temperature difference between the solid and the surrounding atmosphere and the velocity of the gases past the solid surface. Depending on the thickness of the solid and the conditions, it may take a fraction of a second, five minutes or several hours to dry.

Figure 1-1 shows the equilibrium moisture content of some typical cellulosic components of MSW. The moisture content falls off rapidly at humidities below 10%, while above 90% it goes up rapidly. At 100% humidity the moisture content cannot be determined. At 99%, however, it appears to be below 20%. This graph illustrates why moisture determinations are dependent on humidity conditions in the laboratory and in the dryer, and why circulation of fresh air is so necessary to speed drying.

Figure 1-2 shows the time required for drying unstirred RDF in an oven at 110°C. With forced ventilation or stirring, this time can be reduced to less than one hour for small, light (flaky) particles. Also shown is the hygroscopic regain of RDF by exposure to normal ambient humidity. Stirring and forced ventilation can be used to reduce these times substantially.

Figure 1-3 shows the relative humidity of air/water mixtures at various temperatures, including the range above 100°C. A relative humidity (RH) of 100% can exist at any temperature in a steam atmosphere. Unless air or other noncondensible gases are present, the RH cannot be reduced below 100%. On the other hand, at high temperatures very little air is needed to reduce the relative humidity and, hence, the equilibrium moisture content of the hygroscopic material.

Removing the moisture in cellular form requires temperatures high enough to rupture the fibers or to force the water vapor out. We know it takes a long time to cook meat to a dried-out condition even in a 450°F oven, so we must expect that chunks of wood and meat may require at least overnight to dry. They can survive a one-hour residence time in a moving-grate incinerator.

The best way to determine the time and temperature combination required for drying is to periodically determine the weight loss and keep drying until the loss in a time interval is as small as the desired accuracy.

When temperatures above 110°C are used, significant amounts of low-volatile materials other than water vapor may be driven off MSW materials, introducing some degree of error as the temperature is increased. When in doubt, this must be resolved by observing weight loss at different temperatures. Temperatures of 150-170°C often can be used for drying cellulosic materials quickly without volatile losses. If oils or solvents are present in MSW, however, these temperatures would drive them off.

When precise moisture content is desired for cellulosic materials at moisture contents under about 5%, it may be necessary to dry under dessicating conditions.

Combustible Matter

The combustible fraction of a sample is defined as the fractional weight that is removed by complete combustion in the presence of sufficient oxygen, after prior removal of moisture. Therefore, it is the initial weight minus the moisture and ash. It must be distinguished from "combustibles" content, which includes inherent ash and may include moisture.

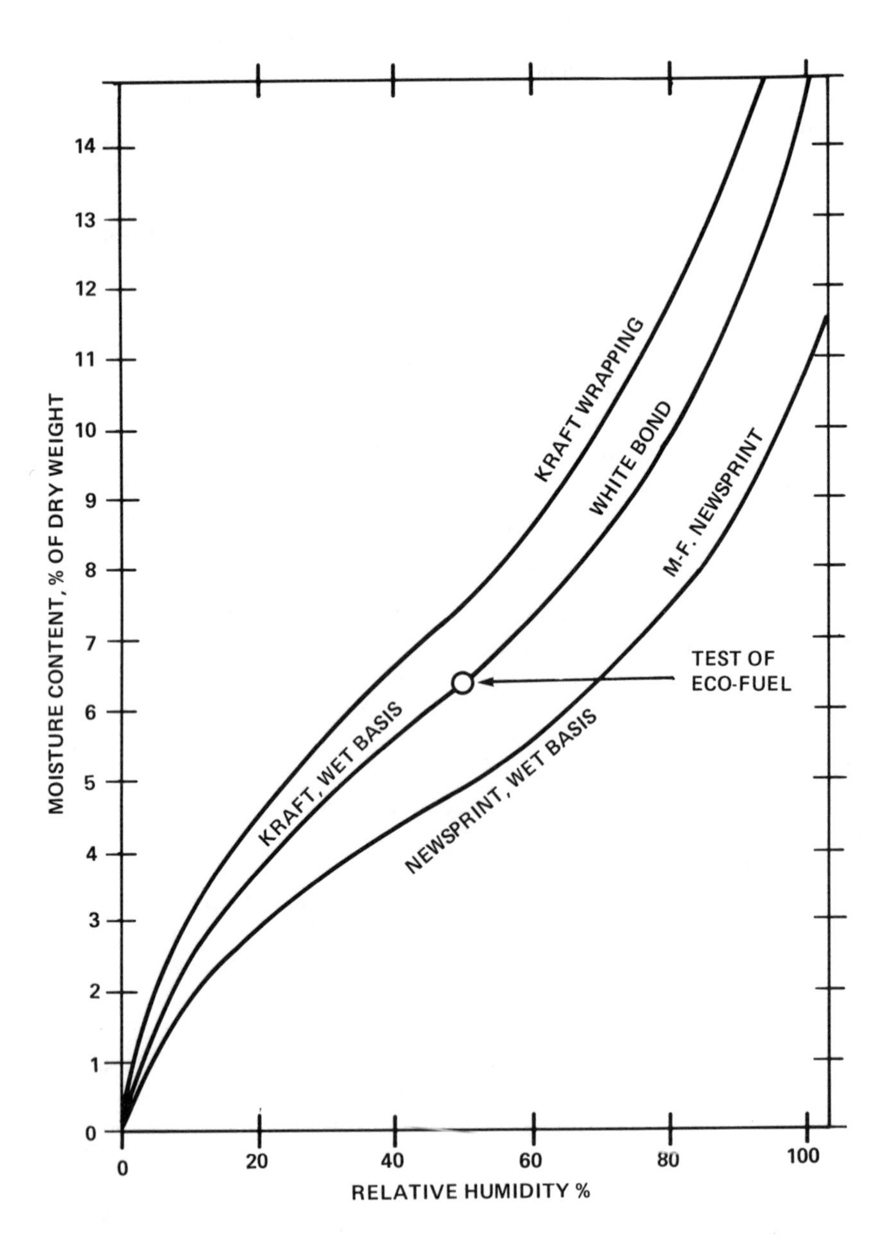

Note: Wet basis is percentage of wet weight.

Figure 1–1. Equilibrium moisture content of cellulosic matter (based on data from Fan Engineering, Buffalo Forge, Buffalo, NY).

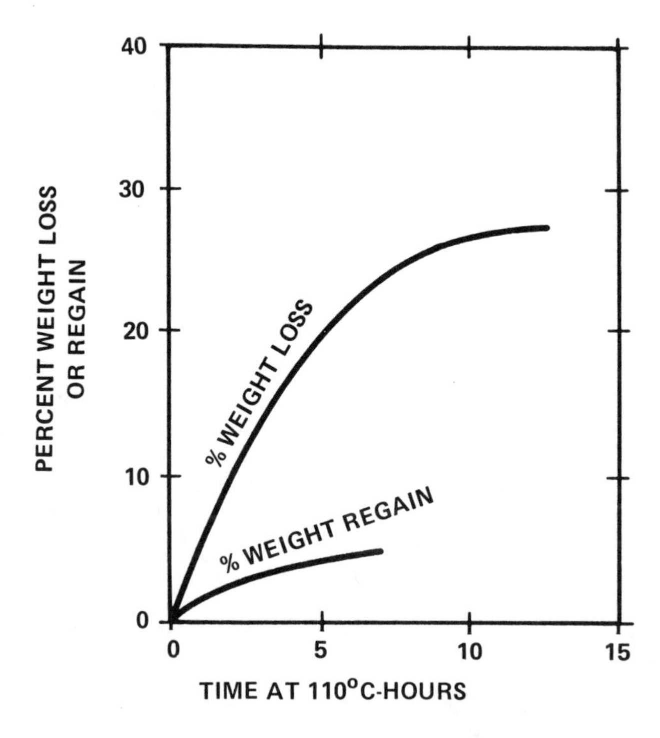

Figure 1-2. Drying and moisture regain of RDF [4].

Gross and Net (Higher and Lower) Heating Value

The higher heating value (HHV) of a combustible sample is determined by the combustion of the sample in a bomb calorimeter [5] under an oxygen atmosphere. The heat produced is absorbed by the stirred water of the calorimeter in which the bomb is placed. The temperature rise of the calorimeter water determines the amount of heat produced by the combustion. The observed temperature rise is corrected for the heat absorbed from the calorimeter environment. The products of combustion cool down and the condensible vapors are condensed, giving up their heat of vaporization to the water bath. Thus, the water vapor produced by the combustion of hydrogen is condensed.

In actual boilers, the water vapor produced by burning the hydrogen is not condensed, so the heat of condensation cannot be recovered. Therefore, the HHV measured by the calorimeter cannot be realized in practice. The lower heating value (LHV) is a more valid indication of the heating value of fuels for combustion in boilers because it is not based on the assumption that the water vapor is condensed by the boiler. To convert HHV to LHV, the latent heat of water vapor at standard conditions (1050 Btu/lb or 2443 J/g), multiplied by the ratio of the sum of the masses of free water in the fuel plus water formed in the reaction, divided by the initial fuel mass, is subtracted from the HHV, thus obtaining the LHV, which is the true

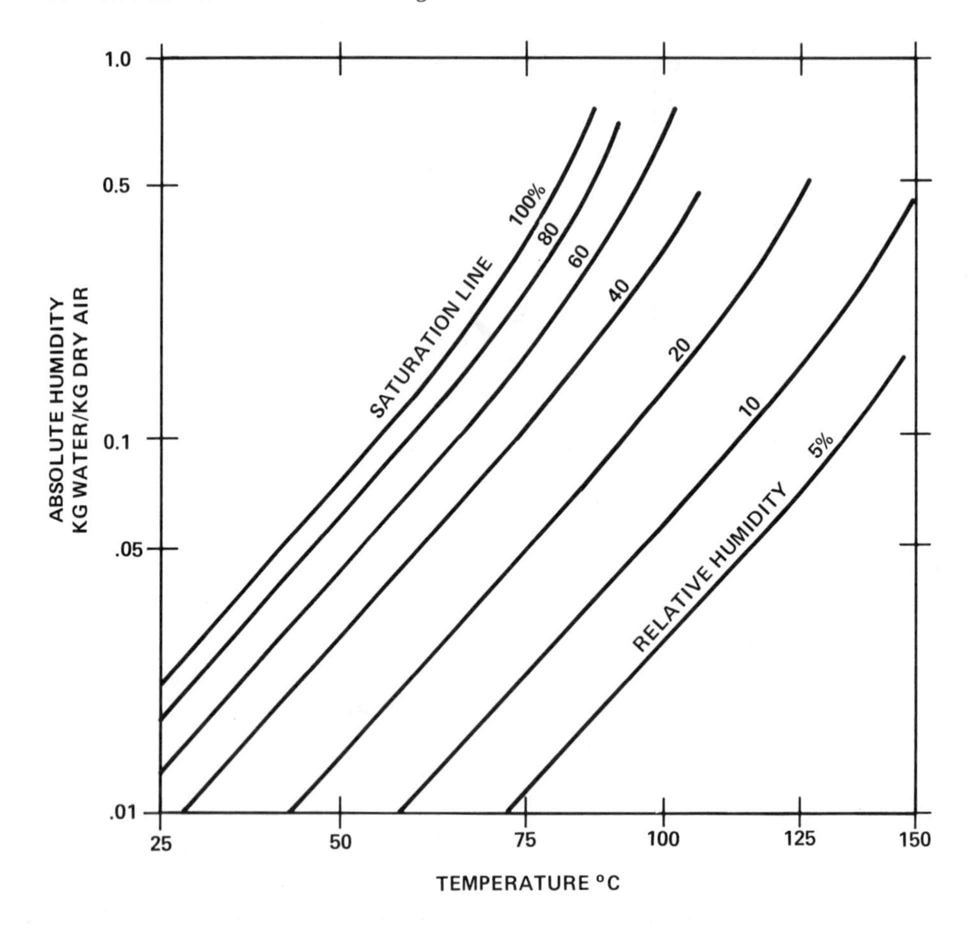

Figure 1-3. Relative humidity of air/water mixtures (data from Proctor and Schwartz, Philadelphia, PA).

potential for heating represented by the fuel properties. Cellulosic fuels have about 5-6% hydrogen content, which results in about 5% difference between HHV and LHV.

The term "available heat" by definition takes into account the effect of fuel moisture on boiler efficiency and thus represents the maximum amount of heat that actually could be delivered to the steam in the boiler. It is not commonly used because it involves not only the fuel but also the characteristics of the boiler.

Available heat is calculated by subtracting from the HHV the heat losses due to the products of combustion leaving the boiler at the stack temperature. The products of combustion include the fuel, air required for combustion of this fuel, excess air actually used, the water vapor in the total air supplied, the water in the fuel and the water generated by combustion of the hydrogen in the fuel. The latent heat is not recovered.

Figure 1-4 shows the available heat for a typical RDF for a boiler operating with a 260°C stack temperature and 50% excess air, for various moisture contents of the RDF. The HHV is shown for comparison.

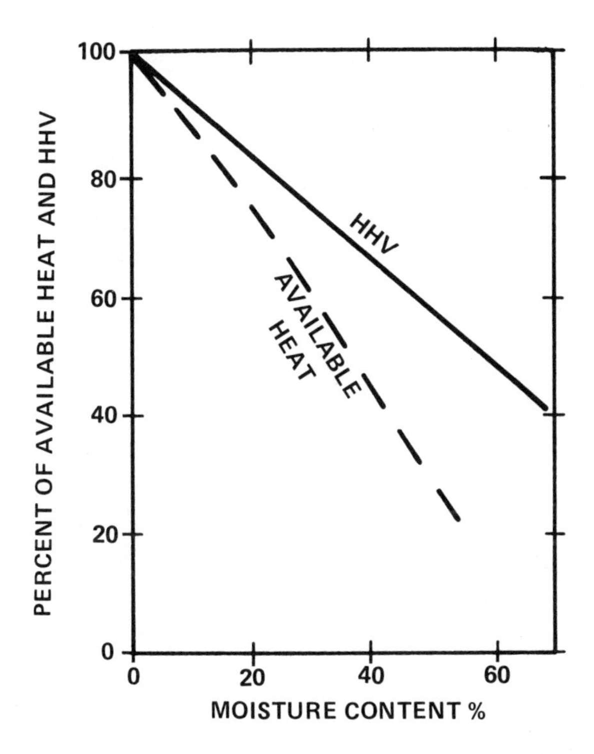

Figure 1-4. Available heat of RDF at various moisture levels.

Methods of Determining the Heating Value of MSW and RDF

There are a number of methods of determining the heating value of samples of MSW and RDF:

- **Use a boiler as a calorimeter**: The boiler must be fully instrumented so that the heat liberated by the combustion can be measured accurately. Some utility and industrial boilers are instrumented well enough to make accurate heat balances, permitting continuous measurement of the heat liberated by the fuels. Normally, however, this is not possible, and heating value must be determined by laboratory analysis of relatively small samples.
- **Use a laboratory calorimeter**: The standard laboratory oxygen bomb calorimeter can accept 1 g of milled refuse. Larger bomb calorimeters accept 2.5 g. The National Bureau of Standards (NBS) has developed a large flow calorimeter [6], which can take 2.5 Kg samples, permitting heating value determinations of coarsely shredded MSW.
- **Calculate heating value from chemical analysis.**

As-received, Dry, and Moisture- and Ash-Free Heating Values

Heating values are expressed on several bases, which must be understood before proceeding further. The as-received heating value (HHV1), is based on the moisture and ash contained in the sample as it is received at the laboratory. The heating value on a dry basis, (HHV2) is higher per unit weight because the moisture has been removed. The moisture- and ash-free heating value (HHV3) is the inherent energy available from the true combustibles in the sample. These are determined as follows:

- **As-received heating value:**

$$HHV1 = HHV3 \times (1-(\text{ash fraction})-(\text{moisture fraction}))$$

- **Dry heating value:**

$$HHV2 = HHV3 \times (1-(\text{ash fraction}))$$

- **Moisture- and ash-free heating value:**

$$HHV3 = HHV1/(1-AF-MF) = HHV2/(1-AF)$$

Higher Heating Value

The (HHV) is defined as the heat released by combustion of a combustible material under conditions in which the water in the products of combustion is condensed.

Although the most direct measurement of heating value can be made with a bomb calorimeter, which gives the HHV of a small sample by combustion in an oxygen atmosphere, the heating value also can be estimated from the elemental analysis or the proximate analysis of a fuel sample.

The proximate analysis of coal can be used to estimate the heating value using Goutal's equation [7]. The same method can be used with refuse fuel samples, provided that the constant, K, can be determined as a function of the volatile matter. For example, for a typical RDF,

Volatile	=	43.60%
Fixed carbon	=	8.17%
Moisture	=	26.60%
Ash	=	21.70%

Goutal's equation is as follows:

$$HHV = 147.6 \times (\%FC) + K \times (\% \text{ volatile})$$

For $K = 80$ (at 45% volatile),

$$\begin{aligned} HHV &= 147.5 \times 8.17 + 80 \times 43.6 \\ &= 4700 \text{ Btu/lb} \end{aligned}$$

- **Heating value, moisture- and Ash-free:**

$$HHV(MAF) = 4700/(1 - 0.266 - 0.217) = 9080 \text{ Btu/lb}$$

The ultimate analysis, which determines the chemical composition of the combustibles, can be used to determine the heating value by multiplying the heating of reaction of each component by its weight fraction, in the same manner as the Dulong equation, which was developed for coal (avoiding the empirical methods used by Wilson [8]), as follows:

$$\begin{aligned} HHV = {} & 14447(\text{Carbon}) + 60958(\text{Hydrogen} - \text{Oxygen}/8) \\ & + 3980(\text{Sulfur}) + 1040(\text{Nitrogen}) \end{aligned}$$

The oxygen content is divided by 8 and subtracted from the hydrogen in order to take into account the portion of the hydrogen that has already reacted with oxygen and hence does not contribute heating value. It is convenient to rewrite this equation as follows:

$$HHV = 144.5\ (\%C) + 609.6\ (\%H) - 76.2\ (\%O) + 40\ (\%S) + 10\ (\%N)$$

- **Theoretical Air Required for Combustion**

The air required for complete combustion of the combustible components of RDF can be calculated from the chemical equations and the atomic weights of the components, and expressed as pounds of air per million Btu of heat released by dividing by the HHV, as follows:

$$\begin{aligned} \text{Theoretical Air (TA)} = {} & 138.17\ (C/12 + H/4 - 0.32 + S/32 \\ & + N/28) \times 10^6/HHV \end{aligned}$$

This form is more convenient to use:

$$\begin{aligned} \text{Theoretical Air (TA)} = {} & [11.51(\%C) + 34.5(\%H) - 4.32(\%O) \\ & + 4.3(\%S)] \times 10^6/HHV \end{aligned}$$

The HHV may be determined by calorimeter, but the determination of TA will be most consistent if the same ultimate analysis is used.

It is convenient to use a tabular form to make these calculations, as follows:

Component	*%WT*	X	*Release*	=	*HHV*	*% WT*	X	*Factor*	=	*Air*
Carbon	26.00	X	144.47	=	3756.22	26.00	X	11.51	=	299.26
Hydrogen	3.80	X	609.58	=	2316.40	3.80	X	34.50	=	131.10
Oxygen	21.21	X	– 76.20	=	–1616.20	21.21	X	– 4.32	=	– 91.63
Sulfur	0.18	X	39.82	=	7.17	0.18	X	4.32	=	9.78
Nitrogen	0.53	X	10.50	=	5.15	–				
Chlorine	0.18									
Nitrogen	0.40									
Moisture	26.54									
Ash	21.70									
Totals:	100.00		HHV	=	4469.10 Btu/lb					348.51

$$\text{Theoretical Air (TA)} = 348.5 \times 10^4/4469.10 = 760 \text{ lb air per million Btu}$$

This value of theoretical air per unit of heat released by combustion, determined in this manner from chemical analysis of samples of MSW and RDF, appears to be independent of the nature of the waste, hence may be used with confidence in estimates of combustion air requirements.

Heating values determined from ultimate analysis appear to be in close agreement with those determined from the same laboratory sample by calorimetry. It is helpful to use the ultimate analysis to check the calorimetric method, to minimize the possibility that large discrepancies will pass unnoticed. There is a chance of error in ultimate analysis as well as calorimetry.

Lower Heating Value

The lower heating value (LHV) is calculated from the higher heating value (HHV), by subtracting the latent heat of evaporation of the water in the products of combustion. This water includes the water content of the fuel and the water produced by the combustion of hydrogen.

The equation for LHV is as follows:

$$\text{LHV} = 144.5(\%C) + 609(\%H) - 76(\%O) + 40(\%S) + 10(\%N) - 10.5(\%H_2O)$$

The water produced by the combustion of hydrogen is nine times the weight of hydrogen in the fuel [16 + 2)/2 = 9]. The hydrogen in the fuel is known from the ultimate analysis. For refuse fuels, the hydrogen is about 7.5% of the combustible matter. The latent heat at atmospheric pressure is 1040 Btu per pound. Hence, the loss heating value due to failure to condense this water is about:

$$\text{Loss of heating value} = 7.5 \times 9 \times 10.50 = 712 \text{ Btu/lb combustible}$$

As the combustibles contain 9000 Btu per pound, the loss of heating value is 712/9000 = 7.9% of the moisture and ash-free heating value.

For a refuse fuel containing 25% moisture and 25% ash, the loss due to failure to condense this moisture is 0.25 × 1050 or 251 Btu per pound of refuse. Since the HHV would be 9000 × 0.5 = 4500 Btu/lb, and the hydrogen loss would be 7.9% of this, or 354 Btu, the total loss is 605 Btu, and the LHV is 3894 Btu per pound.

Fuel efficiency, as distinguished from boiler efficiency, may be defined as the ratio of the heat actually available to a conventional boiler (which cannot condense the water vapor in the products of combustion) to the theoretical heat in the fuel. It is, therefore, the LHV divided by the HHV, reflecting the loss of the heat of condensation. The fuel efficiency would be 3894/4600 = 84.6%. This may be compared with the same fuel, dried, having a loss of 712 × 0.75 = 534 Btu/lb, and a HHV of 9000 × 0.75 = 6750 Btu/lb, hence a fuel efficiency of (6750 – 534)/6750 = 92.1%. The hydrogen loss is, conveniently, about the same as the percentage of hydrogen in the fuel, or 7.5%.

Density

The density of MSW and its derivatives must be determined in accordance with a procedure, which then defines the type of density.

Free-fall density is that measured by allowing the material to fall a minimal distance onto a surface, without confining the material on its sides. This is applicable to materials falling on a belt. The weight of the sample, divided by the volume of the pile, gives the free-fall density.

Bulk density is measured by allowing the material to fall freely into a container of prescribed shape, usually a cube, and dividing the weight of the sample that exactly fills the container by the volume. The container may be overfilled and the excess scraped off to achieve complete fillage. Alternatively, the material can be allowed to partially fill the container and a plunger lowered into it to determine the height of material: this is likely to compress the material and alter the bulk density, however.

True density is the density of the material in question in its natural form.

Compressed density can be defined by the degree of pressure applied to a sample placed in a container and compressed by a plunger. The absolute density is approached as the pressure applied is increased to the limit. Cellulosic materials can be compressed to as high as 1.2 times the density of water, or about 1230 kg/m^3, by applying 2-4 × 10.8 N/m^2 (75 lb/ft^3; 30,000-60,000 psi). Table 1-3 lists densities for various refuse components.

Particle Size of Refuse Components

Figure 1-5 shows, in histogram form, the size distribution of the major components of raw refuse. They range in size from fractions of a millimeter for dirt and sand to 50-500 mm for textiles, plastics, glass, metals and paper.

Additional properties, such as work index and lifting or settling velocity, are described, along with methods of measuring them, in appropriate sections of this book.

Table 1-3. Densities of Refuse Components[a]

Component	*Density*	
Refuse Densities	lb/yd^3	
Loose refuse	100–200	
After dumping from compactor truck	350–400	
In compactor truck	500–700	
In landfill	500–900	
Shredded refuse	600–900	
Baled in paper baler	800–1200	
Bulk Densities	lb/ft^3	
Cardboard	1.87	
Aluminum	2.36	
Plastics	2.37	
Miscellaneous paper	3.81	
Garden waste	4.45	
Newspaper	6.19	
Rubber	14.9	
Glass	18.45	
Food	23.04	
True Densities	lb/ft^3	g/cc
Wood	37	0.6
Cardboard	43	0.69
Paper	44–72	0.70–1.15
Glass	156	2.5
Aluminum	168	2.7
Steel	480	7.7
Polypropylene	56	0.90
Polyethylene	59	0.94
Polystyrene	65	1.05
ABS	64	1.03
Acrylic	74	1.18
Polyvinylchloride (PVC)	78	1.25
Resource Recovery Plant Products	lb/ft^3	kg/m^3
dRDF	39	625
Aluminum scrap	15	253
Ferrous scrap	25	400
Crushed glass	85	1362
Powdered RDF (Eco-Fuel)	27	432

[a]Prepared by Cal Recovery Systems, Richmond, CA, for ASTM E-38 (EPA funding).

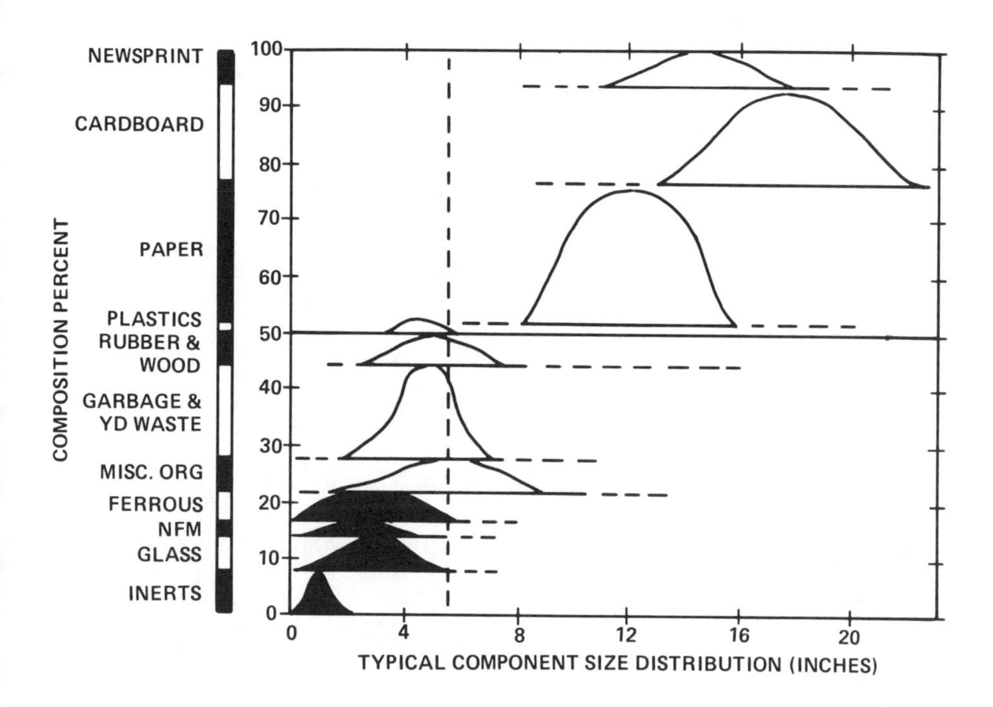

Figure 1-5. Size distribution of MSW.

Chemical Composition

Kaiser [8] investigated the chemical nature of refuse, listing most of the common constituents of refuse along with their chemical compositions, normal moisture and ash, and their heating values. Table 1-4 lists these properties and shows how they can be used to calculate the heating values of the refuse.

Taking the amounts of each constituent into account, the heat contributed by each component can be calculated and the average heating value of the mixture determined. By this method, Btu/lb as-discarded, dry, and dry and ash-free basis can be determined. Table 1-4 shows the individual contributions of each species and group and the resultant heating value of a typical MSW composition. As shown in Table 1-4, the heating value of the representative refuse is calculated to have a heating value of 5019 Btu/lb as received.

SAMPLING AND TESTING OF REFUSE

The first step in determining the properties of MSW and processed streams is to obtain samples that are representative of the lot or flowing stream. The number and size (weight of samples required to achieve acceptable precision in estimating the properties of the lot or stream depend on the variability of the properties and the degree of confidence we desire.

Table 1–4. Conversion of Constituents to Heating Value by Data Analysis*

	As-received Analysis					*Products*		*Ultimate Analysis*				
Constituent	*Wt %*	*% MC*	*% ASH*	*Btu/lb*	*Wt × Btu*	*Wt × MC*	*Wt × Ash*	*C*	*H*	*O*	*N*	*S*
Newspaper	10.33	6.0	1.4	7,974	824	0.62	14.8	49.1	6.1	43.0	0.05	0.16
Brown paper	6.12	5.8	1.0	7,256	444	0.35	6.2	44.9	6.1	47.8	0	0.11
Magazine	7.48	4.1	22.5	5,254	393	0.31	168.0	32.9	5.0	38.6	0.07	0.09
Corrugated	25.68	5.2	5.1	7,043	1,809	1.34	130.0	43.7	5.7	44.9	0.09	0.21
Plastic coated	0.84	4.7	2.6	7,341	62	0.04	2.2	45.3	6.2	45.5	0.18	0.08
Waxed milk cartons	0.74	3.5	1.2	11,327	95	0.03	1.0	59.2	9.3	30.1	0.12	0.10
Paper cartons	2.27	6.1	6.5	7,258	165	0.14	41.8	44.7	6.1	41.9	0.15	0.16
Junk mail	3.03	4.6	13.1	6,088	184	0.14	40.0	37.9	5.4	42.7	0.17	0.09
Tissue paper	2.18	7.0	0.9	7,000	153	0.15	19.6	43.6	6.1	49.0	0.16	0.14
Food wastes	2.52	78.3	1.1	1,795	45	1.97	2.7	49.0	6.6	37.6	1.7	0.20
Citrus rinds	1.68	78.7	0.7	1,707	29	1.32	1.2	48.0	5.7	41.7	1.1	0.12
Cooked meats	2.52	38.7	3.1	7,623	192	0.97	7.8	59.6	9.5	24.7	1.0	0.19
Fried fats	2.52			16,466	415			73.1	12.0	14.3	0.4	0.07
Leather	0.42	7.5	21.2	7,243	30	0.03	8.9	42.0	5.3	22.8	6.0	1.0
Rubber	0.42	1.2	29.7	10,900	46	0.01	12.5	53.2	7.1	7.8	0.5	1.3
Plastics	0.84			15,910	134			78.0	9.0	13.0		
% Combustible 69.60				HHV = 5019		↓	↓	41.7	5.8	27.6	2.8	0.25
% Moisture in Combustibles:						7.42		↓	↓	↓	↓	↓
% Ash in Combustibles:							13.2					
HHV by Ultimate Analysis:	Ultimate Analysis		= 79.4% =					41.7	5.8	27.6	2.8	0.25
	HHV of Elements		=					145	609	−76	40	10
	HHV × % Wt		= 7595.5 =					6046	3532	−2097	112	2.5
Correcting for Ash:	HHV3(MAF)		= 7596/(1–0.132)			= 8751 Btu/lb						
HHV by Components:	HHV1 (as received)		= 5019/(0.696)			= 7201 Btu/lb						
	HHV2 (dry)		= 7201/(1–0.074)			= 7776 Btu/lb						
	HHV3 (MAF)		= 7201/(1–0.074–0.132)			= 9069 Btu/lb						

*Data from Kaiser [9].

The dimensions of the pieces in the sample have a direct bearing on the volume and, hence, on the weight of sample required. Obviously, a sample having the volume of a milk container cannot hold more than one such container. If one milk container is used per family per day, then the daily trash bag will be a good sample volume. However, as some families will discard none, and some two, many sample bags must be collected to achieve the average of one. This primitive example shows that to sample for rare items, large volumes of material must be collected. By comparison, almost any sample of household refuse will contain some paper and, perhaps, on the average of one or more newspapers.

The number of pieces in the sample is more the determining factor than the weight or volume. If we take a ten-pound sample, cut all the pieces in ten parts, then mix them carefully, any one-pound sample will be representative. This illustrates the effect of particle size on the sample size needed to be representative of a material having a given variability of composition.

To be confident that the sample contains roughly the correct number of aluminum cans, it may be necessary to include subsamples for each day in the week, to average extra use on the weekend. This illustrates the need to take time variation into account in sampling.

If part of the refuse is household and part commercial, then the sample should include the average proportion of these two types, or they may be sampled separately and a weighted average taken.

We can sort out all the material in a given truckload and find exactly what it contained, but we do not know how representative that truck was. Many trucks could be sampled totally or we could take small samples from many trucks: which method is best? This basic question is answered by following this procedure: find the errors created by the different aspects of sampling; reduce the largest errors; and allow the small errors to be increased so that the total error is minimized. Find the basic variation that is to be measured and do not try to attain precision better than, say, 10% of the basic variation. In other words, if the moisture content ranges from 15 to 45%, with an average moisture of 30%, if the error in the estimate of the mean were limited to 10% we would be confident that the actual average moisture content was somewhere between 27 and 33%.

It has been stated that a good single sample should contain at least 50,000 pieces. The statement says nothing about the resulting precision, however. To investigate the weight of such a sample, let us consider shredded newspaper, for which we can calculate the relationship between the top nominal size of the paper (95% passing screen size) and the weight of the sample, as shown in Figure 1-6.

Based on a log-normal size distribution, the weight varies to the 1.9 power of the size, or roughly as the square, as would be expected for flakes. This rule is handy for estimates of sample size. For instance, we can compare the sample size needed for raw refuse and shredded refuse to get the same precision. If the raw refuse is minus 24 inch and shredded material is minus 2 inch, a shredded sample equivalent to a 150-kg raw sample would be 1 kg. Using Figure 1-6, we can estimate that 1 kg would contain about 15,000 pieces.

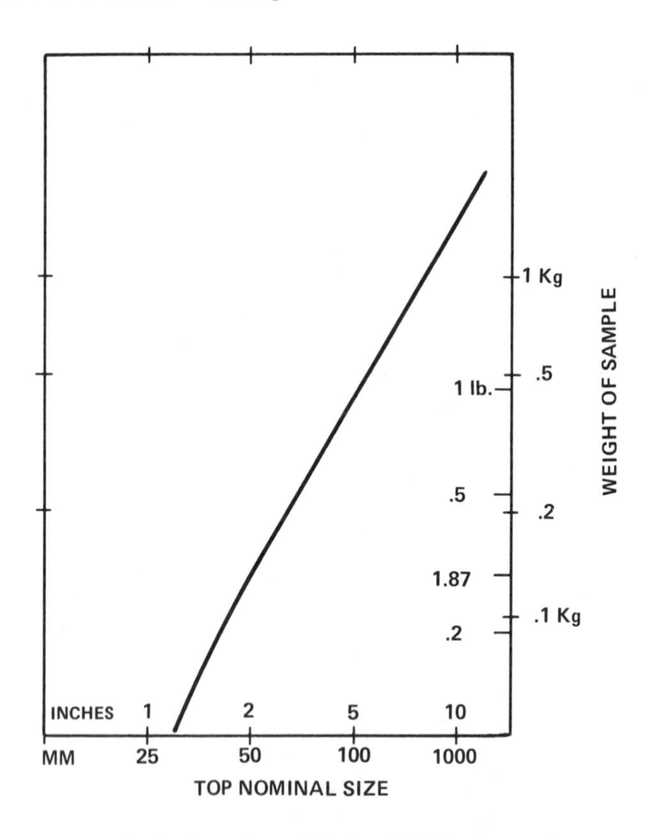

Figure 1-6. Sample weight for 20,000 pieces of paper.

Britton [10] investigated the influence of sample size on the standard deviation in measurement of component fractions of MSW and generated valuable data. A truckload of MSW, weighing 3000 kg, was divided into 66 samples of 44 kg each, which were sorted and the components weighed. The standard deviation was found by calculating the error of deviation from the mean (of the 66 samples) of each sample, and for groups of samples representing 88-, 130- and 260-kg mixed samples. The resulting coefficient of variation (CV) (calculated as standard deviation divided by the mean), is plotted versus total sample weight in Figure 1-7, which shows that the CV is reduced as the sample is increased, up to about 150 kg, after which there is little gain. Points of constant error are plotted as percent precision.

Note that the various components of the refuse have curves that are asymptotic to a value of CV and that above a certain sample weight the error is not reduced much or the precision cannot be improved. This appears to take place in the 50 to 150 kg range; hence, this weight has been selected as optimum for sampling raw MSW.

The highest CV is for wood, a relative rarity, and the lowest is for paper and food, the most common. Garden waste and textiles have high CV, whereas glass is low. There are two factors working here: (1) the probability of a given component being in the refuse, and (2) the concentration of that component in the sample. From

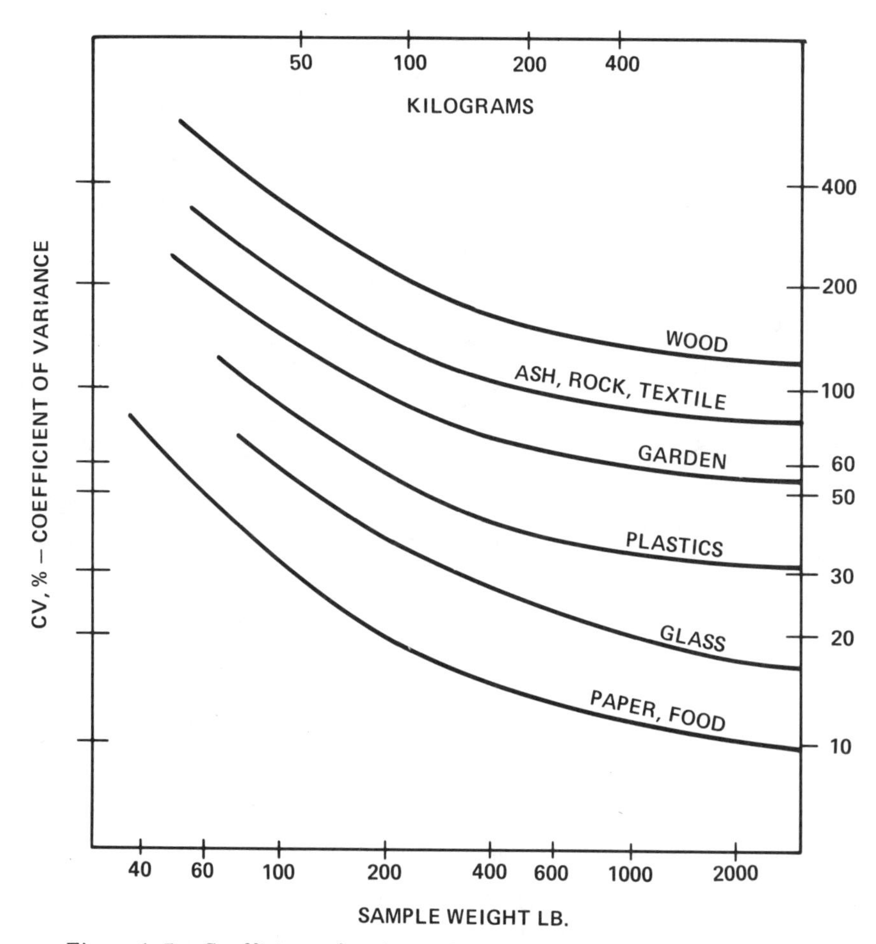

Figure 1-7. Coefficient of variation for various sample weights of MSW (Data from Britton [10]).

theoretical principles, the precision of sampling will be inversely proportional to the concentration of a component in the sample. This is shown in Figure 1-8. If the concentration of combustibles is 50%, 10 samples will give a precision of 10%, but the precision for ferrous metal would be about 30% and, for aluminum, 300%, or three times the standard deviation.

There is some choice whether 10 samples of 44 kg or 2 samples of 220 kg are taken, to obtain a precision of 10% for paper. If the 44-kg samples were not representative due to the large size of the MSW, larger samples would have to be taken. Britton recommended 90- to 130-kg samples of household waste and stated that 3 to 10 samples per 500 tons of refuse would be accurate enough. Commercial waste samples might have to be larger and shredded waste much smaller. It can be argued that the number of samples is not related to the 500 tons, but rather to the period of time over which sampling takes place. Based on coal sampling practice,

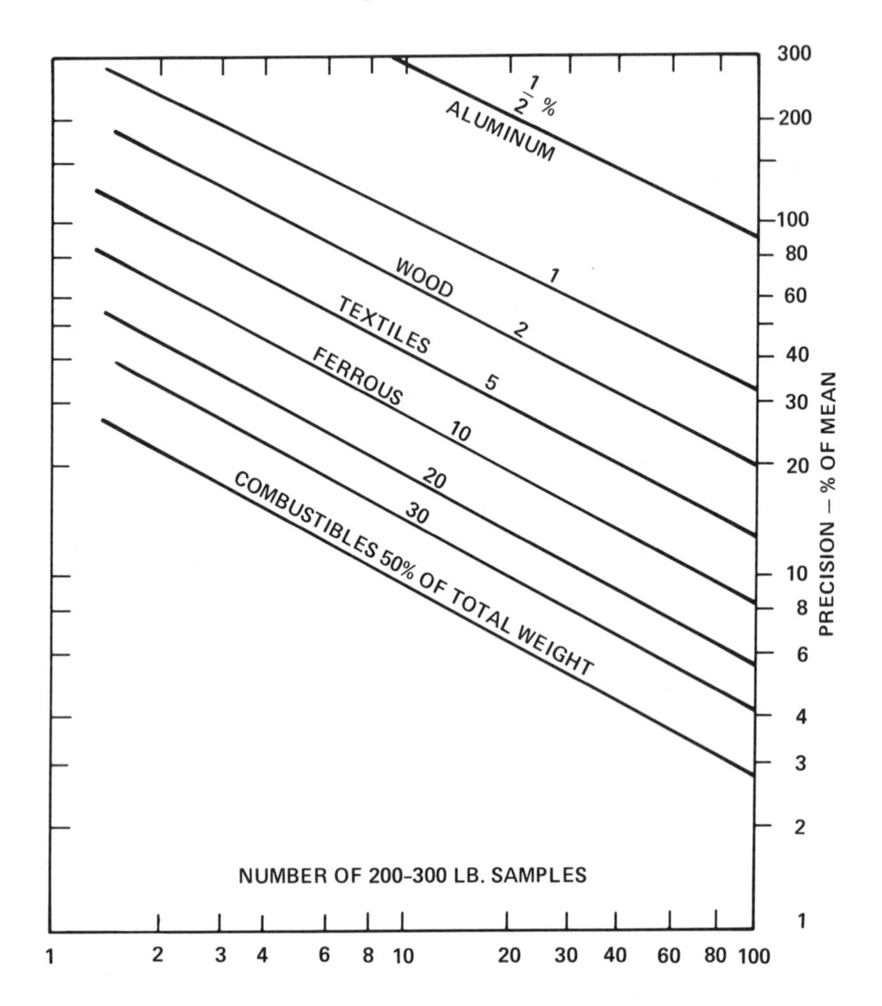

Figure 1-8. Precision versus number of 90–130 kg samples required.

testing should be done during two separate weeks and the results analyzed to determine the mean values and standard deviations of the tested properties or constituents.

A good argument can be made that the smallest sample should be as large as the normal container—hence, a garbage can or bag for household waste and a truck hopper load for a refuse collector or a crane bucket in an incinerator plant—because this is the true sample. The number of such samples would then depend on the variance of the material and reasonable precision levels (10-20% of components of interest at a 90% confidence level generally can be achieved).

One example of a thorough sampling program was the study at the Wayne County Municipal Incinerator, where one truck per day was sampled for five days of one week [11]. Each load was completely sorted to obtain the weight fraction of 16 components, including the significant number of disposable diapers. Component

Table 1-5. Moisture, Ash and Heating Value of Wayne County MSW [11]

Refuse Component	*Weight* (%)	*Moisture* (%)	*Ash* (%)	*Heating Value, Btu/lb*[a] (HHV1)	(HHV2)	(HHV3)	CONTRIB[b]
Newspaper	9.7	15.4	1.17	7,291	8,618	8,720	848
Other	28.6	18.9	6.46	6,455	7,956	8,505	2,432
Diapers	1.5	65.7	1.84	2,873	8,371	8,528	128
Textiles	5.4	13.2	2.07	7,403	8,531	8,711	474
Plastic Film	4.3	25.3	5.1	12,583	16,839	17,744	764
Plastics, Rigid	2.6	7.2	3.22	16,228	17,483	18,065	473
Food	4.1	86.1	6.47	1,116	8,002	8,556	347
Wood	1.7	16.6	2.80	7,108	8,521	8,766	150
Yard Waste	40.6	75.2	17.2	1,970	7,932	9,579	3,889
Sweepings	1.4	39.0	41.2	2,697	4,419	7,519	103
Mean (yard waste)		44.8	10.3	4,971	8,637	9,608	9,608
Mean (no yard waste)		24.0	5.5	5,052	6,952	9,028	9,628

[a]HHV1 = as-received
HHV2 = as dried
HHV3 = moisture- and ash-free
[b]CONTRIB is contribution of each component to HHV3.

samples were taken for laboratory analysis. Moisture was carefully monitored and corrections made to agree with truck weights. The results are summarized in Table 1-5. The heating value of each component was determined from ten samples, along with moisture and ash. Multiplying by the fraction of the component in the mixture, the heating values were calculated, as well as the contribution of each to the total.

The test, run in August, showed about 40% yard waste, not indicative of the annual average. This was subtracted out to derive an estimate of the winter season. It is notable that the yard waste had a very high moisture, hence, low heating value, making little contribution to the total heating value. As the other types of refuse are probably delivered at a fairly constant rate and the yard waste is a highly variable added quantity, the net result is a relatively small change in the heat delivered despite the variable weight.

Another study that generated valuable data was conducted for the Port Authority of New York and New Jersey by Woodyard [12] for a planned refuse-to-power facility in Port Newark to serve Essex and Hudson Counties. The weight flow was estimated by weighing 40-50 trucks per day for 12 days, from which the mean load densities were computed.

The average composition of the refuse was analyzed after a survey following the protocol developed by SCS Engineers for the U.S. Environmental Protection Agency (EPA) [12, 13], sampling newspaper, wood, plastics, glass, ferrous aluminum, corrugated, organics and -3/4-inch fines. Sorting was done on 80-135 kg (200-300 lb) samples. Ten samples per day were sorted by a crew consisting of two supervisors and

Table 1-6. Characterization of MSW by Sorting: Essex County [12]

	Residential, %			*Commercial, %*		
Constituent	*LCL*[a]	*MEAN*	*UCL*[a]	*LCL*	*MEAN*	*UCL*
Newspaper	5.21	6.27	7.33	1.61	2.96	4.31
Corrugated	8.24	11.60	14.96	15.03	23.21	32.39
Plastics	4.83	6.24	7.65	3.03	4.35	5.67
Organics	34.67	38.08	41.49	26.86	35.25	43.64
Wood	1.28	2.34	3.40	3.72	10.31	16.90
Ferrous	4.53	6.33	8.13	2.45	6.96	11.47
Aluminum	1.10	1.36	1.62	0.48	0.74	1.00
Glass	7.76	9.02	10.28	2.20	3.63	5.06
Fines	3.78	5.35	6.92	1.92	3.54	5.12
Miscellaneous	11.22	13.29	15.36	4.14	9.02	13.90

[a]LCL = lower confidence limit.
UCL = upper confidence limit.

six laborers. Trucks were selected for sampling at random and identified as to municipal, municipal contractor, private residential, commercial or industrial. The load was dumped on the floor and a sample taken from it by a front-loader and spread on a 20 × 20-foot area. Thirty-gallon aluminum cans, color coded, were positioned around the area, into which the sorters deposited the various species. The residue was screened and the overs sorted into the appropriate containers. Some of the results of this test are shown in Table 1-6, giving the mean and upper and lower confidence limits within which 95% of samples would be expected to lie. These limits correspond to about two standard deviations above and below the mean: the deviations actually are not symmetrical. The combustible components, including unknown moisture, comprised 64.5% of the mixed refuse.

Another test was run to determine the composition by analytical means, rather than by sorting. To do this accurately, it was decided to have samples shredded at the Monmouth County shredding plant. On one day in each of two weeks, one each of residential and commercial composite samples were prepared by combining 3-yd^3 samples. While these were being shredded, 50-lb samples were collected at 25, 50 and 75% stages of the shredding process. The three samples were mixed and spread over 16 squares. Eight samples of about 80 lb each were then taken for laboratory analysis. The results of proximate analysis by ASTM procedures are given in Table 1-7, with upper and lower confidence limits. An estimated PSD is also shown for comparison with other data. The residential waste has higher ash and moisture content than the commercial waste. These data can be considered to be quite typical of these basic waste types. If moisture and ash are removed from the analysis, the corrected volatile content calculates to 86.9% for the residential and 86% for the commercial refuse, and the carbon content of both types is about 15.5% of the volatile matter, indicating that the combustible material in both wastes is very similar. However,

Table 1-7. Analysis of Shredded Samples of MSW: Essex County [12]

	Residential Waste, %			*Commercial Waste, %*		
Property	*LCL*	*MEAN*	*UCL*	*LCL*	*MEAN*	*UCL*
Moisture	22.16	24.74	27.32	16.90	18.76	20.62
Ash	21.37	23.61	25.85	13.74	16.62	19.50
Volatile	42.09	44.89	47.69	51.97	55.77	59.57
Carbon	5.15	6.77	8.39	7.48	8.84	10.20
HHV1	4347	4898	5149	6031	6279	6527
HHV2	6271	6523	6775	7463	7727	7991
HHV3		8540			9267	
Combustible	76.4%			83.4%		

the moisture- and ash-free heating value of the residential was only 8540 Btu/lb, compared with 9267 Btu/lb for the commercial waste, indicating more oils and plastics in the commercial refuse. The commercial waste had 83% combustibles and 6279 Btu/lb as-received, compared with 76% combustibles and 4898 Btu/lb heating value.

Component and generic composition of the Essex County study, the results of a similar study by Pope et al. [15], carried out in Brooklyn, New York, and the results of the Wayne County study, are listed in Table 1-8 for comparison with RDF-3 data from St. Louis. While the components vary in relative quantity, the combustibles, metals and glass have less variation. Note that the commercial waste has about 10% more combustibles. Not listed, but important, is the lower moisture content.

ANALYSIS OF MSW AND RDF STREAMS

Extensive analysis of raw MSW has been done, as described above, both by sorting constituents and by shredding to obtain samples for laboratory analysis. Due to the effort necessary to carry out these analyses, they have been limited to one or two weeks in duration and to one or two times per year, thus giving little information on seasonal variation.

The material discharged by a primary shredder is a true sample of MSW, size reduced to such an extent that a relatively small number and size of samples are needed to achieve an accurate estimate. Even after magnetic separation, it has not changed its properties except that the troublesome ferrous metal has been removed.

The extensive testing of material streams that has been done at St. Louis [16] and the City of Ames [17] RDF processing plants has generated volumes of detailed analytical data on the properties of these streams, which reflect the MSW fed to these plants.

Table 1-8. Analysis of Municipal Waste and Refuse-Derived Fuel, Percentage Composition

Type MSW: Constituent	*Refuse from Brooklyn, Residential/Commercial [14]*		*New York Mix – 80/30 [15]*	*Essex County [12]*	*St. Louis RDF [16]*	*Wayne County [11]*
Paper	30.0	40.8	32.0	17.9	63.0	54.3
Plastics	8.1	11.6	8.8	6.2	6.5	8.5
Textile	5.3	2.7	4.8			5.9
Wood	3.5	5.2	3.8	2.3	2.3	1.7
Organics	24.7	18.0	23.4	38.0	1.0	5.4
Ferrous	11.0	6.6	10.1	6.3	0.1	5.1
Aluminum	3.8	3.3	3.7	1.4	0.6	0.3
Other Nonferrous	0	0.1			0.4	0.7
Glass	10.1	6.9	9.4	9.0	0.6	10.7
Sand	2.4	1.5	2.2	5.4		2.0
Miscellaneous	1.0	3.4	1.5	13.3	25.6	5.1
Combustibles	71.6	78.3	72.8	64.4	82.5	76.0
Metals	14.8	10.0	13.8	7.7	0.7	6.1
Glass and sand	12.5	8.4	11.6	14.4	0.6	12.7
Miscellaneous	1.0	3.4	1.5	13.3	5.0	5.1

Table 1-9. Variance Analysis of MSW from Eleven Cities [18]

(a) Bureau of Mines, College Park:

	Total MSW				*Combustibles Only*		
	X	s^2	CV		X	s^2	CV
Combustibles	76.2	29.0	7	Paper	67.2	143.0	18
				Corregated	9.4	48	74
Ferrous	6.2	1.1	17	Plastics	6.0	11.7	57
Nonferrous	0.15	–		Fabrics	4.5	35.3	132
				Wood	0.8	1.1	131
Glass	10.15	13.1	36	Yard	4.7	19.1	93
Grit and Dirt	6.5	13.7	57	Moisture	16.3	100.0	61
	100.00	60.00			100.0	259.0	
	s =	7.75			s =	18.95	

(b) Central Wayne County [11]

	As-Received			*Yard Waste-Free*	
	X	s^2	CV	s^2	CV
Newspaper	8.0	0.5	9	13.5	1.5
Other Paper	24.0	10.4	13	40.7	28
Diapers	1.2	0.2	34	2.0	0.5
Textiles	3.5	3.1	50	5.9	8.7
Plastic Film	2.9	8.8	30	4.9	2.2
Plastics, Rigid	2.1	0.2	20	3.6	0.5
Food	3.2	4.1	63	5.4	11.6
Wood	1.1	0.2	37	1.9	0.5
Yard Waste	41.0	14.4	9	–	–
Sweepings	2.3	0.8	39	3.9	2.3
Ferrous	3.0	0.6	26	5.1	1.8
Aluminum	0.2	0.1	122	0.3	0.1
Nonferrous	0.4	0.1	66	0.7	0.2
Glass	3.6	1.3	32	6.1	3.8
Brick	1.2	4.5	177	2.0	12.5
OBW	3.0	5.1	75	5.1	14.6
	100.0	46.3		100.0	88.8
	s = 6.8 × 1.24 = 8.4			s = 9.4 × 1.24 = 11.7	

(c) Union Electric, St. Louis [16]

	Hammermill Out			*RDF Product*		
	X	s^2	CV	X	s^2	CV
Paper	51.0	127.7	22	58.2	106.0	18
Plastics	4.5	7.7	61	4.9	12.3	72
Wood	3.8	13.8	98	3.4	5.9	71
Organic	6.3	37.7	97	4.7	26.1	109
Glass	3.2	7.2	84	2.6	3.4	71
Ferrous	5.6	7.8	50	0.3	1.1	350
Other Metals	0.6	0.1	53	0.5	1.1	210
Miscellaneous	25.0	70.0	33	25.4	50.3	28
	100.0	271.2		100.0	206.2	
	s =	16.5		s =	14.4	

Mean Values of Properties

Table 1-9 lists the weight percentages of the major constituents of MSW and RDF, as reported by various sources, comparing the composition and heating values of samples from St. Louis, Ames, RDF-4 (Bridgeport), the Bureau of Mines summary of analysis of MSW from eleven cities [18, 19] and the analysis of MSW delivered to the Central Wayne County municipal incinerator, reported by Hollander [11]. As terminology and test methods vary somewhat, some manipulation is required to put data of this sort on a common basis. For instance, as the Wayne County test, performed in August, showed 40% yard waste, this had to be corrected to 20% to match the average yard waste reported in the other data.

While there are significant differences in the amounts of various types of paper and plastics, the total amount of combustibles does not vary much, nor do the moisture and ash, in these averages of large numbers of samples. The Bureau of Mines analysis of eleven cities shows almost the same makeup as the other data, except for the relatively high glass content. We note the growth of plastics and variations in metals and glass, which reflect bottle bills. Other differences reflect localities, particularly different economic classes and urban/suburban differences, which affect the amount and nature of food waste.

While the average values of properties certainly have meaning from an overall point of view, the fluctuations or variations in the flow and properties of the material streams in hourly, daily and seasonal operation are major factors in the processing and combustion of MSW and prepared fuels. Therefore, we must study carefully the nature of these variations.

Variability of Properties of MSW

Sampling of MSW and its products reveals a continuous variation in its properties. Data obtained from analysis of samples taken several times per day and continuously over long periods of time give us information on the extent of these property variations.

A summary of mean values and the variability of properties of RDF analyzed in St. Louis and Ames is given in Tables 1-10 and 1-11. The St. Louis data were collected over an entire year, September 1974 to September 1975 [16], and again in the following year (preliminary data are also available from 1971 shredding without an air classifier). Included were complete data from 97 daily samples taken from the major streams and process products, which were analyzed intensively for species (paper, metals, glass, etc.), proximate and ultimate analysis, particle size, heating value, bulk density, elemental composition and ash fusion temperature. Similar data were collected at Ames over a full year, during 1976-77.

Standard Deviations of Refuse Properties

The Ames data are reported in Table 1-11 in terms of the mean value, X, and standard deviation, S_x, calculated from n samples. From this information, the coefficient of variation and the variability about the mean value at a 95% confidence level are

Table 1-10. Mean, Standard Deviation and Relative Standard Error of 97 Samples of RDF Collected at St. Louis in 1976

Property	*Mean*	*Standard Deviation*	*Relative Standard Error (RSE)*
Heating Value	10,636	1372	2.6
Bulk Density	109	24	4.4
Moisture	26.6	7.3	5.6
Paper	58.2	10.3	3.6
Plastics	4.9	3.5	14.3
Wood	3.4	2.4	14.7
Organics	2.6	1.8	21.3
Ferrous Metals	0.3	1.1	66.7
Other Metals	0.5	1.1	40.0
Glass	4.7	5.1	15.4
Miscellaneous	25.4	7.1	5.5
Ash	21.7	4.6	4.1
Fe	0.89	0.56	21.3
Al	1.64	0.78	16.5
Cu	0.04	0.67	50.0
Pb	0.05	0.03	20.0
Ni	0.02	0.02	50.0
Zn	0.07	–	–
Volatiles	43.6	5.1	2.3
Fixed Carbon	8.17	4.1	10.2
Carbon	26.0	2.8	2.2
Hydrogen	3.79	0.46	2.4
Oxygen	21.21	3.7	3.4
Sulfur	0.18	0.1	5.6
Nitrogen	0.53	0.1	3.8
Particle Mean Diameter	7.4	1.9	5.4

calculated. The number of samples used to obtain the mean and standard deviation influences the significance of the mean. When the number of samples is very large, at least 30, the calculated mean may be assumed to be very close to the "population," or "true," mean; likewise, S_X will be close to the population standard deviation. When smaller numbers of samples are taken or the samples are excessively small in weight, the calculated mean and S_X can be substantially different from the population values. This difference, or error in the estimation of the true values, is a function of the number of samples, or "degrees of freedom." The values of X and S_X for a small number of samples can be corrected so that comparison between data based on different numbers and sizes of samples can be properly compared. A simple procedure is available, using the Student's t, described in the Appendix.

Table 1-11. Variability of Daily Values of Characteristics of RDF Discharged from Storage Bin at City of Ames Plant

Characteristic (Percent)	$\bar{X}$ *Mean*	*Variability (about mean at 95% confidence coefficient)* (±)	S_X *Standard Deviation*	*Coefficient of Variation (%)*
Heating Value (kv/kg)	13,050	590	1021.6	7.8
Bulk Density (kg/m^3)	157	10.5	18.1	14.1
Proximate Analysis, %				
Moisture	23.0	2.4	4.2	18.3
Ash	17.4	1.8	3.2	18.3
Volatile	54.6	2.1	3.7	6.8
Fixed carbon	4.95	1.4	2.4	48.6
Ultimate Analysis, %				
Carbon	30.9	1.3	2.2	7.2
Hydrogen	4.8	0.2	0.3	6.8
Oxygen	22.9	2.3	3.9	17.1
Sulfur	0.43	0.1	0.2	44.8
Chlorine	0.24	0.06	0.11	45.1
Nitrogen	0.42	0.06	0.11	25.4
Ash Analysis, %				
SiO_2	48.19	2.3	4.1	8.4
Al_2O_3	11.75	1.3	2.3	19.5
Fe_2O_3	4.3	7.7	1.3	31.0
TiO_2	1.45	0.15	0.26	17.74
P_2O_5	0.79	0.16	0.28	34.9
CaO	12.71	0.93	1.7	12.7
MgO	2.38	0.18	0.31	13.1
Na_2O	4.37	0.35	0.60	13.7
K_2O	1.8	0.14	0.24	13.6
Ash Fusion, °C				
Reducing Atmosphere				
Initial Deform	1,106	72	42	3.8
Softening	1,143	31	18	1.6
Hemispherical	1,158	41	24	2.0
Fluidization	1,177	65	37	3.2
Oxidizing Atmosphere				
Initial Deform	1,149	41	23	2.0
Softening	1,165	50	29	2.5
Hemispherical	1,181	64	37	3.1
Fluidization	1,197	79	46	3.8
Particle Size				
Geometric Mean, mm	12	1.2	1.4	13.0

Standard Error

The standard error, E, corrects the standard deviation for the number of samples so that results of various sampling programs can be compared. The standard error is proportional to a probability function and inversely proportional to the square root of the degrees of freedom (number of samples minus one). For large samples, the function "Z" applies. For small numbers, especially under ten, the Student's *t* function gives more accurate estimates.

$$E = S_x \frac{t}{\sqrt{n-1}} \tag{1-1}$$

The relative standard error, E/X, can be used to compare data from various sources, in preference to the coefficient of variation, CV, which does not take into account the number of samples.

Confidence Interval

When relatively few samples have been taken, the error in the estimate is relatively large. By calculating the standard error, we can determine the range of values of the measured mean within which we can expect to find the true mean value. In 95 samples out of 100, we can be confident that the calculated mean value will be within a range of ± the standard error for that number of samples. This range is called the confidence interval.

As an example, we can take the Central Wayne County data which are based on five large (truck) samples, from which subsamples were taken to estimate the properties of this MSW. In this case, we have the opportunity to compare the real differences that exist between the five truckloads themselves. These are important data because they give an indication of the degree of variation to which a real plant is subjected as trucks come in, one at a time, with no mixing of their contents. As these trucks number only five, however, we do need to use statistical analysis to improve the estimate of the degree of variation that will take place as hundreds of trucks come in over days and weeks.

By calculating the confidence interval, the mean value can be compared with the means of other data. Also, a correction can be made to the measured standard deviation and the coefficient of variation to compensate for the error resulting from the small number of samples, giving an improved estimate of how wide the variations would be in an actual stream of this material and what fraction of the material is likely to exceed this estimate. For most types of variation in refuse, the mean is "central," that is, the range is roughly equal on both sides of the mean.

The range of the confidence interval is the standard error:

$$\text{Range} = \frac{S_x}{\sqrt{n-1}}t \qquad (1\text{-}2)$$

where n = the number of samples
t = the Student's "t," found from tables

As an example, at the Wayne County plant five sample trucks were sorted into components. From each of these sortings, five samples of each component were analyzed to obtain moisture, ash and heating value. The heating values were then calculated based on the weight composition of the five MSW samples, and the mean and standard deviation calculated.

The mean HHV1 was 4650 Btu/lb, with a standard deviation of 242. The CV was, therefore, 242/4650 = 5.2%. The value of t is found in the tables to be 2.776 for a 95% certainty and (5–1) degrees of freedom. The range or, in other words, the variability about the mean, is therefore $2.776/(\sqrt{4})$ times S_x, or 1.2415 X 242 = 300.

The mean HHV1 at a 95% confidence interval is, therefore,

$$\text{HHV1} = 4650 \pm 300 \text{ Btu/lb}$$

This means that it is estimated that 95 out of 100 samples will have a HHV1 within the range of 4350 to 4950. The relative standard error (RSE) is 300/4650 = 6.45%.

The analysis of 30 samples of Essex County residential waste gave a mean HHV1 of 4900 Btu/lb, with a standard deviation of 1075, and a standard error of 1075 X 2.04/5.477 = 400 Btu/lb, giving a range of 4500 to 5300. The CV was 1075/4900 = 21.9%, and the relative standard error was 400/4900 = 8.16%.

By comparison, 14 samples of RDF taken from the storage bin at Ames had a mean HHV1 of 6800 Btu/lb and a standard deviation of 456.6. Here, t = 2.2, and the standard error = 456.6 X $2.2/(\sqrt{14})$ = 268 Btu/lb. At 95% confidence level, the HHV1 was therefore 6800 ± 268, or the range was from 5632 to 7068 Btu/lb. CV = 456.6/6800 = 6.7%. The relative standard error was 268/6800 = 3.9%.

Variations in Heating Value

Table 1-12 shows the HHV of MSW determined by the Wayne County tests, compared with data from 97 daily samples taken during the St. Louis demonstration and 60 samples of Eco-Fuel from Bridgeport. HHV values are listed as the following: as-received (AR), dry basis (DB) and moisture- and ash-free (MAF), with the S_x values for each. There is a substantial difference in S_x for each type of HHV.

Table 1-12. Comparison of Heating Values and Standard Deviations of MSW, RDF-3 and RDF-4

	MSW		*RDF-3*				*RDF-4*	
	Wayne County		*St. Louis*		*Ames*		*Bridgeport*	
	X	S_x	X	S_x	X	S_x	X	S_x
As-received (HHV1)	4,648	421	4,573	589	5,103	447	7,828	319
Dry Basis (HHV2)	8,719	409	6,232	602	7,189	376	8,070	310
Dry Ash-Free (HHV3)	9,586	340	8,845	544	9,480	248	9,252	282
Number of Samples	25 Samples (from 5 trucks)		97 Weekly Composites		22 Weekly Composites		60 Trailers	

The relationships of the ranges at 95% confidence level of HHV1, HHV2 and HHV3 for several types of MSW and prepared RDF are given in Figure 1-9 showing the substantial improvement in variability that results from removing inert materials and moisture from MSW while producing a RDF. Values of CV are about three times the ranges shown.

Seasonal Variations

Figure 1-10 shows the substantial variation in as-received heating value (HHV1) of daily RDF samples collected at St. Louis during the course of an entire year. Both moisture and ash content exhibit a seasonal swing of seven to eight months rising and four to five months falling, causing a reverse variation in the heating value. The width of the band shows the range of measured values. Note that at any given time the range is less than the annual range. The seasonal variation of moisture ranges from 10 to 43%, or ± 16.5% about the mean of 26.5%, whereas at a time when the moisture averages 26.5%, the range is from 12.5 to 35%, or 26.5 ± 11.25%. This indicates that the annual S_x is about 45% greater than the momentary S_x.

The values of moisture- and ash-free HHV3 (Figure 1-11) also show a seasonal variation, despite elimination of the variations in moisture and ash. This variation is the result of increasing the ratio of yard waste to plastics yard waste, displacing the high heating value of the plastics component.

Figure 1-12 (top) shows the seasonal variation of monthly averages of HHV1, HHV2 and HHV3 for Monroe County. The variation of HHV1 is extremely small, indicating that the heating value of the combustible fraction is nearly constant.

Figure 1-12 (bottom) shows the high variability of ash content reported for Monroe County RDF. The data from 1981-82 has been repeated to emphasize the cyclical trend.

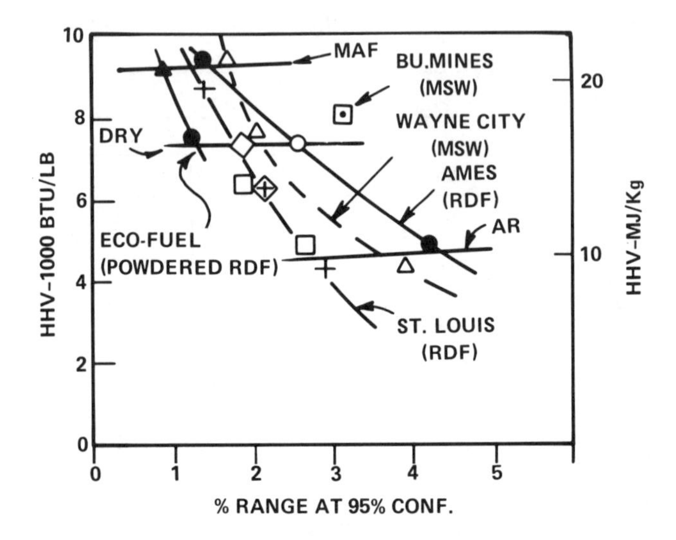

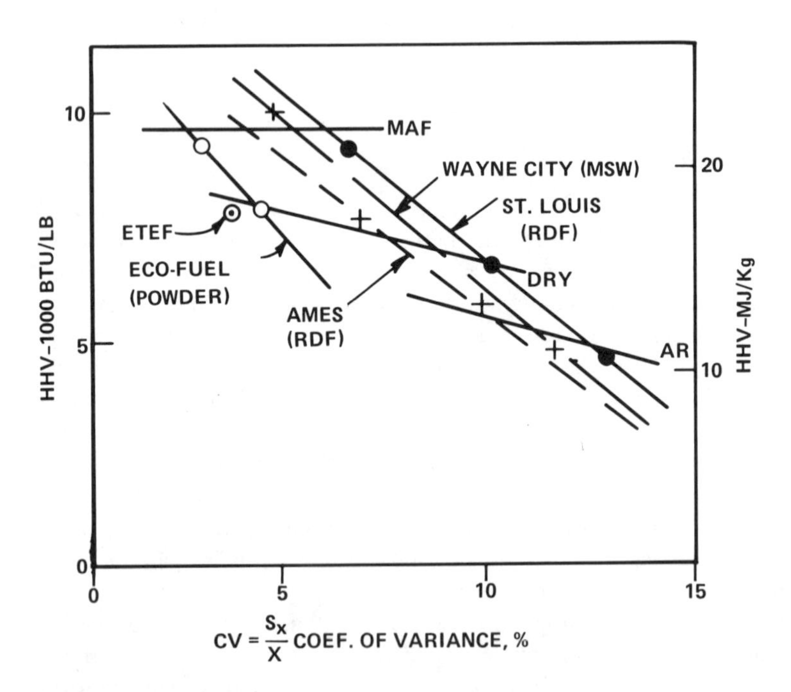

Figure 1-9. Higher heating value versus range and coefficient of variation for MSW and RDF.

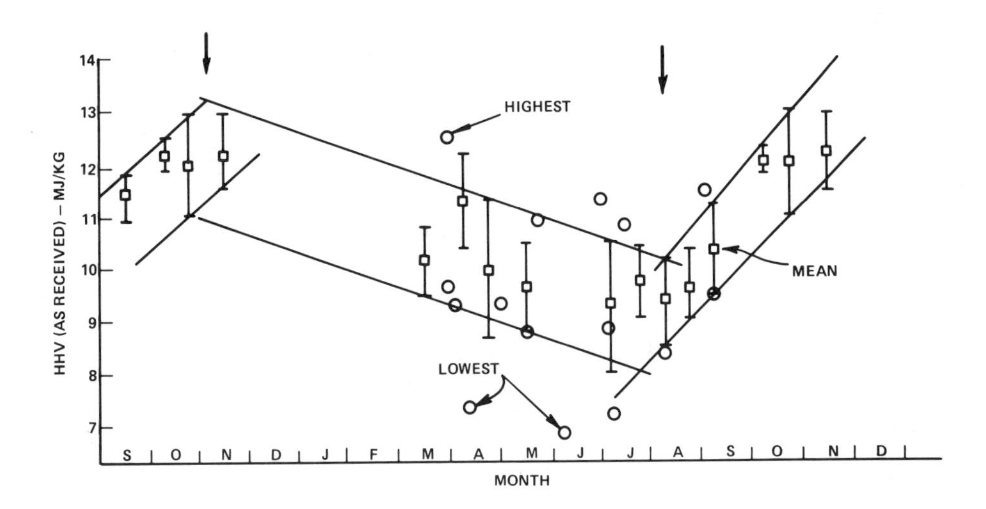

Figure 1-10. As-received higher heating value – monthly average and range of one standard deviation from the mean (St. Louis).

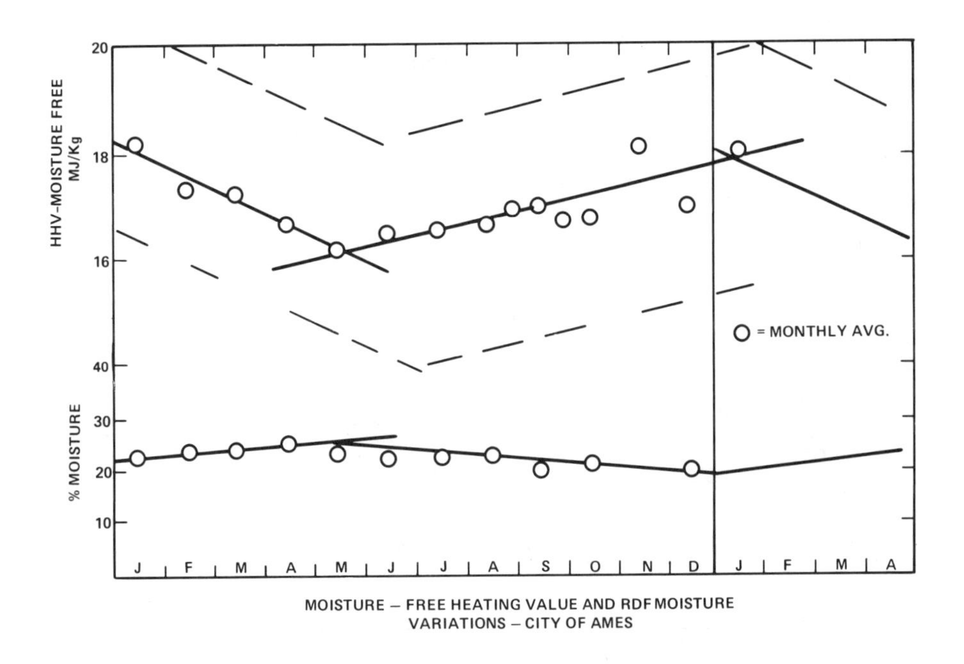

Figure 1-11. Moisture-free HHV and percent moisture variations (Ames).

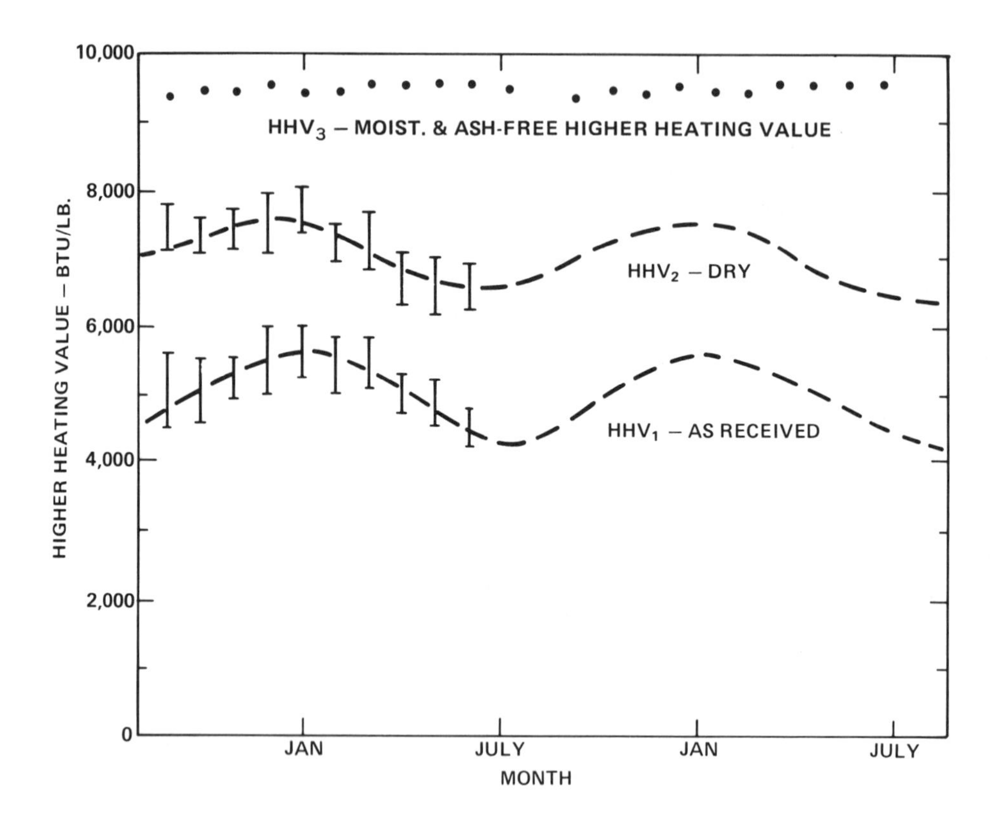

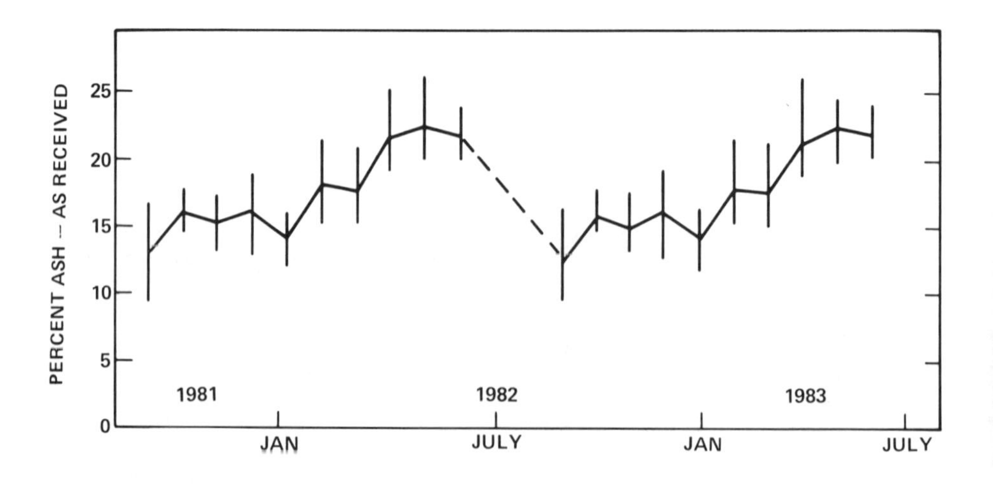

Figure 1-12. Moisture- and ash-free higher heating value monthly average and range of one standard deviation from the mean (Monroe County, NY).

The Effect of Mixing on Variance of RDF

Mixing, grinding or homogenizing would be expected to reduce the variances caused both by moisture and ash and by variations in plastics content. During collection, transportation, storage and retrieval, a great deal of mixing takes place that makes the moisture and other properties more uniform. Additional mixing naturally takes place as a result of handling the refuse, and as a necessary function of feeding MSW to a mass-burning incinerator, as well as processing it in an RDF plant.

Figure 1-9 shows the range in HHV1, HHV2 and HHV3 for various MSW and RDF types. As would be expected, the most highly processed RDF-4 (Eco-Fuel) has less variance than RDF-3 (St. Louis and Ames). The data for Wayne County show the same trends; however, being based on only two weeks, perhaps they should not be compared to the others.

To study the sources of variance in properties, it is convenient to plot one property against another to determine whether there is a correlation between them. Surprisingly, a strong correlation can be shown between HHV3 and ash, as shown in Figure 1-13. A similar, but weaker, correlation is found between HHV3 and plastics content.

A large number of sample data from Ames were found to correlate to a line showing a zero ash intercept of 7700 Btu/lb dry, which is roughly the heating value of cellulose. The mean values of large numbers of samples from Chemung County [20], Ames, St. Louis and East Bridgewater, each having a different ash level due to the MSW and the method of processing, all fall on the same line as determined by the Ames data. This is a remarkable correlation.

On the other hand, data from tests at the BuMines pilot plant [21] show a different relationship between HHV3 and ash for samples collected by the three cyclones. Cyclone #1, which has the lightest product, including thin plastics, has the lowest ash and highest heating value, and correlates to a line having the same slope as the Ames data. Cyclones #2 and #3 do not correlate as well, having more variable ingredients. Bridgeport Eco-Fuel, which has a primary trommel to remove glass, shows low ash and a relatively higher heating value.

There are several explanations for these correlations. First, the degree of processing determines the extraneous inerts content: most of the products have about the same range of heating values. Secondly, the plastics content influences heating value in a way that is related to inherent ash. As plastics have about double the heating value of cellulose and generally have a high inert content (fillers), they significantly influence the RDF heating value and ash content. An additional factor is the inherent ash of the paper types in MSW, which varies from 2% for newspaper to more than 20% for coated papers, which have a high variance. This accounts for the fact that even a finely ground and well-mixed powdered RDF retains a degree of variability, albeit small.

At this point these principles may be brought out:

- The seasonal variation in moisture and ash, mainly due to yard waste, greatly affects the heating value of RDF and is a major factor that must be taken

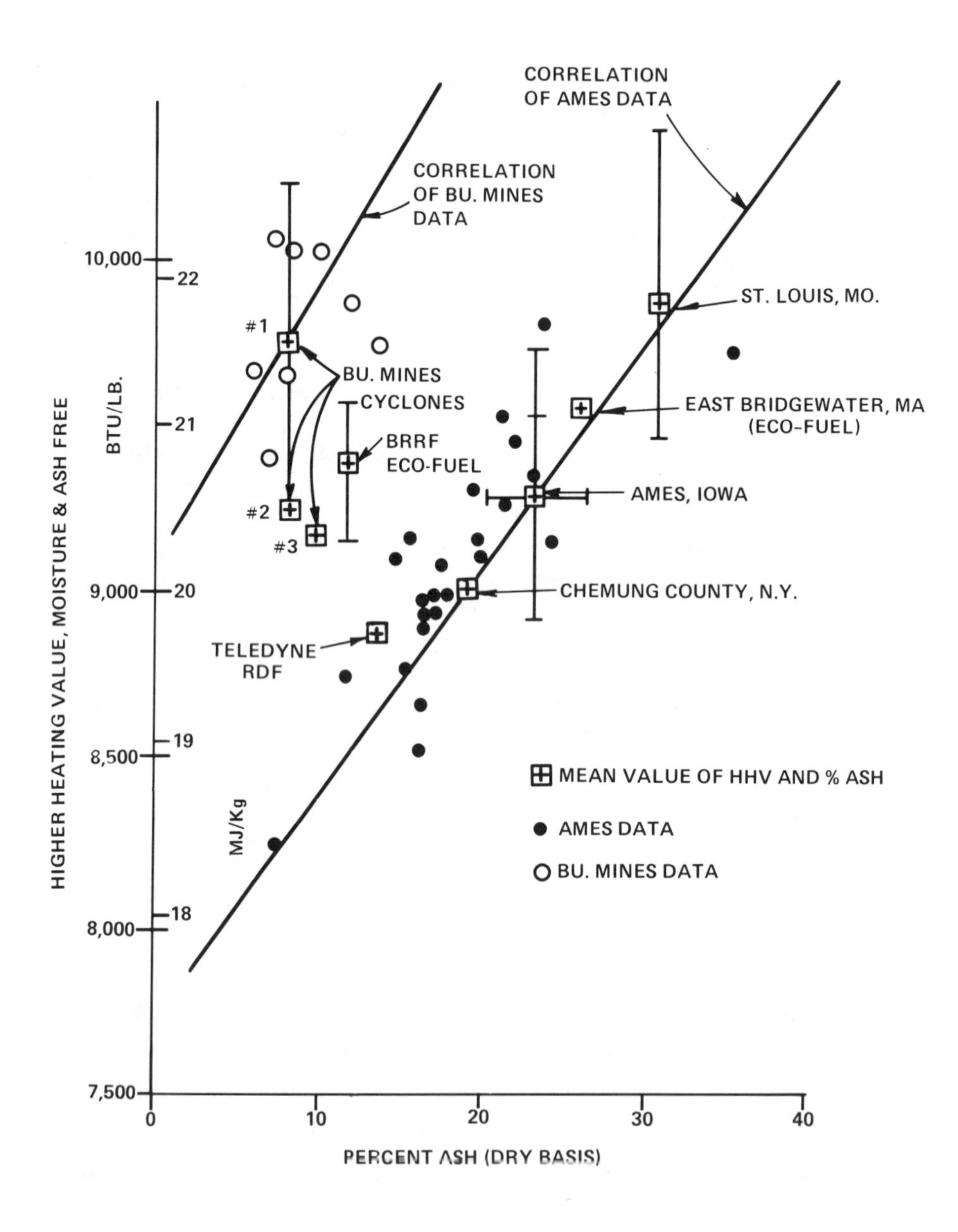

Figure 1-13. Correlation of moisture- and ash-free higher heating value versus percent ash (data from various sources).

into account when analyzing data. Moisture and ash content affect the processing and burning of MSW and RDF.

- Processing MSW reduces the variance of the product by physically removing or mixing the more variable elements.
- Drying MSW and RDF can remove the greatest component of variance.
- Ash is also highly variable and seasonal because yard waste contributes a major portion. Removal of the dirt associated with the yard waste would significantly reduce variance of a fuel product.
- Standard deviation is an important parameter in describing properties or mixtures of materials.
- Relative standard error can be used to compare data from tests using different numbers of samples.
- The square of S_X, called the Variance (VAR), is an additive property. When there are many components of a mixture, each having its own variance, the standard deviation of a property of the mixture can be found by taking the square root of the sum of the separate variances.
- The variances of the total mixture, the combustible content of MSW and the individual constituents of household refuse collected in different localities are remarkably similar, but great differences exist between household refuse and commercial and industrial wastes.
- The number of samples required to attain a given precision of estimate depends on the size of the particles and the concentration of the property of interest.

REFERENCES

1. Smith, F.A., and F.L. Smith. "The Quantity and Composition of Community Solid Waste: Comparative Estimates," USEPA USWMP (September 1973).
2. Niessen, W.R., and S.H. Chansky. "The Nature of Refuse," in *Proc. 1970 Nat. Incinerator Conf.*, ASME, New York (1970).
3. Hollander, H., ed. *Thesaurus of Resource Recovery Terminology* (Philadelphia: American Society for Testing and Materials (1982), STP 832.
4. Hecklinger, R. "Determination of Fuel Characteristics of RDF by Macroanalysis," *Proc. 7th Mineral Waste Symposium* (U.S. Bureau of Mines, 1980), pp. 84–90.
5. Kirklin, D.R., J. Colbert, P. Decker, S. Abramowitz and E.S. Domalski. "Test Procedures for the Determination of the Gross Calorific Value of Refuse and Refuse-Derived Fuels by Conventional and Large Bomb Calorimetry, 1979 Summary," NBSIR 81-2278, National Bureau of Standards, Washington, DC (1981).
6. Domalski, E., K.L. Churney, M.L. Reilly, D.R. Kirklin, A.E. Ledord and D.D. Thornton. "25 Gram Capacity Combustion Flow Calorimetry," NBSIR 82-2457, National Bureau of Standards, Washington, DC (1982).
7. Keator, F.W. *Mechanical Engineering Laboratory Methods* (New York: D. Van Nostrand Co., 1947).
8. Wilson, D.L. "Prediction of the Heat of Combustion of Solid Wastes from Ultimate Analysis," *Environ. Sci. Technol.* 13 (June 1972).

9. Kaiser, E. "Chemical Analysis of Refuse Components," in *Proc. 1966 Nat. Incinerator Conf.* (New York: American Society of Mechanical Engineers, 1966).
10. Britton, P.W. "Improving Manual Solid Waste Separation Studies," *J. San. Eng. Div., ASCE* (October 1972).
11. Hollander, H.I., J.K. Kieffer, V.L. Eller and J.W. Stephenson. "A Comprehensive Municipal Refuse Characterization Program," in *Proc. 1980 Nat. Waste Processing Conf.* (New York: American Society of Mechanical Engineers, 1980).
12. Woodyard, J.P., J.C. Anderson, M. Neisser and S.S. Passage. "Estimating Solid Waste Quality and Composition," in *Proc. 1982 Nat. Solid Waste Processing Conf.* (New York: American Society of Mechanical Engineers, 1982).
13. SCS Engineers. "Municipal Solid Waste Survey Protocol," Contract 68-03-44-2486, MERL, U.S. EPA, Cincinnati, OH (1979).
14. Klee, A.J. *Quantitative Decision-Making*, Vol. 3 in the series *Design & Management for Resource Recovery*, P.A. Vesilind, Ed. (Ann Arbor, MI: Ann Arbor Science Publishers, 1980).
15. Pope, Evans and Robbins, Consulting Engineers. "Study of Municipal Solid Waste Quantity, Composition and Heating Value," Port Authority of New York and New Jersey (January 1979).
16. Fiscus, D.E., et al. "St. Louis Demonstration Final Report," EPA-600/2-77-155a, MERL/ORD/U.S. EPA, Cincinnati, OH (1977).
17. Even, J.C., S.K. Adams, P. Gheresus, A.W. Joensen, J.L. Hall, D.E. Fiscus and C.A. Romine, "Evaluation of the Ames Solid Waste Recovery System, Part I," Engineering Research Institute, Iowa State University, Ames, IA (1977).
18. DeCesare, R. "Pilot-Scale Studies on the Composition and Characteristics of Urban Refuse," U.S. BuMines RI 8429 Washington, DC (1980).
19. Sullivan, P.M., M.H. Stanczyk and M.J. Spendlove. "Resource Recovery from Raw Urban Refuse," RI 7760, U.S. BuMines, Washington, DC (1972).
20. Lawler, S.P. "A Statistical Analysis of the Chemung County Americology Air Classifier on Tests Made October to November, 1975," American Can Company, Greenwich, CT, unpublished results.
21. Schultz, H., P.M. Sullivan and F.E. Walker. "Characterizing Combustible Portions of Urban Refuse for Potential Use as Fuel," RI-8044, U.S. Bureau of Mines, Pittsburgh, PA (1975).

CHAPTER 2

Transport, Storage and Handling

In the process of disposing of refuse, it must be collected, transported, handled, stored and retrieved. To facilitate collection and transportation, it is generally compacted into compactor trucks, which transport it to landfill or a facility for processing or burning. Transfer stations may be interposed between the collection trucks and the disposal facility.

Within the facility, the refuse must again be moved, either to storage or to feed the process or furnace. The refuse and the processed materials are handled by various types of machines and conveyors, which move it between storage and processing operations. Special equipment is required to retrieve these materials from storage and to feed them to processing operations and combustion devices. Processing systems must be fed properly to assure good operation.

The size of containers that can be used for transport is limited by the weights that can be transported in accordance with local regulations and by permissible or practical dimensions, which generally cannot exceed 8 feet in width, 30-40 feet in length and 10 feet in height.

The ratio of weight to volume, or density of MSW can be decreased by a ratio between three and five times its free density before the container or vehicle weight reaches the practical or legal transport weight.

The rate at which MSW can be transported depends on the weight of the container and the time required for the vehicle to collect or pick up a load, deliver it and return.

When storing MSW, the volume of the storage required depends on the rate of storage, the storage time required and the rate of retrieval, as well as on its density. Stored in a pile, the density of MSW can increase by as much as two times. When retrieving it from a pile, its density will be reduced by a factor of two or more.

When moving MSW and its products, the area needed for transport depends on the density of the material and the velocity at which the transport equipment can move it. This, in turn, depends on the particle size and nature of the material.

In summary, the density, volume and time or velocity control are required so that the dimensions of plant and equipment required to store and handle the refuse and its products can be sized properly. These four parameters can be defined as

$$\text{Storage volume} = \text{length} \times \text{width} \times \text{height}$$
$$V = L \times W \times H$$
$$\text{units} = \text{cubic feet } (ft^3), \text{ cubic meters } (m^3), \text{ cubic yards } (yd^3)$$

Storage weight = volume $\times$ density
$W = V \times d$
$W = L \times W \times H \times d$
units : pounds (lb), kilograms (kg), tons (2000 lb), tonnes (1000 kg)

Flow (volume) = (flow cross-sectional area) $\times$ velocity
$Q = A \times V$
units : cubic feet per second (ft^3/sec), cubic meters per second, (m^3/sec), or per minute (m^3/min)

Flow (weight) = (flow volume) $\times$ (density)
$W = Q \times d$
$W = A \times V \times d$
units : pounds per second (lb/sec), kilograms per second (kg/sec), tons per hour (TPH), tonnes per hour (tph)

PROPERTIES OF MSW AND ITS PRODUCTS THAT INFLUENCE HANDLING

The following properties of MSW and its products influence the design, dimensions and performance of the equipment that can be used to handle them:

- Density: free-fall, loose, compacted
- Compressibility and rebound
- Particle size and shape
- Flowability
- Stickiness
- Abrasiveness
- Lint and dust content
- Tendency to mat
- Angle of repose
- Angle of slide
- Presence of strings, rags and wires

The bulk density is the weight of material, in a free form, contained within the given volume. This is many times less than the true density due to the large fraction of voids between the solid materials. For instance, metal cans, plastic and glass bottles and cardboard boxes have true densities from 70 to 300 lb/ft^3, but bulk densities of less than 2 lb/ft^3 and voids of more than 95%.

Due to the large void volume of refuse, it is economically essential to compact refuse. As soon as pressure is applied, the cardboard boxes and plastic bags yield and their void spaces are reduced. Additional pressure breaks bottles and flattens plastic and cardboard containers. Some of the compaction achieved will be lost on release of pressure.

The size of the particles determines the size openings that must be provided for feeding and retrieving materials. MSW can contain particles up to 8 feet in length and 4 feet in width. Shredded MSW can be handled by much smaller cross sections, as small as 6 inches in diameter.

The flow characteristics determine whether the material will slide down chutes or whether it will tend to jam whenever it has to pass through converging passages. Materials that are stringy or contain rags or wires are likely to cause trouble when handled by rotating equipment, which will tend to wind these materials around shafts. Materials that are abrasive will tend to wear equipment, especially where high velocities are used and direct impact takes place. Linty materials tend to plug screens and cause housekeeping problems. Dusty materials create environmental and safety problems.

The capacities of transport, storage and handling equipment are directly sensitive to the densities of the materials involved. Figure 2-1 shows the approximate ranges of densities of municipal solid wastes in the raw form, compacted, released and stored. The raw waste can be compacted into truck bodies, raising the density from 100 to 300 lb/yd^3 to anywhere from 400 to 900 lb/yd^3. When released from the compactor it tends to relax to about 350 to 400 lb/yd^3. In storage it can reach up to 900 lb/yd^3, averaging closer to 600 lb/yd^3.

Figure 2-2 shows the effect of the height of storage on the average density of municipal refuse and the local density of the refuse at various heights, as determined by Scott and Holmes [1]. Note the linearity of the curves on logarithmic coordinates. By storing 4 meters deep, the average density is increased from 200 to 300 lb/yd^3; increasing the depth to 15 meters only increases the average density to 350 lb/yd^3.

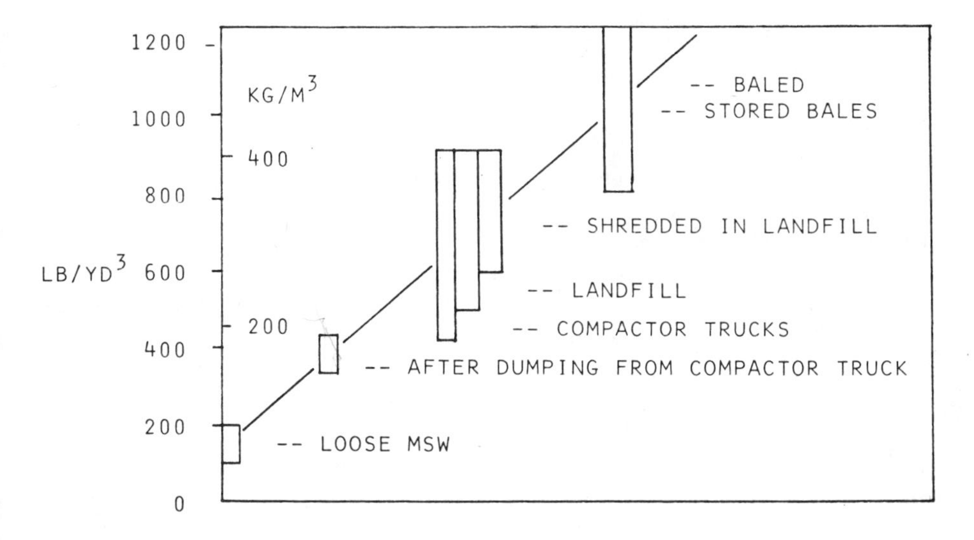

Figure 2-1. Densities of raw and compacted MSW.

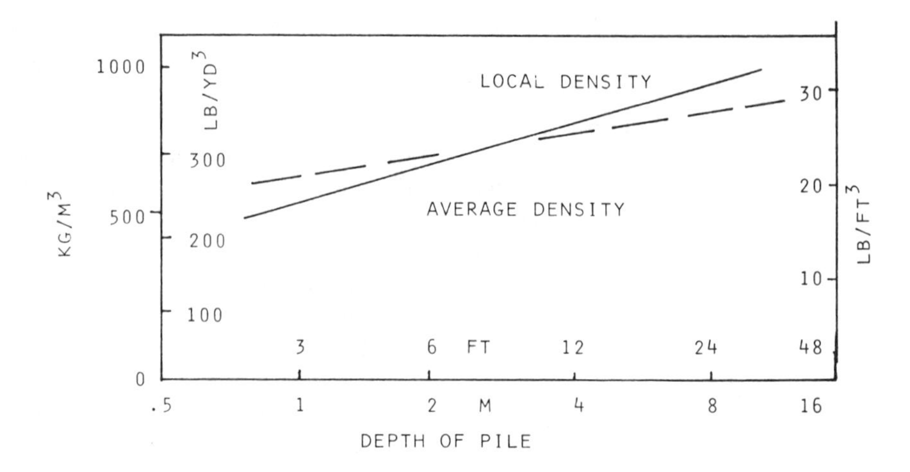

Figure 2-2. Densities of MSW as they vary with pile depth [1].

The influence of compaction pressure on refuse density is shown in Figure 2-3, along with points showing typical ranges achieved by front-, rear- and side-loader trucks. Transfer trailers can be loaded to roughly the same densities, depending on the material.

Stored refuse has essentially no "flowability." When released, especially when pushed out of a compactor body, it expands somewhat but can retain a vertical angle of repose. To retrieve it from storage and move it, it is necessary to expand it into a flowable condition, which is in the range of 10 to 15 lb/ft^3 or 275 to 400 lb/yd^3 (150 to 250 kg/m^3).

Discharging refuse directly from a compactor truck may not release the density much. Allowing it to fall freely will, on the other hand, release it to a loose condition.

COLLECTION OF MSW

The process of collecting municipal refuse can be analyzed by separating activities into unit operations. Models can be developed that make it possible to optimize the entire process. Once the operations have been identified and evaluated, the necessary labor and equipment can be determined for given waste quantities, locations, pickup, haul and disposal site conditions. The alternate ways to handle the collection process can be considered and evaluated to determine the optimum method.

The effectiveness of the process can be expressed as the ratio between the minimum cost of operation divided by the sum of the minimum cost and the cost of the various inefficiencies that have been identified. When costs can be reduced to labor time, cost of labor per hour and the daily cost of equipment and mileage, the analysis can be reduced to a study of time, taking into account the weight of the material handled.

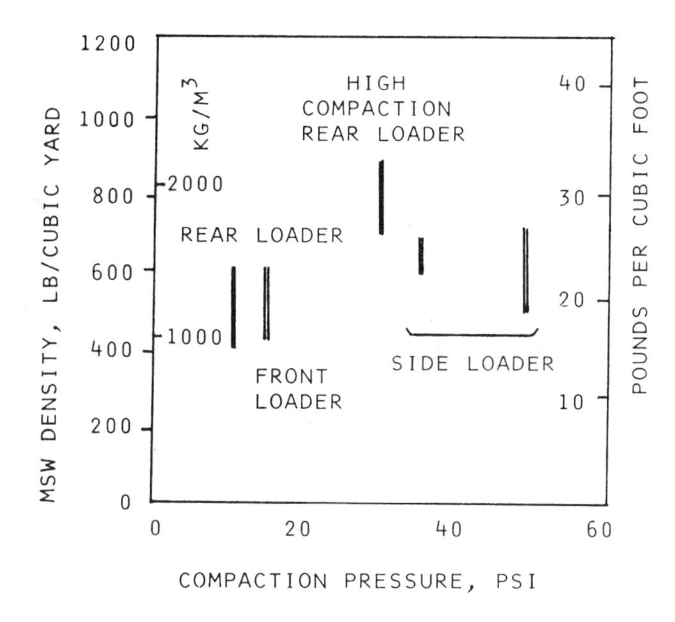

Figure 2-3. Effect of compacting pressure on refuse density.

The quality of the data collected will determine the value of the results of the studies. Rough estimates can be very useful to put together the overall picture, from which the major factors can be discerned. After the preliminary picture has been evaluated, it becomes more evident where more exact data are required.

The distance between the points of collection and the disposal site, the amounts of waste picked up along the route and accumulated until the vehicle is full, the time required for the round trip and the number of workers required determine the cost of collection. The system must be able to handle the peak loads by such methods as using all the equipment and working overtime, storing the waste and diverting the excess to landfill.

Wastes are generally handled by a one- to three-man crew, who lift or hoist it into side-, rear- or top-loaded collection vehicles. Hydraulic systems are used for loading and compacting the refuse to achieve maximum weight capacity.

Municipal wastes are generally picked up at street curb or alley and dumped into a hopper, which is hydraulically discharged into a "packer truck" body. Industrial or commercial refuse may be manually or mechanically loaded; the driver may need a helper to bring the loaded containers to the vehicle and return the empties. In the hauled container system, the collector may have a helper to attach and detach chains and cables used in loading and unloading the containers.

Side- and rear-loaders, which range up to 32 yd^3 in capacity, are generally used for household collection and can compact the refuse to 500–900 lb/yd^3. Commercial packer bodies may be rear- or front-loaders holding up to 50 yd^3.

Stationary compactors, which have capacities up to 75 yd^3, compact the refuse as it is received and then load transfer trailers having capacities up to 65–80 yd^3. The latter have bodies 40 feet long, 8 feet wide and 73 feet 6 inches high (the maximum practical size for shipping) and payloads up to 25 tons. They need heavy-duty rigs to handle them, which are equipped with hydraulic drives to discharge the trailers at the dump site, using push plates or live-bottom drag chains.

In stationary container systems, the waste is dumped in relatively small increments into loose containers or compactor bodies, where it is stored until collected. At that time it is dumped into a top-loader, which transports it to the disposal site.

In the case of hauled container systems, each container picked up has to be hauled to the disposal site after dropping off an empty one.

The efficiency of a collection system can be evaluated by comparing the actual with the ideal performance. Considering time as the major factor, the efficiency can be expressed as the theoretical time divided by the actual time, as follows:

$$\text{Efficiency} = \frac{\text{Ti}}{(\text{Ti} + \text{Wn} + \text{Wun})}$$

where Ti = round trip time to the individual spot

Wn = necessary nonproductive time, such as checking in and out, driving to the first and last pickup points, traffic congestion, equipment downtime for repairs, fueling and maintenance

Wun = unnecessary time for unauthorized coffee breaks, etc.

When weights are known, the efficiency can be expressed as the weight actually carried divided by the weight that potentially could be carried.

In container systems, the container utilization factor, F, the ratio of actual capacity to "rated" capacity, usually expressed in volumetric units, can be included:

$$\text{Efficiency} = \frac{\text{F} \times \text{Ti}}{(\text{Ti} + \text{Wn} + \text{Wun})}$$

The utilization factor introduces other considerations, such as the size of container supplied as it relates to the amount of refuse delivered to it and the regularity of daily delivery. Leaving a larger container and collecting it less frequently so that it can be collected more nearly full allows the utilization factor to increase at the expense of tying up more capital in the container and making the pickup schedule more complicated.

The number of trips per day, the number of men per vehicle, and the size of the containers and vehicles contribute to optimizing the collection system design.

The number of people served per cubic yard of truck capacity, per truck and per worker is shown in Figure 2-4, based on these data [2]. A city with a population of 200,000 seems to achieve the maximum efficiency, with a drop-off to a lower level for larger cities. The number of workers per truck reaches a minimum of 2.4 at the peak, increasing and decreasing with larger and smaller cities, as the result of using the size and type of trucks that are best suited to the city.

Note that with the geometric (logarithmic) scales used for the population axis, the number of people served per truck and per crew member is a linear function, reflecting the fact that people occupy an area, whereas trucks travel a linear distance.

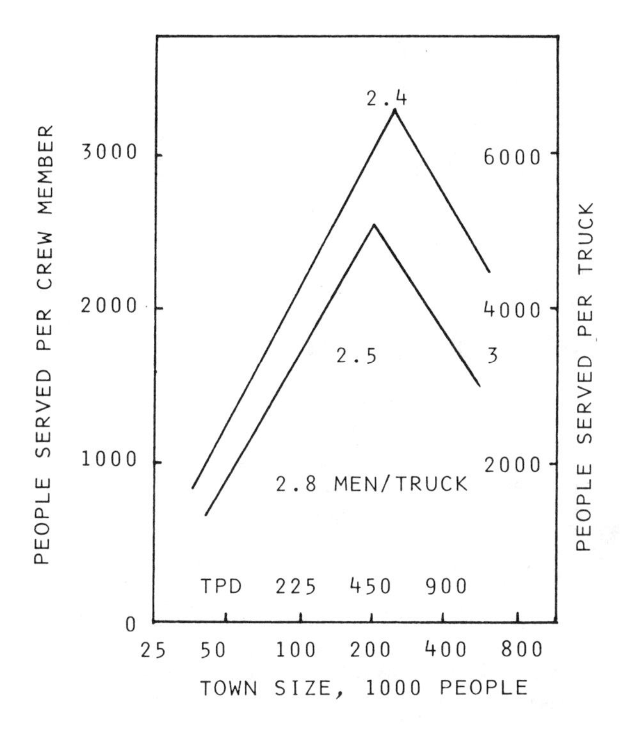

Figure 2-4. Relationship between crew size and population served [2].

In small cities the population density is low, requiring long runs between houses. As the density of the city increases, the runs shorten between houses and the pickup per mile traveled increases. However, ultimately the trip to the disposal site increases, causing a loss of production.

Side-loading trucks can be handled by three to seven men, whereas rear-loaders are generally served by two to four men. Figure 2-5 shows the trend observed in the percentage of trucks having various crew sizes. Side-loaders usually have a single worker, as compared with three for rear-loaders.

The percentage use of rear-loaders and side-loaders is shown in Figure 2-6. Side-loaders increase to 25% at a population of 750,000 but fall off in larger cities. As the city increases in size, the percentage of 20-yard side-loaders decreases while the percentage of 25- and 30-yard trucks increase, crossing at a population of 725,000 and reaching 700% at a population of 400,000. Twenty-yard rear-loaders decrease in number to 50% at a population of 700,000, but peak at 75% at a population of 400,000, above which 25- or 30-yard units take over. The optimum combination seems to take place at a population of 200,000.

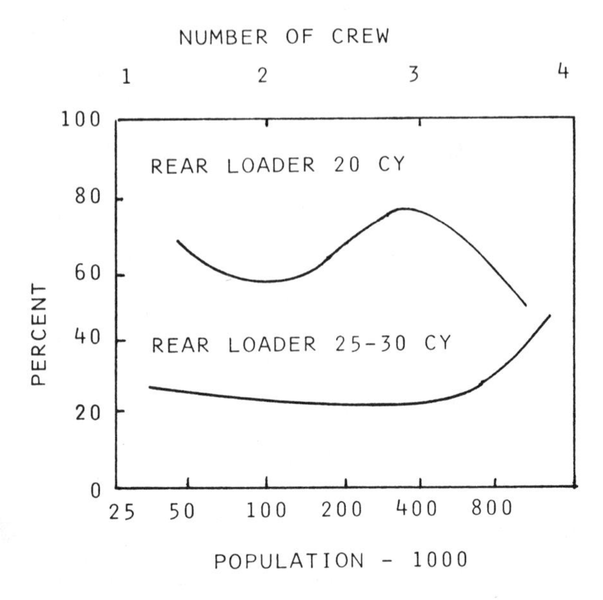

Figure 2-5. Crew size and type of truck used by various towns.

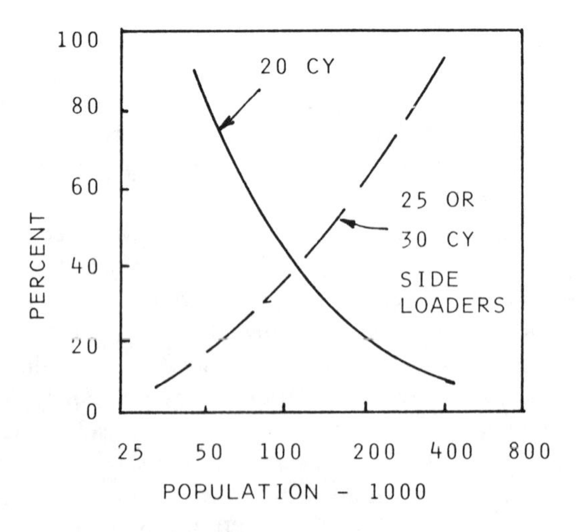

Figure 2-6. Percent of side- and rear-loaders used.

Estimating Solid Waste Quantity and Ton-Miles to Disposal

A study performed by SCS Engineers for a proposed Essex County, New Jersey energy recovery facility illustrates a number of principles that apply to refuse collection in urban areas [3].

The refuse produced by twenty towns in Essex County was measured as it was brought to several landfill sites in Hackensack Meadows, close to the planned processing site at Port Newark. The number, volume and origin of randomly selected trucks were recorded by observers for a period of two weeks. Randomly selected trucks were weighed to obtain the weight per cubic yard of truck volume.

The average production of refuse per capita was found to be about 4.5 lb/day per person, with a standard deviation of over 25% between towns. A population of 800,000 people produced about 7800 (TPD) on average. During each week, the daily capacity had a relative standard deviation of about 30%. The weekly averages varied from 7350 to 7520 TPD, with daily peaks over 2000 TPD.

Figure 2-7 is a map of the area showing the daily tonnage collected from each town. Essex County is pie shaped, with the processing plant located at the apex, not at the center. Such a configuration is not unusual for counties associated with populous urban areas. Essex County encompasses about one-quarter of the total area, two quarters being counties and one water. This configuration results in the tonnage being a fraction of that which could be collected in a more ideal circular area with a central "downtown" processing plant.

By tabulating the distances of the towns from the Port Newark site, the daily tonnage of the cities, the accumulated tonnage, the ton-miles and, finally, the miles per ton, graphs can be drawn showing these relationships with distance from the plant.

Figure 2-8 shows how the accumulated amount in tons per day falls off versus distance. A radius of 13 miles is necessary to collect 1800 TPD, whereas 1000 TPD is collected at half that radius.

Figure 2-9 shows the miles per ton versus the capacity of the disposal facility. Note that the zero intercept is 3 miles and that this has doubled to 6 miles at 7500 TPD.

Variation in Daily, Weekly and Annual Refuse Disposal

The design of a disposal facility requires knowledge of the hourly, daily, weekly, monthly and seasonal variations in refuse intake, as well as the total annual quantity.

Figure 2-10 shows the variations reported throughout an entire year of operation of the East Bridgewater, Massachusetts plant, serving Brockton and a number of other towns. The monthly receipts varied from 6200 to 9200 tons, averaging 250-370 TPD, a variation of ± 20% for months. February was low, and peaks occurred in May and October, representing yard waste peaks, while tire inspections resulted in large quantities of tires.

In January, the daily total ranged from 750 tons to 500 tons, with a three-day average of 450 tons, which was the ultimate capacity of the incinerator. In May, a peak of 550 tons was recorded for one day, with a three-day average of about 420 TPD.

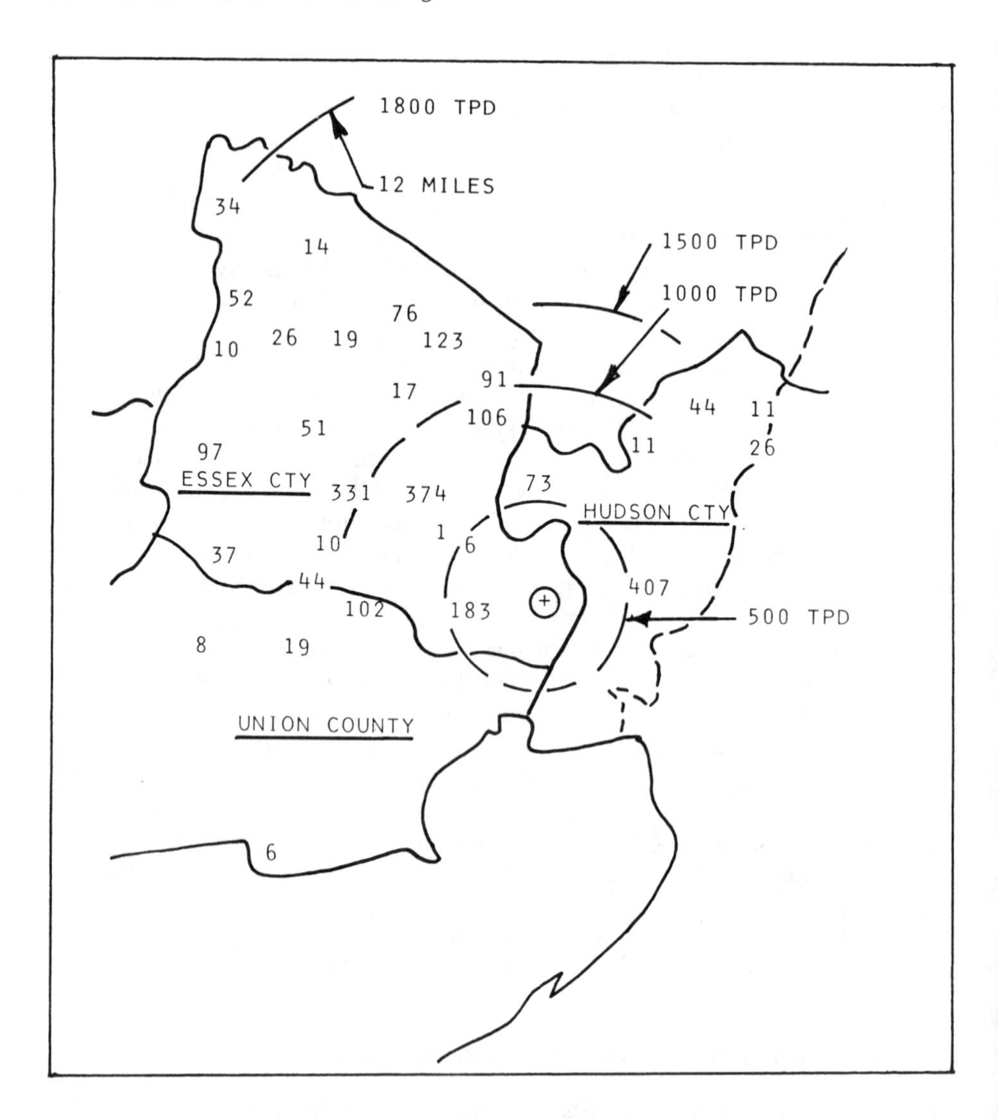

Figure 2-7. Map of Essex County study area, showing daily tonnage collected from each town.

Data on daily capacity can be plotted as shown in Figure 2-11, in cumulative and histogram form. The mean value is about 350 TPD, with a 550 TPD maximum for one day. Leaving out five days brings the capacity down from 500 TPD to 400 TPD.

Transfer Stations

Transfer stations are used to receive refuse from packer trucks and compact it into transfer trailers to reduce the cost of transportation to distant landfills or processing

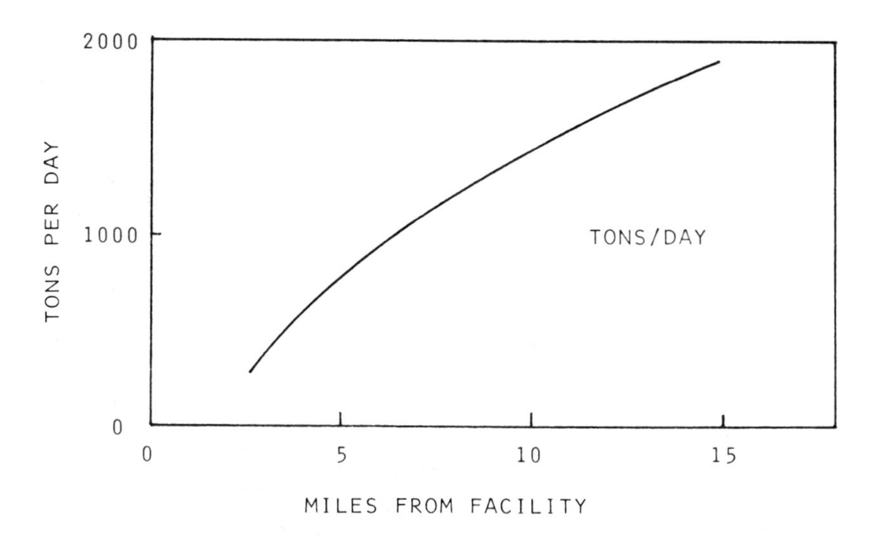

Figure 2-8. Increased TPD handled as radius of service is increased (Essex County study).

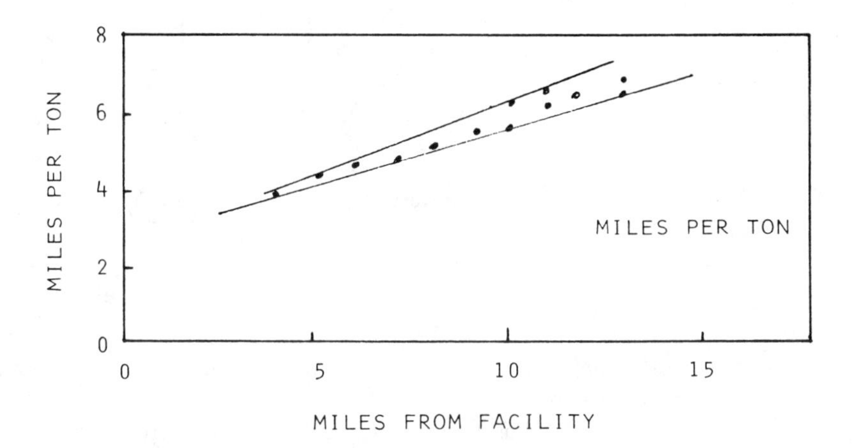

Figure 2-9. Increase in miles traveled per ton of refuse with the increase in radius of service (Essex County study).

plants. They may be constructed for the purpose or may be converted from obsolete municipal incinerators, which could not meet current codes and, hence, have been shut down. Transfer stations often fit into resource recovery plans.

In a transfer station, one or more compactors is installed to receive the refuse. The refuse carrier may dump it on the floor for a loader to move, or it may be dumped directly into the compactor, which delivers it to a transfer trailer or container, or

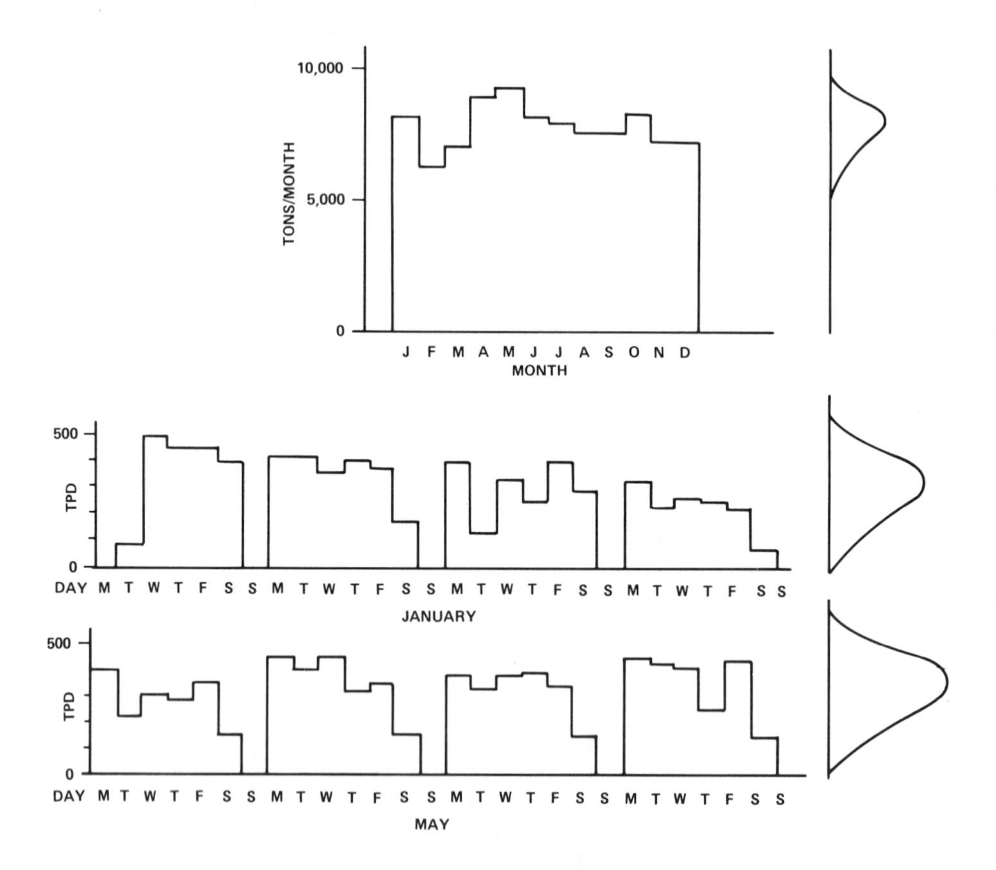

Figure 2-10. Variation of refuse delivered to disposal site (data from East Bridgewater, MA).

directly to an open trailer. As each transfer trailer or container is filled to allowable weight (20 tons or other amount, depending on local regulations) it is hauled away and an empty container put in its place.

Transfer trailers can be used as a means of storage, thereby permitting a steady rate of delivery to the disposal site, even though the compactor operates at higher and lower rates during the day.

Studies of the pickup patterns of the carriers should precede the design of the transfer station to determine the rate of arrival curve during abnormal, as well as normal, days. Adequate apron area must be provided so that the carriers do not lose too much time waiting. A means of inspecting the refuse as dumped and of removing OBW and objectionable items is desirable. A separate section is usually required for local citizens to bring their refuse to avoid interference with the trucks. These areas are also used as metals, glass, paper and antique or junk recycling centers.

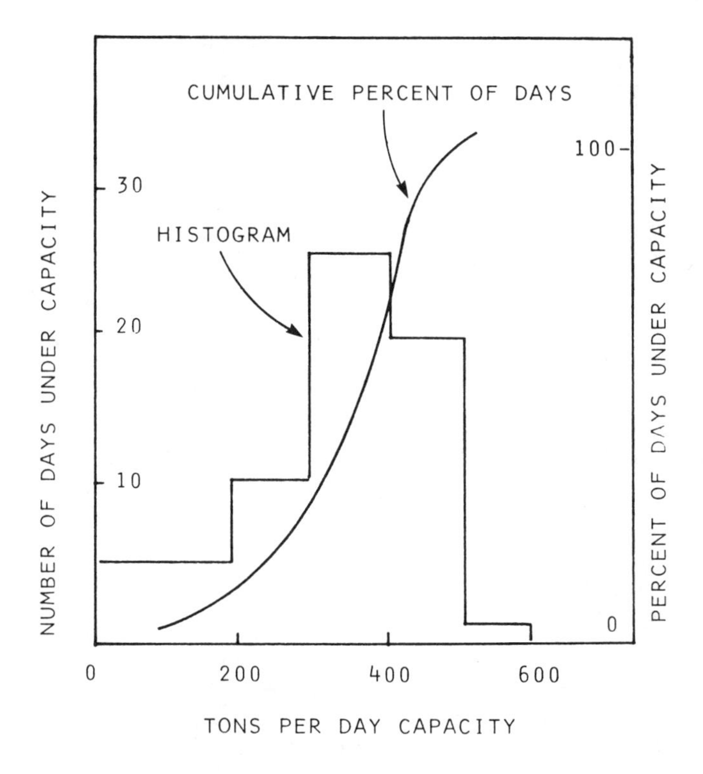

Figure 2-11. Daily capacity in cumulative and histogram form.

Spare compactor equipment must be provided in case of breakdown. Planned maintenance minimizes the amount of spare equipment required. The spare equipment can be used during peak periods, provided that the necessary preventive maintenance is carried out during slack periods. When working close to the capacity of the installed equipment, spare parts inventories and maintenance crews have to be kept at appropriate levels. When many transfer stations are included in a large system, systems analysis can be applied to optimize and control the flow of refuse under normal and abnormal conditions, and centralized parts inventory and maintenance crews can be used, especially if components are standardized.

The benefits of a transfer station derive from the reduced cost of transporting a heavier load of refuse with a smaller crew, so that the collection crews can return for another collection. The benefits will be highly dependent on site conditions and the length of the trip to the disposal site. An example will illustrate the principles involved. Table 2-1 [4] shows the costs incurred without the transfer station as compared to the costs of operating the station.

Table 2-1. Figuring Transfer Station Savings – An Example [4]

Without Transfer Station		*With Transfer Station*	
Truck Operation			
34 trucks, 2 trips/day = 68 trips @ 40 miles = 68 X 40 X $1.44/mi =	$3,907	3 trucks, 77 trips = 77 X 40 = 680 mi/day @ $1.44 =	$ 986
Labor			
3 men/truck, 7-hr trip = 3 X 68 = 240 X $6.00/hr =	1,440	7 men/truck, 7-hr trip = 77 hr/day = 3 men = 3 X 8 X $8.75/hr =	210
Transfer Station Operating Costs	none	@ $3.00 per ton =	1,020
Total Daily Cost	$5,347		$2,216
At 252 days per year, the savings are		$1,347,400 –558,432	
Net Savings		$ 788,970 per year	

RECEIVING, HANDLING, STORAGE AND FEEDING

Receiving facilities must be designed to receive MSW collection or transfer vehicles at the rate at which they will arrive, generally during an 8- to 10-hour period, 5 or 5½ days per week. Processing may take place from 8 to 24 hours per day and 5 to 7 days per week. As refuse is received in many cases only 5 days per normal week and 4 days in a holiday week, the full week's material may come in 4 days. Also, storage may be needed for 3 days if the plant must produce heat and/or power continuously. The size of the storage area is determined on this basis.

Additional storage must be allowed for times when the plant is down due to outages, maintenance, breakdown or modifications. An alternate solution is temporary diversion to landfill.

The imbalance between the rate at which refuse is received and the rate of processing must be absorbed by some form of storage, which must take into account holidays, long weekends, bad weather and other factors, as well as the daily factors.

The rate of arrival of refuse trucks determines the space required by these trucks while they wait to discharge, the method of handling the discharge and the storage requirements. Figure 2-12 shows a typical graph of the rate of arrival of refuse trucks.

The average rate of receipt during these 10 hours was about 24, with a peak of 42. The input to the plant, averaged over 20 hours, would then be about 12, making the peak to average ratio about 3.5. If the trucks averaged 6 tons per load, the plant capacity was 6 X 12 X 20 = 1440 TPD.

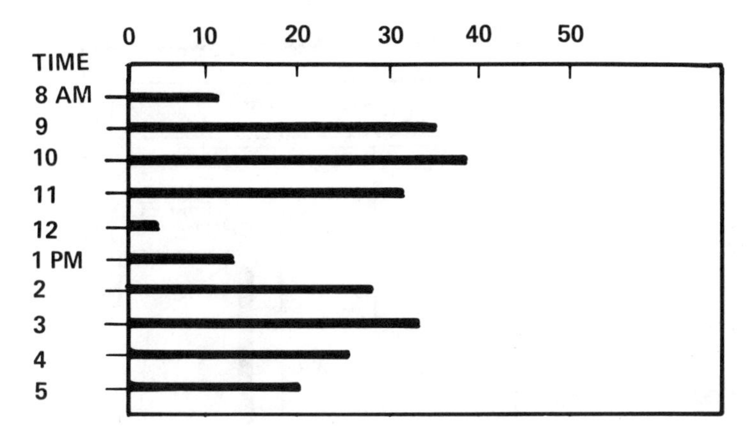

Figure 2-12. Rate of arrival of trucks during the day.

The daily average would not be a constant. Typically, it is light in summer and especially heavy in early spring and fall, when the leaves come down. The 1440 TPD is likely to exceed 2000 TPD at times and for several days, so storage could not accommodate the excess.

Storage Pits for Incinerators

Incineration plants have generally used storage pits into which the packer trucks dump directly. Figure 2-13 shows a receiving pit of the type long used with mass-burning incinerators, in which a crane is used for mixing and charging the incinerator hopper. Pits of this type can be filled level with the trucking floor, or, by closing the doors, the crane can load the refuse up to the level of the hoppers. Their capacity depends on the average density of the MSW, which may range from 300 to 400 lb/yd^3, depending on the height of storage and the nature of the refuse.

The design of storage pits for incinerators is reviewed briefly here [1].

A 1000-TPD plant requiring two days of storage would need the following storage volume, assuming an average storage density of 300 lb/yd^3 (11 lb/ft^3):

$$V = 2(\text{days}) \times 1000(\text{tons}) \times \frac{2000(\text{lb/ton})}{11(\text{lb/ft}^3)} = 365{,}000 \text{ ft}^3 \qquad (2.1)$$

A refuse storage pit 40 feet deep and 40 feet wide would require a length of 225 feet. A tipping floor, which could be piled 12 feet high, would require a storage area of 30,000 ft^2, 100 × 300 feet or 50 × 600 feet.

Tipping Floors

Refuse processing plants generally have been provided with "tipping floors," on which the refuse is dumped by the packer trucks or transfer trailers, after which front-loaders are used to pile the refuse in storage or feed it to the plant.

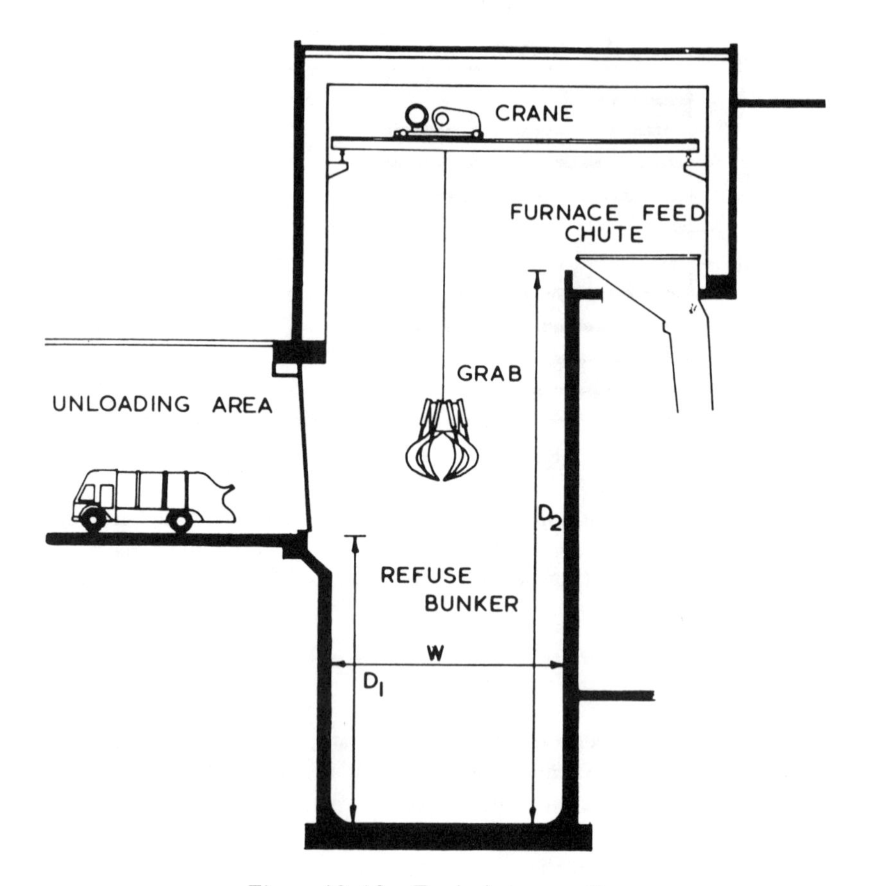

Figure 12-13. Typical storage pit.

Many arrangements have been tried for tipping floors, resulting in various degrees of efficiency. Systems analysis can be applied to determine the most effective layout, based on a set of objectives, and the unit operations that are available to achieve them.

Figure 2-14 shows the layout of a typical tipping floor, which can be used to store MSW up to a height that depends on the front-loader's discharge capability. Large 7- to 12-yard loaders can discharge at 12-14 feet in height. The storage capacity of the floor then depends on the average refuse density, which may vary from 300 to 400 lb/yd^3.

The tipping floor using front-loaders allows first-in, first-out handling, as well as inspection of the wastes during the process of handling it bucket-by-bucket. In addition, the loader can spread out the waste for inspection in a thin enough layer for pickers to inspect it and remove objectionable materials. Loaders with suitable attachments can be used to facilitate removal of OBW.

The average receiving rate at a plant having a maximum design capacity of 1000 TPD is 125 TPD, based on 8 hours, and 42 TPH, based on feeding the plant

Figure 2-14. Typical tipping floor.

over 24 hours. The layout must permit loads to be received during the approximately 8-hour day, yet permit continuous feed to the process. Generally, this requires handling of the material shortly after it is dumped and delivering it either to storage or directly to feed the plant.

Time studies can be made to determine the floor space and distances traveled under real conditions to ensure that the loaders and men actually can handle the refuse at the rates of arrival, as well as allow for looking it over, picking unwanted materials out, mixing dry and wet and other operations. For instance, it might be found that within the 100- to 200-foot range of distances required for storage, as described above, the loader can go for a load and spread it in half a minute, transfer a load from its point of discharge in half a minute and deliver a load to the feed conveyor in half a minute. During off-hours, when no refuse is coming in, a single loader may pick up a load from the pile, spread it on the floor and push another load into the receiving conveyor in one minute. If the loader handles 1.5 tons per load, it would be handling 90 TPH.

When refuse is being received, one loader may be able to pick up a load and deliver it to the storage pile in less than one minute, and thereby handle about 90 TPH. With trucks coming in at a peak rate of 40 trucks per hour ($40 \times 6 = 240$ TPH), three loaders would be needed. For the average of 24 trucks per hour ($24 \times 6 = 144$ TPH), two loaders will handle the refuse, one feeding the plant and the other storing. When the peaks come, they can shift their functions, the one feeding the plant having some spare time.

Inspecting and Picking the Refuse

It is generally advisable to employ pickers to inspect the refuse and remove oversize and otherwise objectionable items. When pickers are used on the floor, the loader operator can spread the bucket load on the floor in a layer thin enough for the pickers to inspect it and remove the objectionable objects. After inspection, the loader returns and pushes the load into the receiving conveyor. By making the receiving conveyor long enough, three or four picking areas can be arranged so that the time for picking can be several minutes, even at full plant capacity. The number of pickers would depend on the waste material and the equipment available to transport the OBW to the container.

Other arrangements provide picking stations along conveyors. If the conveyors can be stopped and restarted safely and interruption of flow to the plant does not upset its operation too much, this procedure has merit.

Feeding the Plant

The way the refuse is fed to processing and burning plants requires careful attention because irregularities in feed interfere with plant performance. MSW is extremely difficult to feed in a controlled manner due to the variations in moisture and density.

Feed to mass-burning incinerators is certainly a critical factor in the control of combustion and emissions, not to mention control of the rate of steam and power output of a connected plant. The design of the feeding system must be such that the operator can keep up with the demand of the system while still having enough time to pile and mix MSW, travel from pile to conveyor or hopper, and take rest breaks or change crew.

When a mass-burning incinerator plant is fed with a crane and grapple, the operator fills the hoppers under instruction from the incinerator operator or keeps the hopper full while an automatic control varies the feeding mechanism that draws from the hopper. Mass-burning incinerators also can be fed by a conveyor, which supplies the hopper, in which case the incinerator controls the operation of the conveyor and the front-loader operator feeds the conveyor as required. This arrangement was used in the East Bridgewater plant to feed a stoker-fed incinerator.

A crane holding 12 yd^3, or about 2 tons, can feed a 200-TPD incinerator while using only about 30% of this time for feeding, or a 400-TPD plant with 60% feeding.

When feeding a processing plant, the front-loader operator dumps loads onto the primary conveyor in as uniform a manner as possible, generally filling the empty part of the conveyor. When feeding a shredder, a current control can start and stop the conveyor to prevent overload and keep a fairly steady feedrate. When feeding a trommel, the process operator or a mechanical device can control the speed of the feed conveyor to keep the input rate fairly constant.

A front-loader with a 10-yd^3 bucket could feed a 50-TPH (1000-TPD) shredding module with 1.5-ton loads brought every two minutes. If the feed conveyor is long enough to hold three loads side by side, they can be brought one after the other in three minutes, allowing three minutes for other duties.

If a plant fed by front-loaders working on a tipping floor had to receive for 8 hours per day and process for 24 hours, it would have to have a loader committed to piling two-thirds of the incoming refuse during the 8 hours, in addition to the loader that feeds the plant.

Front-loader Capacity and Horsepower

Front-loaders are widely used for off-the-road uses and are available in a wide range of sizes. For handling MSW, the buckets can be increased in volume compared with those supplied for earth and stone. The buckets designed for earth have capacities ranging from 1 to 12 yd^3, for which the flywheel horsepower provided ranges from 65 to 700 hp. Buckets designed especially for MSW can carry 50% more volume.

Innovative Systems for Feeding MSW

In an effort to reduce the labor required to operate pit-and-crane and front-loader conveyor systems, several other approaches have been tried with various degrees of success.

The Hamilton, Ontario refuse-to-energy plant has a single pit into which the refuse is dumped by the trucks. The pit has a "live bottom" with four pan conveyors, each of which is speed controlled to feed a shredder. In the original concept the conveyors ran along the bottom, then up at a 45° angle to the shredder, relying on "fallback" to even out the feed. This arrangement has been changed to separate the horizontal conveyor from the inclined conveyor so that they can be run at different speeds, allowing the horizontal conveyor to be used for storage and the inclined conveyor for feed control.

The refuse-to-energy systems at Akron, Ohio and at the Occidential Chemical Co. in Buffalo, New York were provided with storage hoppers into which the trucks could dump directly. At the bottom of these hoppers, hydraulic rams were provided to control the feed from the hoppers. It was found that MSW bridges too severely to allow large amounts of storage to be held in this hopper: the material has to be discharged as-received or storage limited to manageable levels. These systems were also provided with secondary storage of shredded refuse.

The second-generation RDF system planned for Union Electric in St. Louis was designed to store the MSW received from transfer stations in compactor-containers so that they could be discharged as needed to feed the plant, eliminating any other forms of storage other than a small surge bin. This design was not implemented. The containers were to be discharged by hydraulic pushers, conventionally supplied with transfer trailers, into a conveyor feeding the shredding line. A system of this design would have to be provided with an conveyor inspection station for removal of OBW and objectionable materials.

Storage and Feeding of Shredded Refuse

To provide storage for all-day operation as well as to take care of surges in production or consumption and provide controlled feed to boilers or other processes, shredded refuse has been stored in silos, bins and on the floor of a storage room. In any case, the method of retrieval is critical to the effectiveness of the system.

Ferrous and glass products may also require storage: the containers used for shipping are the most convenient devices for this purpose.

When retrieving shredded refuse to feed boilers, the feed generally must be fluffed into a reasonably consistent density to achieve optimum handling characteristics. A small surge bin has sometimes been provided in an effort to smooth out the differences between the irregular rate of retrieval and the desired steady, controlled rate of feed to boilers. When feeding utility boilers at 10–20% of capacity, the feed control is not as critical as, for instance, feeding a dedicated boiler at 100%.

Atlas Bins

Many Atlas storage bins have been installed to store single- and double-shredded RDF and feed it to burners. Coarse-shredded RDF, however, has generally contained an excessive amount of textiles and wires, which wrap around moving parts and interfere in other ways with reliable operation. This bin has the advantage that up to four separate burners can be fed simultaneously from one bin, while maintaining separate control of each feed. Figure 2–15 shows a photograph of an Atlas bin. The bin is shaped conically to hold a maximum amount of material without bridging. A ring is provided around the periphery of the bin floor, which revolves, pulling four sweep chains around the bottom edge of the pile that pass over trenches in the floor containing drag conveyors. These conveyors are speed controlled to adjust the feedrate at which they discharge into a pneumatic conveying system.

Feedrates from the Atlas bin are sensitive to the density and flow characteristics of the RDF. A given speed setting can give a wide range of feedrates, depending on the height of the pile in the silo, its moisture content and other properties. By using a weigh scale to control the speed, these variations can be minimized.

The Atlas bin is a last-in, first-out device, requiring running out the entire contents periodically so that the core material does not compost into material that can spontaneously combust or decompose to such an extent that it will not retrieve.

Atlas bins have been built for RDF facilities with capacities up to 100,000 ft^3 and with discharge rates up to 90 TPH (based on 4 lb/ft^3 density). The discharge density has been found to range from 4 to 10 lb/ft^3 during operation.

Figure 2-15. Atlas storage bin.

When equipped with automatic controls, the discharge rate (at constant rate) was held within ± 5% during a 15-minute test period, as reported by Atlas. Unless a weigh scale is used to adjust controls, pile height and diameter, off-center loading and other variables may be needed inputs to the controller.

Storage Bins with Auger Dischargers

A number of storage bins, perfected for the wood industry, have screws or augers on the floor of the bin that ream out material and drop it on a transverse conveyor, which carries it to the burner or process. These systems are subject to considerable wear when used with MSW containing abrasive materials. The tendency of rags and wires to wind up on rotary devices is another drawback to the use of screws.

"Live-bottom" bins have the entire bottom covered with screws, which must be turned very slowly to control discharge rate or be operated intermittently. They have been used for small surge bins, but when loaded with moist RDF have not been found to discharge smoothly. This arrangement causes the entire contents of the bin to move toward the discharge wall unless the screws are set in pairs to drive to the opposite walls.

Moving auger bins (Figure 2-16) have one or two augers that discharge to a central or side conveyor. They have the advantage of being first-in first-out, minimizing decomposition of material that is retained too long. To serve the entire bin these

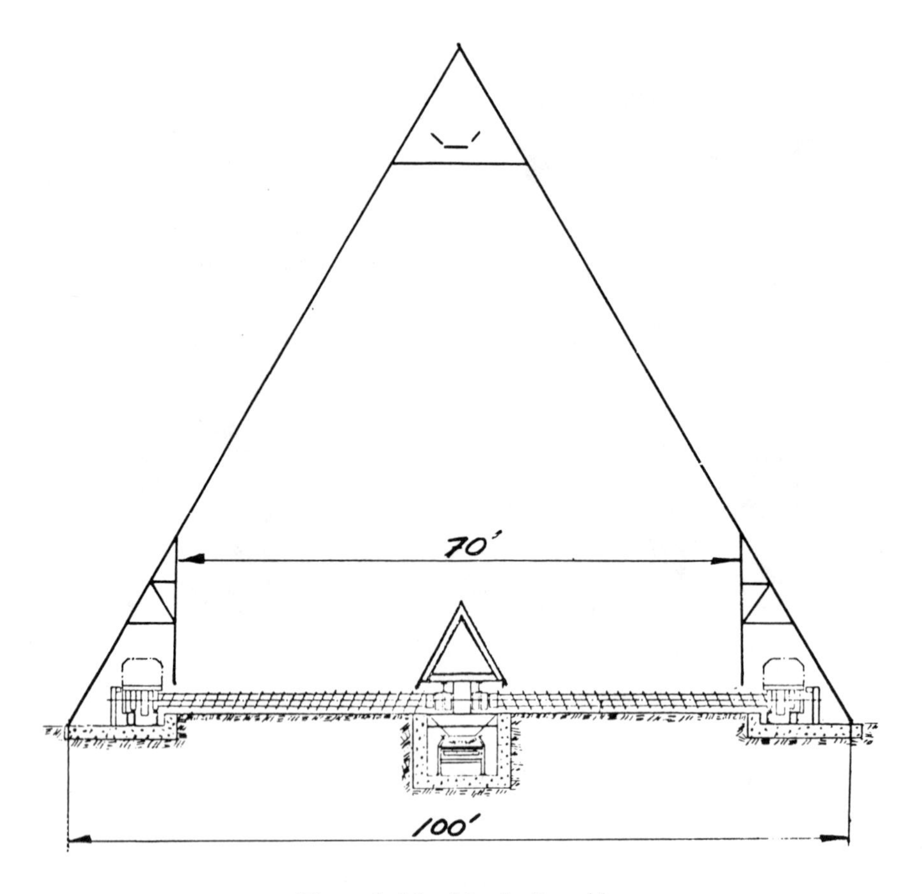

Figure 2–16. Live-bottom bin.

augers are driven laterally. This arrangement requires coordination of the lateral speed with the discharging speed and a reversal of direction at each end. A Miller-Hofft bin of this type was installed at the St. Louis Demonstration Plant. It operated satisfactorily after the drive horsepower was increased as needed for RDF and the height of storage in the bin was restricted well below its potential capacity. The discharger has two counter-rotating screws, which ream material out of the pile and screw it to the discharge. As might be expected, the screws are subject to wrapping with rags and wires. Double-shredding is necessary to reduce rags to a size that does not create serious wrapping. The Wennberg retrieval system (Figure 2–17) uses a screw for retrieval that tapers gradually.

The Koppers bin (Figure 2–18) uses vertical screws to lift material away from the pair of screws at the hopper bottom. This controls the density of the RDF where it enters the discharge screws and offers more consistent flow characteristics. Figure 2-19 shows the results of a performance test in feeding double-shredded RDF. The relationship between screw speed and capacity is linear and within a ± 5% tolerance.

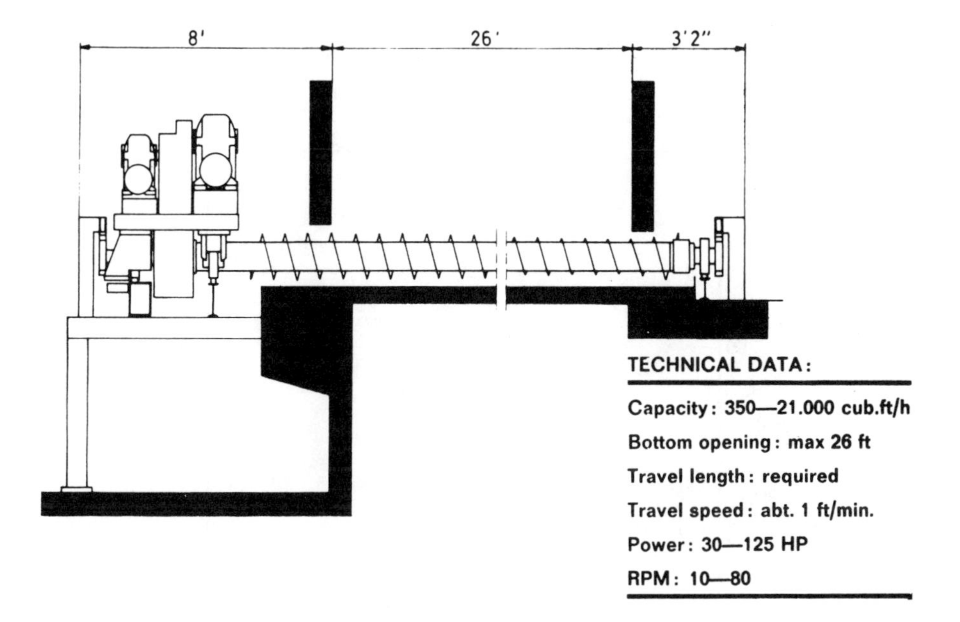

Figure 2-17. Wennberg bin.

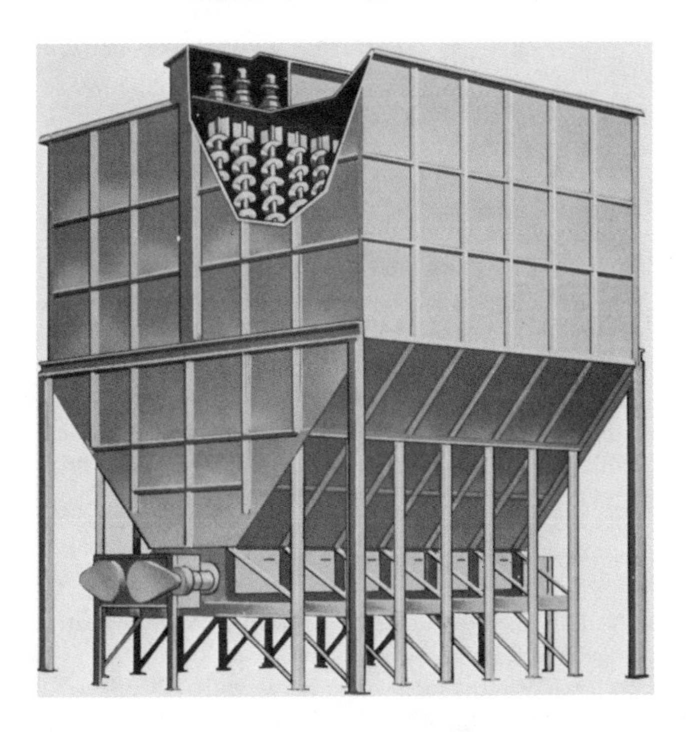

Figure 2-18. Koppers bin.

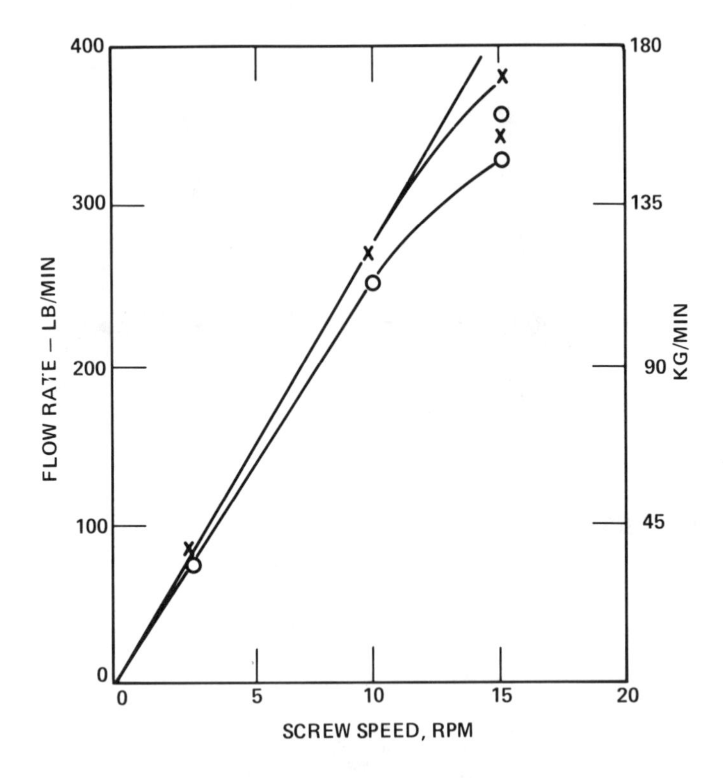

Figure 2-19. Performance of a Koppers live-bottom bin.

Screw Conveyor Capacities

The theoretical capacity of screw conveyors is easy to calculate. Single-pitch screws move their contents in one turn and double-pitch in two turns. Actual capacity depends on the amount of "fillage." A full screw loses efficiency due to interference with infeed material. The actual density of the material handled and the percent fillage determines its actual capacity because a screw is basically a volumetric device. The maximum displacement capacities of twin screws are shown in Table 2-2. Actual capacities are closer to 25% of the maximum theoretical displacement.

Mechanical Conveyors

The term mechanical conveyor refers to conveyors on which materials are carried bodily, as distinguished from pneumatic conveyors, which use air or gases to convey materials. The major types of mechanical conveyors are listed in Table 2-3, along with the types of materials for which they are best suited and comments regarding their use. Several types of conveyors are pictured in Figure 2-20.

Table 2-2. Maximum Capacities of Twin-Screw Conveyors

Diameter (in.)	*$ft^3/hr/rpm$*	*Maximum ft^3/hr*	*Maximum RPM*	*Maximum ($4\ lb/ft^3$)*	*TPH ($10\ lb/ft^3$)*
6	4.9	585	120	1	3
9	17.8	2,100	120	4	10
12	43.0	5,100	120	10	25

Table 2-3. Major Characteristics of Mechanical Conveyors

Type	*Suitability*	*Comments*
Pan, Hinge, Apron	MSW, heavy metal, stone and glass	Rugged, resistant to impact, subject to slow hinge and bearing wear; up to 100 ft/min
Drag	Damp or wet wood, RDF, bagasse	Rugged, not good for stringy materials, up to 100 ft/min
Bucket	Dry, dense, free-flowing solids	Rugged, not good for wet or dusty solids; up to 60 ft/min
Belt	Shredded MSW, wood, RDF, light metals	Not good for impact or dust unless protected; up to 450 ft/min
Screw	Dry, dense, free-flowing solids	Good for dusty but not wet or stringy material. Speed: up to 150 rpm
Vibrating	Dry or damp dense solids	Rugged, resistant to impact, good for stringy materials

Hinge conveyors are made of steel plates, which are hinged directly together. Axles in the hinges are connected to rollers, which run on tracks and are driven over head and tail pulleys by the mechanical drive.

Apron conveyors have pans of various shapes that are supported by chains located outside of the material stream. This, in turn, has wheels that run on rails and are driven over the head and drive pulleys by the drive motor. These conveyors can be run horizontally or inclined up to about 40° to the horizontal. Cleats can be used to assure movement of the load, and side walls can be provided to prevent loss of material. Side skirts can be used to permit moving beds of material up to about 8 feet, although 4 feet is the practical limit due to risk of personnel or equipment falling in.

Bucket conveyors are similar to apron conveyors except that the buckets are open only on the top and can be carried vertically or up steep inclines.

Drag conveyors have flights or crossbars attached to drive chains, which slide on a smooth floor and thus drag solid materials along from the feedpoint to the drop-off. The chain is driven by sprockets outside of the material flow.

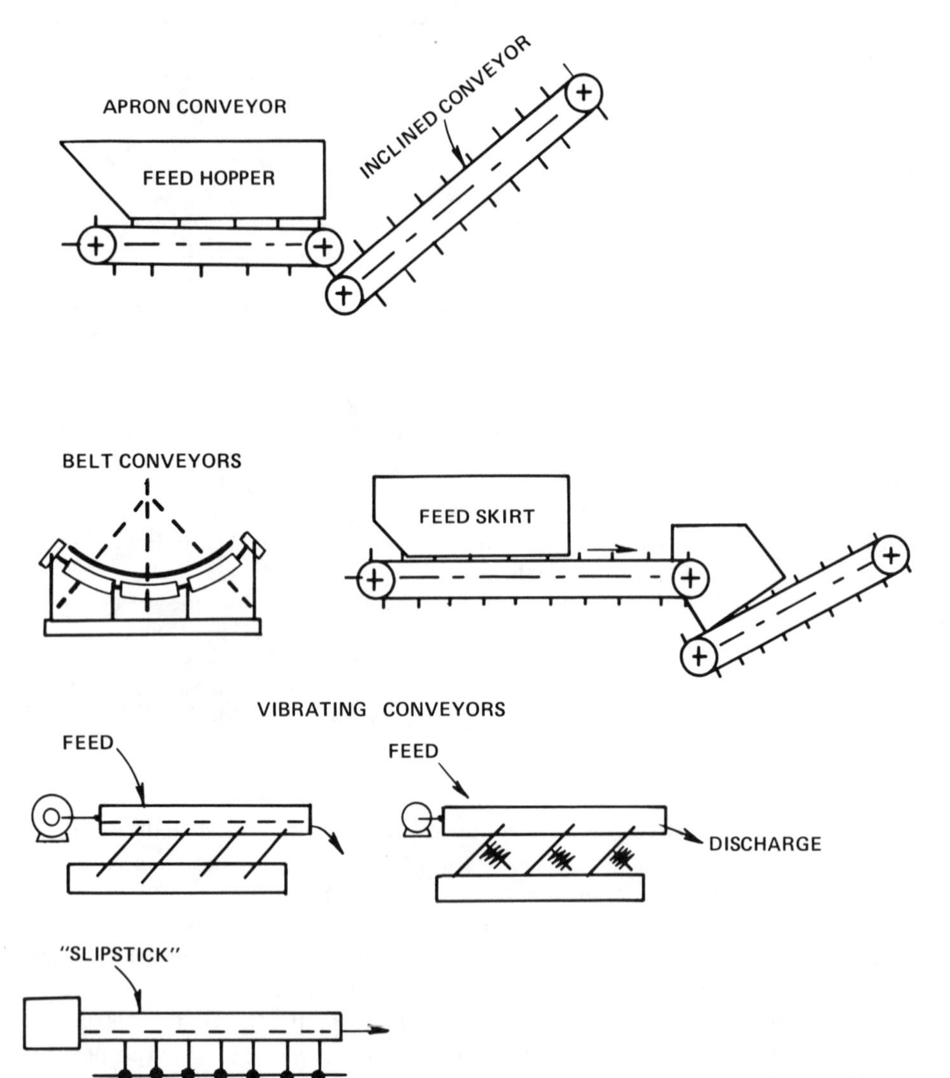

Figure 2-20. Three typical mechanical conveyors.

Vibrating conveyors have solid pans that vibrate or oscillate in such a manner as to move the solid material. They may also have perforated bottoms that can screen the material.

Belt conveyors carry the material above a belt, which can be curved by supporting it with angled rollers, thus providing a concave shape to carry the material. Some belt conveyors have flexible walls and pockets to contain solid or fluid materials.

Pneumatic conveyors use a stream of air to convey suspendable materials from one end of a pipe to another.

Capacity of Mechanical Conveyors

The capacity of a mechanical conveyor is primarily limited by volumetric considerations because the weight generally can be handled by the equipment. The volume carried by a conveyor is the product of the velocity of the material and the cross-sectional area perpendicular to the direction of the velocity vector.

The manner by which the material enters or leaves the conveyor system (at the transfer points) may be more important than the conveyor itself in determining its capacity and satisfactory performance.

When material falls onto a conveyor, it seldom falls with a velocity in the direction of the conveyor equal to the conveyor velocity. Therefore, the conveyor must provide a means of imparting its velocity to the material so that the material is reliably carried away from the feed region. Flytes or drag bars provide this function in metal conveyors. Belt conveyors normally rely on friction between the belt and the material, although cleats can be added to assist in accelerating the material.

The capacity of a conveyor, subject to the limitations caused by the transfer points and feeding methods, is

$$C = V \times D \times B \times W \times 60 = V \times D \times A \times 60 \qquad (2\text{-}1)$$

where C = capacity (lb/hr)
V = velocity (ft/min)
D = bulk density (lb/ft^3)
B = burden (height of bed) (ft)
W = width of material on bed (ft)
A = sectional area perpendicular to velocity (ft^2)

Conveyor Velocities

The velocities that can be used vary with the type of conveyor. Metal belt conveyors are used at velocities of 10–100 ft/min, whereas flexible belt conveyors are used at velocities of 100–450 ft/min.

Vibrating conveyors can generally be operated at 400-600 vibrations per minute (cpm), with strokes up to 1 inch, which can produce velocities of 20-60 ft/min. Various mechanical actions are available, most of which deliver a horizontal, as well as a vertical, velocity component.

The axial component of the stroke multiplied by the number of strokes per unit time gives the maximum velocity that could be imparted to the flowing material. Springy and fluffy materials will absorb some of the pan motion, reducing the actual capacity. Also, wall friction can seriously interfere with capacity, especially when the materials are moist or sticky.

Vibrating conveyors are usually used horizontally or with a downward slope. These conveyors generally cannot support a high burden and are not as effective for light materials.

Bulk Density

Second to material velocity, the bulk density of the materials handled is the primary component in the equation of capacity. Table 2-4 lists typical bulk densities of materials found in MSW, the approximate weight of each constituent and the resultant contribution to total bulk density. The average is the bulk density of the mixture, in a loose form.

Table 2-5 lists the bulk densities of some intermediate and final products of resource recovery plants.

Table 2-4. Bulk Densities of MSW Components and Mixtures

Component	*Weight (%)*	*Density (lb/ft^3)*	*Volume (ft^3)*
Food Wastes	9.5	18	0.53
Paper	43.1	5.1	8.45
Cardboard	6.5	6.2	1.05
Plastics	1.8	4	0.45
Textiles	0.2	4	0.05
Rubber	–	8	–
Leather	1.5	10	0.15
Garden	14.3	6.5	(2.20)
Wood	3.5	15	(0.23)
Glass	7.5	12.1	0.62
Cans	5.2	5.5	0.95
Nonferrous	1.5	20	0.15
Ferrous	4.3	10	(0.22)
Dirt, Ceramics	1.1	30	(0.04)
TOTAL	100.0		15.09 = 180 lb/yd^3
Total, not including (items)			(12.40) = 220 lb/yd^3

Table 2-5. Bulk Densities of Various Products

Product	*Bulk Density (lb/ft³) (kg/m³)*
Bureau of Mines, College Park, MD	
Refuse as-received	7.7
Primary shredder discharge	6.5
Magnetic product	8.3
Feed to air classifier	4.5
Air classifier heavies	12.0
Heavies over 3/4-inch screen (trommel)	18.8
Heavies under 3/4-inch screen	41.1
Combustibles,	
light fraction	1.8
medium fraction	2.7
heavy fraction	4.3
Clean coarse glass (5% moisture)	85.0
Clean fine glass (12% moisture)	90.0
Other RDF materials	
Densified RDF (pellets)	25-35.0
Powdered RDF (Eco-Fuel II)	25-35.0
Fluff RDF	5-15

Cross-sectional Area

The third factor in conveying capacity is the cross-sectional area of the material perpendicular to the direction of flow. Metal and rubber conveyors having integral side walls can carry material up to the level of the walls.

Metal conveyors having high stationary walls can carry material up to the height of the walls, provided that flytes on the conveyor can apply forces greater than those created by material friction against the sides.

Flat belt conveyors can put material in a pile, provided that the pile does not reach to the edge of the belt and spill off. To avoid this, belt conveyors commonly use troughing rollers to curve the belt, thereby increasing the cross-sectional area of the material carried and reducing the chance of side spillage.

Transfer Points

The points at which materials are transferred from one device or conveyor to another are called transfer points: they can be a major source of problems.

The acceleration zone, in which the material falling onto a conveyor accommodates itself to the belt speed, is critical. At the point at which the feed material falls onto the belt, at a velocity less than that of the belt, the height of the material burden on the belt will be greater than it will be farther down the belt, where the material is moving at belt speed. The excessive burden at the feedpoint would overflow the belt unless skirts and guides are provided to contain the material. In spite of such skirting, material will tend to work its way under the skirts and spill from the belt. This can be minimized by providing flexible skirting material and a pneumatic dust control system. The lighter the material, the greater the problems at the feed and discharge ends due to air entrainment of the material.

Steep chutes can help direct the materials flowing off of a belt toward the new direction of motion, thus reducing the degree of acceleration that must be provided by the next conveyor.

Transitions must allow for expansion of the material because contraction of cross-sectional areas will tend to cause jam-ups. Where material velocity is expected to decrease, larger cross sections must be provided.

Materials falling off the end of a belt conveyor have a trajectory that depends partly on gravitational effects and partly on aerodynamic forces. Dense materials have trajectories that can be predicted by calculation. On the other hand, light, flaky materials like RDF carry air with them and develop air currents that affect the behavior of the stream, especially the outer edges.

At the point at which the material leaves the belt, it follows a trajectory that depends mainly on the belt velocity and falls to the next conveyor or piece of equipment. The way in which it falls determines how it will pile on the next belt or enter the next piece of equipment. The "splash" of the material after this fall creates velocities that will drive the material through openings in the skirting. For these reasons, the shaping of this transition must be carefully worked out.

Layout and Angle of Inclination of Conveyors

Layout of conveyors is of major importance in plant layout and design because it determines the location of the pieces of equipment, thus freezing the plant configuration and making it hard, if not impossible, to make field corrections when the material flow is found to be adverse.

The equipment served by conveyors generally requires feed from a higher elevation than the discharge. This elevation difference is provided by the conveyors. Hence, the ability of conveyors to elevate material is essential to their application. The angle of inclination that can be used determines the length of the conveyor and, hence, the physical dimensions of the plant.

The angle of inclination that can be used for a given conveyor and type of material is limited by the frictional forces that the conveyor can apply to the material to accelerate it to conveyor speed, and also by the angle at which "rollback" occurs. Some materials have a clear, consistent "angle of repose," which limits rollback, whereas other materials will roll at any angle. Table 2-6 [5] gives some of the typical angles used for different materials.

Table 2-6. Angle of Inclination of Conveyors [5]

Material	*Range of Angles that Can Be Used (°)*
Sand	15–20
Coal, Gravel	17–20
Stone	20–26
Wood Chips	22–28
Paper	22–25
Densified RDF	1–25

A steep angle reduces the length of conveyor needed to reach a given height. When steep angles are used, it usually is necessary to provide cleats of some sort on the belts to increase the grip of the conveyor surface. By using a steep conveyor angle, the material can be made to fall back so as to help in leveling, although this effect is not reliable.

HOUSEKEEPING

The subject of material handling cannot be treated without a discussion of plant housekeeping. When linty or dusty materials are carried, windage effects at seals and transfer points will raise the lint and dust into the moving air and cause them to be deposited elsewhere in the plant and be carried in the surrounding air while waiting to alight.

Dust and lint are harmful to the workers if they inhale these materials. There is a greatly increased hazard of fire and explosion when dust and lint are present in the air and on surfaces. If there is a puff due to even a minor explosion, the dust and lint deposited around the plant can easily be disturbed, become airborne and contribute to a conflagration or cause secondary explosions. In serious cases there is enough dust around to create pressures in the building sufficient to blow off roof and wall sections. Generally, roof and wall construction is made to relieve this type of weak explosion to prevent more harmful pressure buildup.

Prevention of dust and lint accumulation requires careful attention to the design of conveyor skirts and transfer points. In addition, dust collectors are usually provided at points known to raise dust and designed to remove it. It is easier said than done, yet all steps taken are rewarded by enhanced safety and reduction of effort by cleanup crews.

DUST COLLECTION

To minimize the dust collection system, the openings through which the materials enter and leave the transfer points must be kept to a minimum, yet not interfere with the flow of materials.

Curtains may be hung at the entrances and exits of transfer points. If they are placed incorrectly or are too stiff, they may interfere with material flow and also not seal well. The materials flowing off the incoming belt are allowed to maintain their natural direction of flow while carrying a minimum amount of air with them. The leaving curtain is placed as far downstream as possible to allow the materials to acquire the speed of the belt before having to contend with the curtain. In some cases, crude rotary valves may be justified to seal off excessively dusty atmospheres at transfer points.

MAINTENANCE

Conveyor systems are essential to the reliable operation of refuse processing plants and, with good design and care, will not be the reasons for shutdown. Unfortunately, too often their malfunction has been the cause of interruption of flow.

Assuming the transfer points have been designed properly and the materials handled are not too low in density to be carried properly, it remains to be certain that preventive maintenance can be carried out on the conveyor working parts and that they are protected from abrasive damage from dirt, glass and metals.

These abrasive materials are ever-present in MSW and the intermediate products, if not in the final product. All moving parts of conveyors are prone to damage from sand and dirt. Good design attempts to prevent the sand from getting to the moving parts and to remove it where it tends to collect so that it does not build up over a period of time and cause problems. Belt cleaners are available to remove most of the sand that follows the belts. The sand tends to continue to fall off the belt on the return pass, causing poor housekeeping and itchy necks.

DESIGN OF BELT CONVEYORS

The CEMA handbook [5] is a general source of information on the application and design of conveyors, including drag chains, metal pan and rubber belt conveyors. Unfortunately, this handbook does not include information on the properties of MSW fractions and by-products, or on the application of standard equipment to these materials. Consequently, it is not surprising that in the first and second generations of recovery facilities this lack of information results in misapplication, underdesign and a variety of problems with the conveyor systems.

The difficulties arose from a lack of knowledge of the extreme range in material densities encountered with MSW and its intermediate products, ranging from 3 to 35 lb/ft^3, and overlooking the variations in feedrate that can readily occur.

The conveyors used to feed MSW to the plant can easily be overloaded by the feeding device, be it an uncontrolled garbage truck or a front-loader under the control of an operator. The front-loader may pick up a volume of refuse weighing as little as 1000 lb when it is dry and light, and as heavy as 2500 lb when it is wet and heavy. As the operator makes this deposit in a width of conveyor about the same as that of his machine, this tends to determine the loading rate on the conveyor. In addition, he may overlap his load over the adjoining one, creating the effect of a 5000-lb load, and perhaps miss a portion of the conveyor to further compound the fluctuations.

Table 2-7 shows the range of conditions affecting the design of a primary conveyor designed for 75 TPH, when the feed density varies from 10 to 30 lb/ft^3, and the speed is adjusted to feed this rate despite the density variation, while maintaining a constant height of load of 5 feet.

If the conveyor were designed for 10 lb/ft^3, the increased density of 30 lb/ft^3 would cause overload of the motor. However, it would create severely increased torque loads on the headshaft. If the motor overloads were used as protection, so that the 2.79 hp/ft load was permitted at the higher density, the torque would increase to 30/10 = 3 times the torque at the lower density. This would be destructive to the drive system if this possibility were not taken into account.

Due to the propensity for overload of the primary conveyor, it is necessary to provide not only motor overload protection, but also devices to protect the conveyor mechanically. Shear pins must be replaced when they fail, requiring costly downtime. A hydraulic drive permits the use of a fluid relief valve to limit the torque applied to the conveyor and the speed when it is overloaded. A variable-volume, constant or maximum horsepower pump can be applied effectively, permitting speed control over a wide range of conditions. As the power is proportional to the feedrate, setting constant power is a good way to maintain fairly constant weight flow into the system. This is not necessarily the correct parameter of feed, but it is better than none at all.

Rubber conveyor belts are likewise sensitive to the density of the material because this determines the belt loading, along with the shape and height of the material piled on the belt.

The use of continuous skirt boards with belt conveyors increases the power draw of the belt and the torque applied through the entire system in an unpredictable manner. When the correct idler angle, belt width and correct material density are used, there should be no need for continuous skirt boards.

Table 2-7. Design Parameters for Primary MSW Conveyor[a]

Design Feedrate, 75 TPH			
Width of Conveyor, 84 in.			
Depth of Loading, 60 in.			
Density of Material on Conveyor	10	30	lb/ft^3
Necessary Conveyor Speed	7	2.3	ft/min
Chain Pull	13,140	19,404	lb
Headshaft Torque	157,680	232,848	lb-ft
Brake Horsepower (per foot)	2.79	1.35	hp/ft
Torque (based on 2.79 hp/ft)	157,680	480,365	lb-ft

[a]Prepared by the Rexnord Company.

The Conveyor Equipment Manufacturers Association (CEMA) has developed standards for design of conveyors for handling various materials. Refuse-derived materials are not listed in the CEMA standards, making it necessary to perform tests to confirm the applicability of these methods.

Figure 2-21 shows the cross section of a belt conveyor with angled rollers, as well as the standardized recommended cross section on which CEMA tables are based. The dimension "c" is left clear to keep material from approaching the edge too closely. From these dimensions, the material cross section and the volumetric capacity of belt conveyors can be calculated, the latter as a function of belt speed and belt width, for various optional roller angles and angles of surcharge.

The cross-sectional area of conveyors of various belt widths, with 35° rollers and a 5° angle of surcharge, is listed in Table 2-8. By multiplying the volume by the material density, the capacity of the conveyor in TPH can be calculated. Values for typical materials with their characteristic angles of surcharge and densities are listed for 5-20° angles of surcharge for the range of belt widths.

Some fundamental research has been carried out by NCRR on belt and vibrating conveyors handling shredded MSW, air classified lights, densified RDF pellets, the air classified heavy fraction, the magnetic fraction (ferrous) and a blend of dRDF and coal [6]. These properties were measured as listed in Table 2-9:

- **Angle of maximum inclination**: the angle, in degrees to the horizontal, at which the empty belt will successfully elevate material fed to it. This depends on the mass flowrate and belt speed.
- **Angle of repose**: the angle between the horizontal and the sloping line from the top of the pile to the base, determined by measurements of materials being stockpiled after free fall.

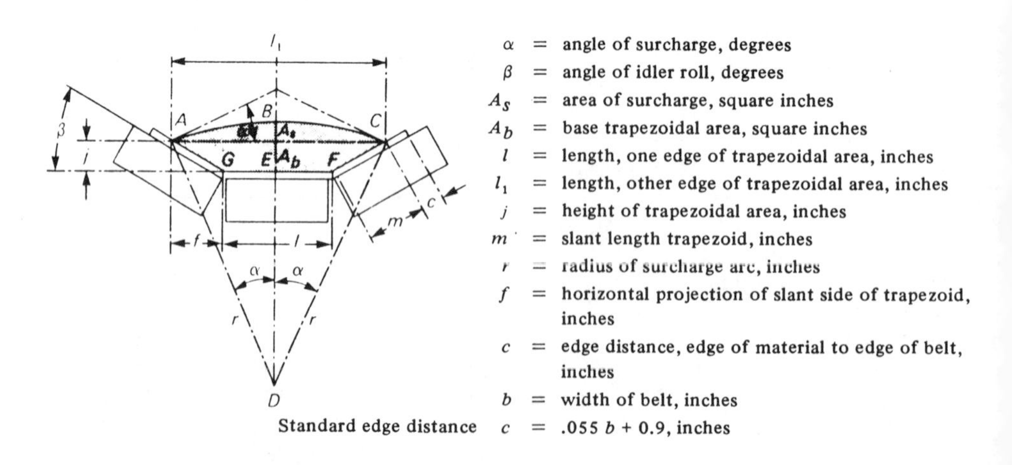

Figure 2-21. Area of load cross section.

Table 2-8. Conveyor Capacities (TPH) at 100 ft/min Belt Speed [5]

		Angle of surcharges, °				
		5		*10*	*20*	
Material Density (lb/ft^3)		*3*	*5*	*5*	*10*	*25*
Belt Width (in.)	*CS Area[a] (ft^2)*					
18	0.16	1.45	2.4	2.6	6.3	15.7
24	0.31	2.79	4.7	5.1	12.2	30.5
36	0.75	6.76	11.3	12.4	29.4	73.5
48	1.38	12.5	20.8	22.0	54.2	135
60	2.21	20.0	33.0	36.4	86.3	215
72	3.23	29	48	53	126	315
84	4.44	40	67	73	173	433
96	5.84	52	87	96	227	567

[a]At 5° surcharge and 35° roller angle.

Table 2-9. Characteristic Angles of Refuse Products

Solid Waste Fraction	*Angle of Repose*	*Angle of Slide Steel Belt*		*Angle of Maximum Surcharge (20°)*	*(35°)*	*Angle of Maximum Inclination*
MSW (–100 mm)	25–52	29	30	55	54	19
RDF (–50 mm)	29–49	31	35	51	65	21
d-RDF	27–46	33	34	–	49	30
Heavy Fraction	30–59	27	28	48	59	28
Ferrous Fraction	N/A	17	32	–	52	28
d-RDF/Coal	40–45	22	24	–	40	27

- **Angle of slide**: the angle to the horizontal of an inclined flat surface at which a small amount of material will start to slide downward due to its own weight.
- **Angle of surcharge**: the angle to the horizontal that the surface of the material assumes while the material is at rest on a moving conveyor belt. It is not an intrinsic property of the material because it depends somewhat on the conveyor velocity, configuration and loading. The Rexnord Company has determined that the angle of surcharge is about 5° for shredded MSW and RDF, with a bulk density ranging from 3 to 5 lb/ft^3; 10° for materials having 5 to 10 lb/ft^3; and 20° for more dense materials.

- **Maximum angle of surcharge**: the angle to the horizontal determined after the fall of material onto a static belt and before leveling out under agitation and vibration.
- **Loose bulk density**: determined by measuring the dimensions of the cone of material developed by free fall of the material.
- **Vibrated bulk density**: determined by the volume occupied by the material in a container after vibrating or tapping the container. A 1-ft^3 container was used.

Rates of Spillage from Belt Conveyors

Belt conveyors are simple in design and operation, but operating experience in resource recovery plants shows that they are subject to spillage, jam-ups, blowback and rollback.

The materials carried by conveyor belts can spill from the sides of the belts, particularly in the feed region, and in the region where they leave the belt and are transferred to another belt. Sand and other fine materials are released from the belt as it passes over the head pulley and runs over idler rollers on the return under the conveyor. These discharges of material cause cleanup, odor, sanitation and maintenance problems. Pans can be installed under the conveyor to catch the idler discharges and prevent them from contaminating the air, equipment and floor below.

The relationship of belt spillage, idler angle, loading and belt speed was determined by NCRR tests of a 7.6-meter (25-foot)-long belt, 0.5-meter (18 in) wide, at speeds between 0.2 and 2.4 m/sec (40–480 ft/min), with two sets of belt support rollers set at 20° and 35°. The spillage was reported either as the actual weight or as a percentage of the total weight carried by a 25-foot length of belt.

Data from tests of a complete conveyor [6], including the feed end, the main length of the conveyor and the discharge end, bring to light some of the factors contributing to spillage. The distribution of spilled material was recorded by dividing the belt into four sections along the length of the belt, from the tail pulley toward the head pulley, the first two at the feed end, a long section representing unskirted belt and one at the discharge end.

Results for RDF conveyed at 3 TPH are shown in Figure 2-22, at belt speeds of 0.75, 1.5 and 2.25 m/sec (150, 300 and 450 ft/min). There was little difference between the results at 1.5 and 2.25 m/sec. At 0.76 m/sec (150 ft/min), most of the spillage occurred at the feed section, accumulating 30 lb/hr within 7 feet. After this section, the spillage falls to about 0.4 lb/hr/ft, with a slight jump at the head pulley where the material exits. The spillage at the entry was reported to consist mainly of fines, inerts and inorganic particles, which became wedged between the belt and its skirting and then worked their way out. This spillage is caused to a large extent by the impact of the material onto the belt, resulting in a splash, or transverse velocity after impact, and by the high level of material in the zone before it accelerates to belt speed. The light materials are restricted by the side walls in the transfer area and, being of larger size, cannot pass through the clearances between the skirt and the belt.

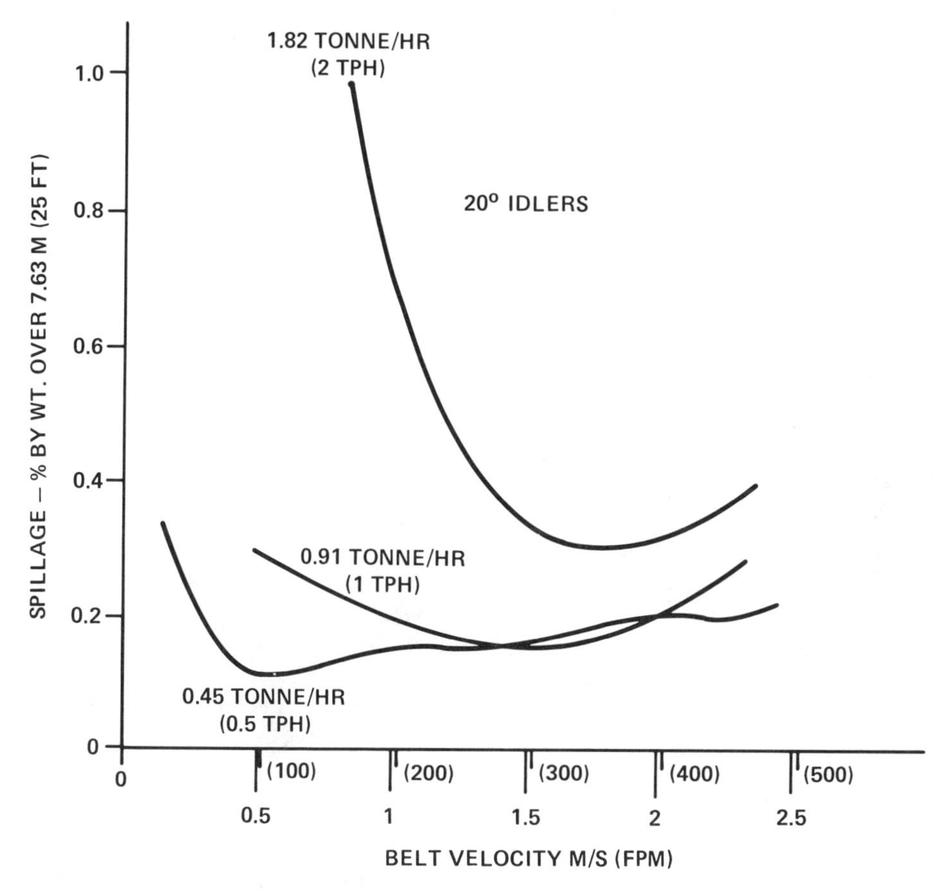

Figure 2-22. Effect of belt speed on the spillage of RDF [6].

After the entrance section, spillage at 0.75 m/sec results from crumbling and troughing of the RDF along the sides of the conveyor. At 1.5 m/sec (300 ft/min), the spillage is much less. Only 2 lb/hr occurs in the feed section. The spillage along the belt was only 0.1 lb/hr/ft of length, due to the reduced height of the material on the belt.

Most of the spillage occurred in the exit section, resulting from the higher velocity of the material leaving the head pulley, hitting the opposite wall and sides of the conveyor cover or being released from the belt as it passed over the head pulley. The total spillage of 8 lb/hr is only 0.13% of the 6000 lb/hr carried by the belt, a very small percentage loss, although this quantity would be significant in the course of a day.

Figure 2-23 shows the relationship between total spillage and belt speed at 1, 3 and 5 TPH for shredded MSW, RDF and ferrous metal, with 20° and 35° troughing roller angles. The spillage varied with belt speed at different TPH loading rates. The following relationships can be seen:

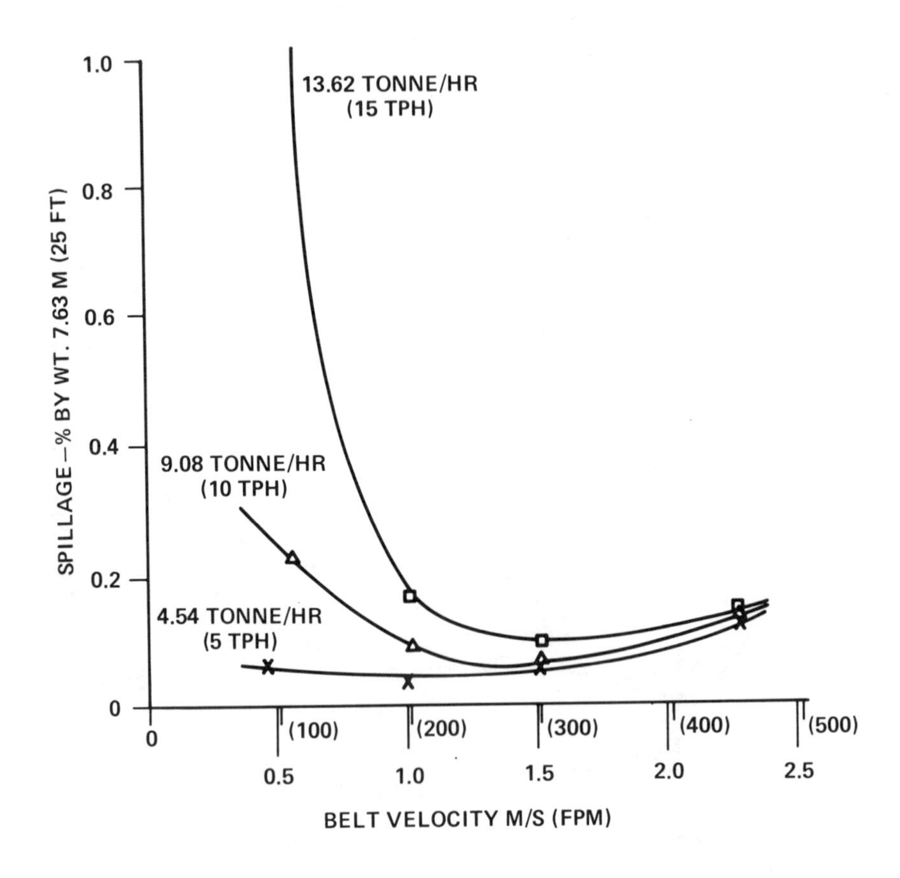

Figure 2-23. Effect of belt speed on the spillage of ferrous metal [6].

- As the belt speed is increased for a given loading, the amount of spillage decreases, levels off, then increases again.
- The heavier the feedrate at a given speed, the greater the spillage because the bed height increases.
- The 35° idlers show less spillage than the corresponding 20° idlers because the belt can hold more material when the side edges of the belt are raised.
- The more uniform and dense materials, such as dRDF, have less spillage than the fluffy, light, heterogeneous fractions.

Two factors seem to influence spillage: at low velocity, spillage is the result of a high burden; at high velocity, spillage increases linearly with velocity. By subtracting from total spillage the spillage resulting from high belt speed, the remainder will be the spillage caused by overloading the feed end of the conveyor.

This information indicates that these are the minimum speeds at which an 18-inch belt conveyor, with this configuration of feed and discharge end, must be operated to carry these loads with minimum spillage. At higher feedrates, higher belt speeds are needed to reduce the burden enough to prevent spillage. Speeds higher than these result in increased spillage but may be justified if the spillage is not serious or can be effectively collected.

The feed end spillages can be minimized by shaping the skirting and feed enclosures properly; however, due to the splashing of material and the height of the material in the accelerating zone, they cannot be avoided entirely. The head pulley and outlet losses can be reduced by good skirt design, but damp fines will persist in following the belt. Lint and dust can be prevented from entering the room by using the dust control system to maintain negative pressure in these regions.

Spillage along the belt is best controlled by using the 35° roller configuration and not loading the belt too high with material. Due to the tendency for the material to lose its angle of repose as the belt gallops over the pulleys, it is conservative to design belts carrying MSW and RDF on the assumption that the material attains a flat profile.

It was also confirmed in this study that as the inclination of the belt increases, spillage increases due to the tendency of material to fall outward. Figure 2-24 shows that at 18° the spillage is not significant, except at high belt speeds (above 1.5 m/sec). Renard and Khan [6] determined that shredded MSW could not be conveyed at inclinations above 19° without backsliding and irregular acceleration of the material at the point at which the material falls onto the belt. At this point, acceleration of the material must take place reliably to avoid jam-ups.

If the waste fraction contains a significant amount of spherical or cylindrical pieces, which are likely to roll down the belt, spillage can be expected to increase more rapidly as the speed is increased because a reduced burden will not restrain these materials as well as a heavier one.

DESIGN OF VIBRATING CONVEYORS

The carrying capacity of vibrating conveyors depends on the frequency of vibration and the amplitude of stroke, and is also influenced by the nature of the material being conveyed.

They are used for feeding and discharging shredders, air classifiers and fuel feeders and have the advantage of being simple in construction, while being resistant to damage from heavy and abrasive objects. They tend to level the materials fed to them and, hence, will deliver a more constant feed than they receive.

The theoretical principle of their operation is illustrated in Figure 2-25. The pan is caused to oscillate by a reciprocating drive, while the pan moves on hinged supports. The pan follows the sinusoidal path shown. Particles located above the pan are lifted during the rising cycle; however, due to the inertia and air resistance of the particles, they do not fall all the way back to the pan. As the pan has a forward component, the particles are progressed axially during each cycle. To assist movement of the material, the pan can be sloped. The frequency of the pan depends on the spring

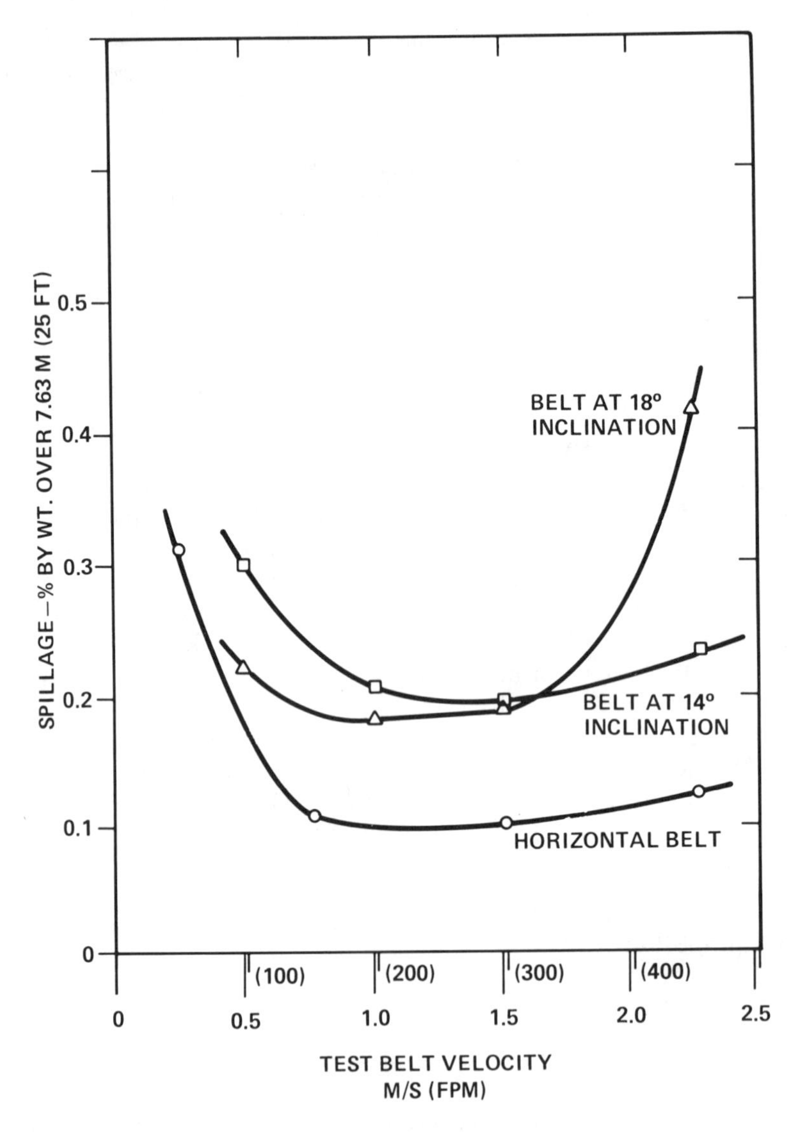

Figure 2-24. Effect of belt inclination on spillage [6].

rate of the springs and the weight of the pan. By using suitable values, a wide range of frequencies can be provided to suit the properties of the materials being handled.

While a significant amount of data is available on the performance of vibrating conveyors with dense materials, the performance of light, damp, flaky and soft materials contained in MSW cannot be readily predicted: only tests with the real materials can produce reliable data.

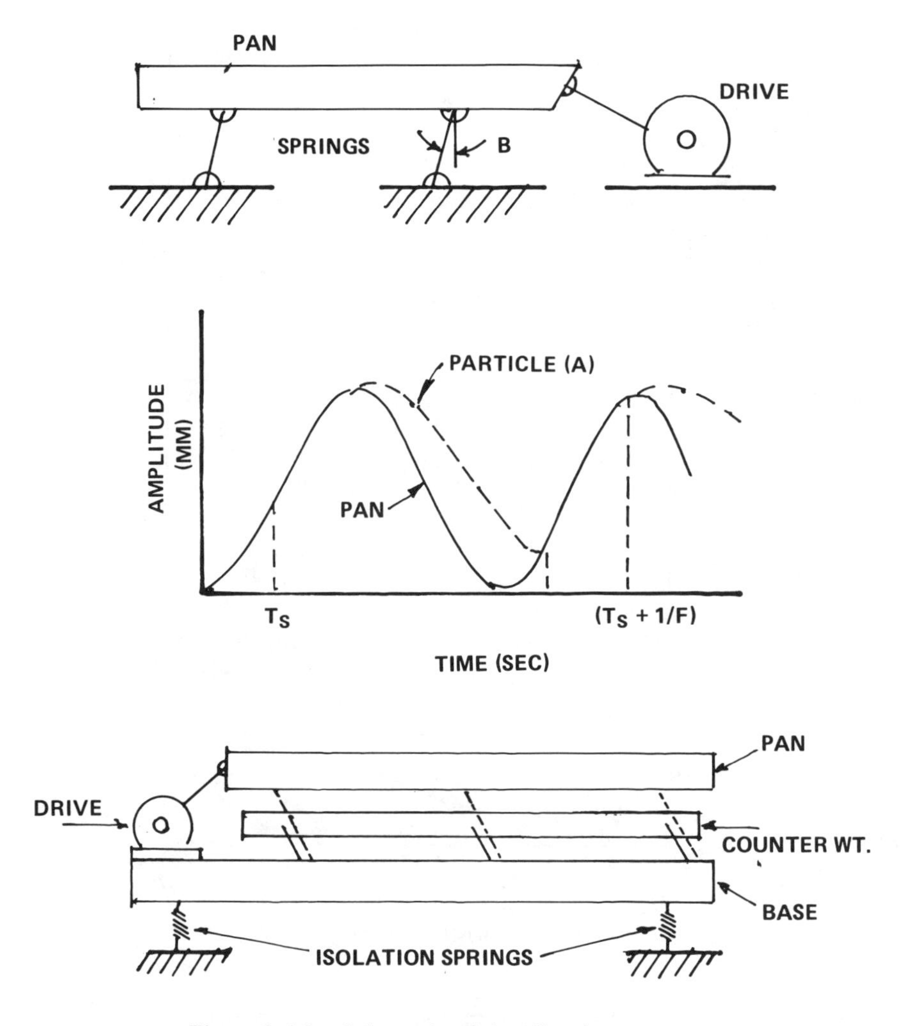

Figure 2-25. Schematic of the vibrating conveyor.

Renard and Khan [6] tested the performance of a vibrating conveyor with refuse components, using a 0.6-meter (2-foot)-wide by 4.6-meter (15-foot)-long pan conveyor driven at 430 and 545 cycles per minute (cpm). The eccentric driving cams could be changed from 12.7 mm (0.5 in) to 22.2 mm (0.875 in). The tests determined, for the two values of stroke, the maximum carrying capacity and conveying speed versus frequency for the various materials.

Figure 2-26 shows a typical set of performance curves at 13-mm (1/2-in.) stroke. The MSW and RDF did not convey well until frequencies greater than 550 cpm were used, after which a steep rise in capacity is indicated. The denser materials conveyed well at all frequencies, indicating the use of this device as a feed control.

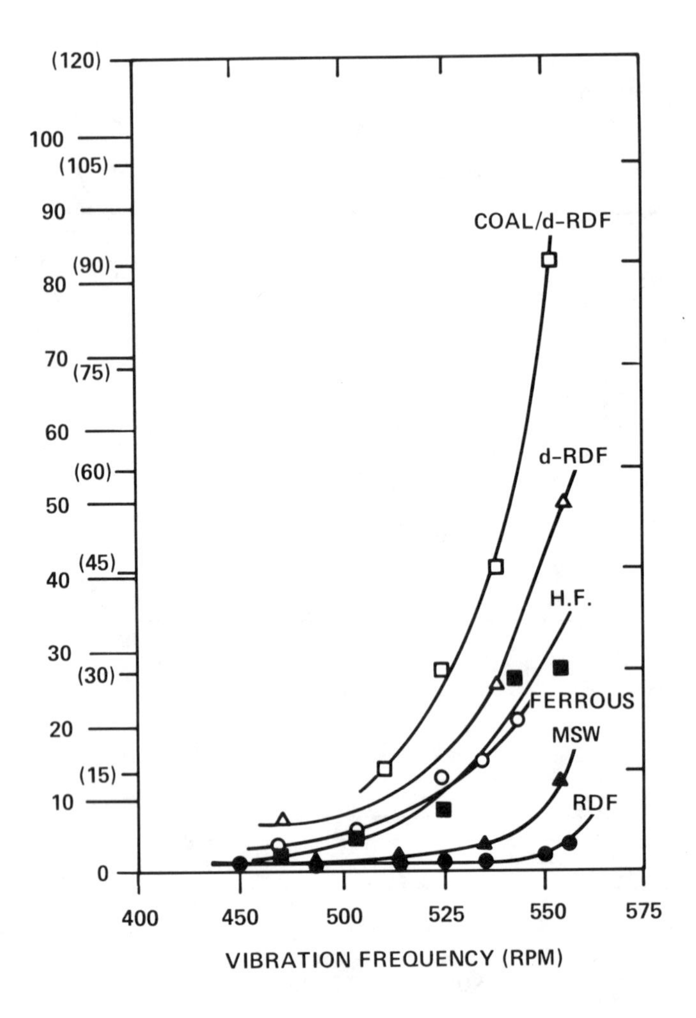

Figure 2-26. Carrying capacity of vibrating conveyor, 13-mm (1/2-in) stroke [6].

The following conclusions were drawn from the tests:

- Carrying capacity increases with frequency and stroke up to a saturation level that was not determined.
- The conveying speed increased with frequency, ranging from 0.2 m/sec (40 ft/min) with the short stroke to 0.5 m/sec (100 ft/min) with the longer stroke, at 540 cpm.
- The height of burden influenced the conveying speed. The combustible light fractions were substantially influenced by burden, whereas the dense materials were not significantly affected.
- Dust generation increased with amplitude.

The capacity of the conveyor was limited by the 8-inch-high sides and can be calculated by multiplying material density by width, burden height and velocity.

For instance, the limiting velocity with RDF was:

$$C\,(1\text{ inch}) = 46\,(\text{ft/min}) \times 2(\text{ft}) \times (1/12)(\text{ft}) \times 3(\text{lb/ft}^3) = 23\text{ lb/min} = 0.7\text{ TPH}$$
$$C\,(3\text{ inch}) = 40 \times 2 \times (3/12) \times 3 = 60\text{ lb/min} = 1.8\text{ TPH}$$
$$C\,(6\text{ inch}) = 20 \times 2 \times (6/12) \times 3 = 60\text{ lb/min} = 1.8\text{ TPH}$$

Here we see that the capacity increased as the burden was increased from 1 to 3 inches, but not as the burden was increased to 6 inches, indicating a tendency to plug up. MSW, on the other hand, shows a positive change with loading:

$$C\,(1\text{ inch}) = 72 \times 2 \times (1/12) \times 8 = 96\text{ lb/min} = 2.9\text{ TPH}$$
$$C\,(6\text{ inch}) = 50 \times 2 \times (6/12) \times 8 = 400\text{ lb/min} = 8\text{ TPH}$$

All the denser materials had positive slopes, indicating stable operation despite increased burden.

Vibrating conveyors are most efficient when used with dry, dense materials, and least effective with damp flaky materials. Nonetheless, some resource recovery plants have used vibrating conveyors under shredders rather than metal or rubber belt conveyors because the latter might not be able to withstand the impact of high-velocity objects that can be thrown from the shredder. A 6-foot-wide conveyor was tested while carrying 25 TPH of shredded MSW with a burden of 24–30 inches. At 8 lb/ft^3 this corresponds to a velocity of about 60 ft/min. To sustain this velocity, the pan must be sloped about 5–10°, with a stroke of 3/4–1 inch. With sticky or damp materials, it may be necessary to provide cleats or other devices on the pan surface to provide positive forward motion to the material.

PLANT OPERATION AND MAINTENANCE

This book would be incomplete if it dealt only with the mechanical function of the equipment and the mechanical systems. No project can be a success without the effort and devotion of people, and the concept and design of processing plants must reckon with people from start to finish. Somewhere in the equations that describe functioning plants will be people: how they work and what they do to make the plant, concept, design, construction and operation work successfully.

The "people function" is especially important in transport and handling of refuse and its products. People must decide what is wanted to be able to figure out what needs to be done and how to do it. They must learn, train others and supervise.

As the scope of this book is machine oriented, the mention of people must be brief. These are the major factors that should be mentioned:

Each person who deals with refuse must pay attention to it, if only to avoid poison ivy and broken glass. Does it require muscle? Should it be mixed before feeding? Should something be pulled out?

As refuse enters a conveying system, its potential for trouble persists. Those who find a jam-up and have to pitch in and remove the offending material learn the need to watch and warn others. Those who have to shut down and restart equipment after getting everything running so nicely develop a concern about the activities (or lack thereof) of those who might help avoid these misfortunes.

The following factors can help to avoid annoying, inefficient and destructive happenings:

- Good design: simple, responsive, not accident-prone
- Accessibility: to view, reach, inspect and maintain
- Maintenance-free: large safety factors; protected from dust, water, steam; sturdy equipment; reliable cooling systems; wear parts easily replaced; standard parts; wear-resistant materials and designs
- Watchable: direct view from safe, accessible locations; inspection ports and doors; television cameras
- Trouble-sensing devices
- Communication: control room to tipping floor and plant; radio to maintenance personnel; radio to process monitors; refuse trucks, receiving, weighing; container handling

A well-designed plant will have minimum need for people, but sufficient help for them so they can do what has to be done.

Safety is the result of good design, reliable equipment and lack of temptation to do what is unsafe. What must be done should be done safely and not under pressure or emergency conditions.

To achieve good plant design requires broad, continuous and thorough communication between people with the necessary ability, knowledge, experience and training, and the ability to hear, listen, reflect and think about what others think, say or do. Management of the human effort can be a process far more complex than that of inanimate machinery.

REFERENCES

1. Scott, P.J., and J.R. Holmes. "The Capacity and Principal Dimensions of Refuse Storage Bunkers in Modern Incineration Plants," paper presented at the ASME National Incineration Conference, 1974.
2. "Refuse Collection Practices," *Am. City and County* (1980).
3. Woodyard, J.P., J.C. Anderson, J. Wood and S.S. Passage. "Estimating Solid Waste Quantity and Composition for Resource Recovery: A Case Study of the Hackensack Meadowlands Survey," 1982 ASME SWPC, New York (1982).
4. Hessler, "Transfer Stations," *Solid Waste Management* (December 1980).
5. CEMA. *Belt Conveyors for Bulk Materials* (Boston: CBI Publishing Co., Inc., 1979).
6. Renard, M.L., and Z. Khan. "Design Considerations for Municipal Solid Waste Conveyors," *Seventh Annual Res. Symp.* EPA-600/91-81-002c, USEPA/MERL, Cincinnati, OH (1981).

CHAPTER 3

Size Reduction

Recovery of valuable materials from MSW can be accomplished by processing it to release, separate and remove the desired components. To do this effectively, the materials in MSW require size reduction for these reasons:

- Plastic bags, bottles and cans must be broken to release their contents.
- Textiles, wood and other miscellaneous large objects must be cut to small sizes for further processing.
- Size reduction reduces the tendency of materials to tangle.
- Storage and retrieval of waste-derived materials become more efficient after size reduction.
- Magnets cannot effectively remove large pieces of metals from a mixture of other large objects.
- Air classifiers are not effective in separating objects that are large in proportion to the cross-sectional areas, objects that can tangle and clumped materials. Size reduction expands the refuse and creates aerodynamic shapes.
- Combustion is slow when objects are large and dense, and inefficient when some objects are large and others small.

Prior to the size reduction step, it is desirable to pick the refuse for objectionable items, such as rugs, tanks, containers of liquid combustibles, furniture, refrigerators, and so forth, and worthwhile to salvage good corrugated cardboard cartons and paper that can be compacted and sold. It also helps to screen out glass and other materials that should not be shredded, as well as small-sized combustibles that do not need to be shredded. The machinery that handles raw MSW must have feed apertures larger than the largest object. After size reduction, the openings and apertures in the system can be reduced accordingly. To avoid jamming, openings should be two to four times the size of the largest particles encountered.

Size reduction of materials can be accomplished by one of more of these actions:

- **Impact (dropping, smashing or crushing)**: effective in breaking brittle materials, such as stone, glass and wood, and in opening cans and breaking metal objects.
- **Shear (cutting or tearing)**: effective in size reducing paper, cardboard, plastics, light metals and wire.
- **Attrition (wearing and grinding)**: effective in reducing materials to lint and powder by surface abrasion.

- **Densification**: effective in improving storage, retrieval and burning properties of refuse fuels.

Most of the machinery used in waste recovery operations has been developed for other purposes, although a few machines have been developed specifically for refuse processing.

These are some of the machines now in use, in order of severity of duty and ruggedness of construction:

- **Slitters**: for paper products only
- **Bag breakers**: open plastic bags and cardboard boxes
- **Chippers**: convert wood into chips
- **Granulators**: convert wood and plastics to granules
- **Pelletizers**: densify materials into pellets
- **Briquetters**: form materials into briquets
- **Cutters**: cut wood, light metals, rubber, plastics, paper
- **Flail mills**: cut paper and wood, smash glass and metals
- **Hammermills**: reduce most materials from metals to paper
- **Ball and rod mills**: grind feed to powder
- **Cascade mills**: size reduce most materials

TYPES OF SIZE REDUCTION MACHINERY

The devices capable of size reducing the materials in MSW range from automobile shredders, which can handle anything, to granulators that can only handle materials that will not damage the sharp knives and will pass through the discharge holes without plugging them. The basic types are shown in Figure 3-1.

Cutters

Cutters, such as chippers, shears and granulators, use sharp edges to shear materials, after which they are passed through apertures.

Chippers are high-speed rotary devices (up to 1800 rpm) designed to reduce logs or heavy members of wood to chips by allowing the wood to enter a cavity in a rotating cylinder, sufficiently for the sharp cutting blades to shave off a chip as the moving blade passes a stationary blade. These machines are as unsuited to cutting paper or rubber as they are ideal for wood. The cutters can be damaged by metals and other hard objects in the feed.

Granulators are medium-speed rotary devices (up to 100 rpm) having sharp knives in the rotor that engage in a scissor-like manner with angled stationary knives to cut plastics and other dense, uniform materials. The feed material is kept in suspension above a grate so that the cutters keep cutting the material until it will pass through the holes in the discharge grate. These machines are designed to minimize fines, as are the chippers, and rely on the sharpness of knives. They are not suited for sticky or abrasive materials. Air movement is required to assist the material in passing through the grate.

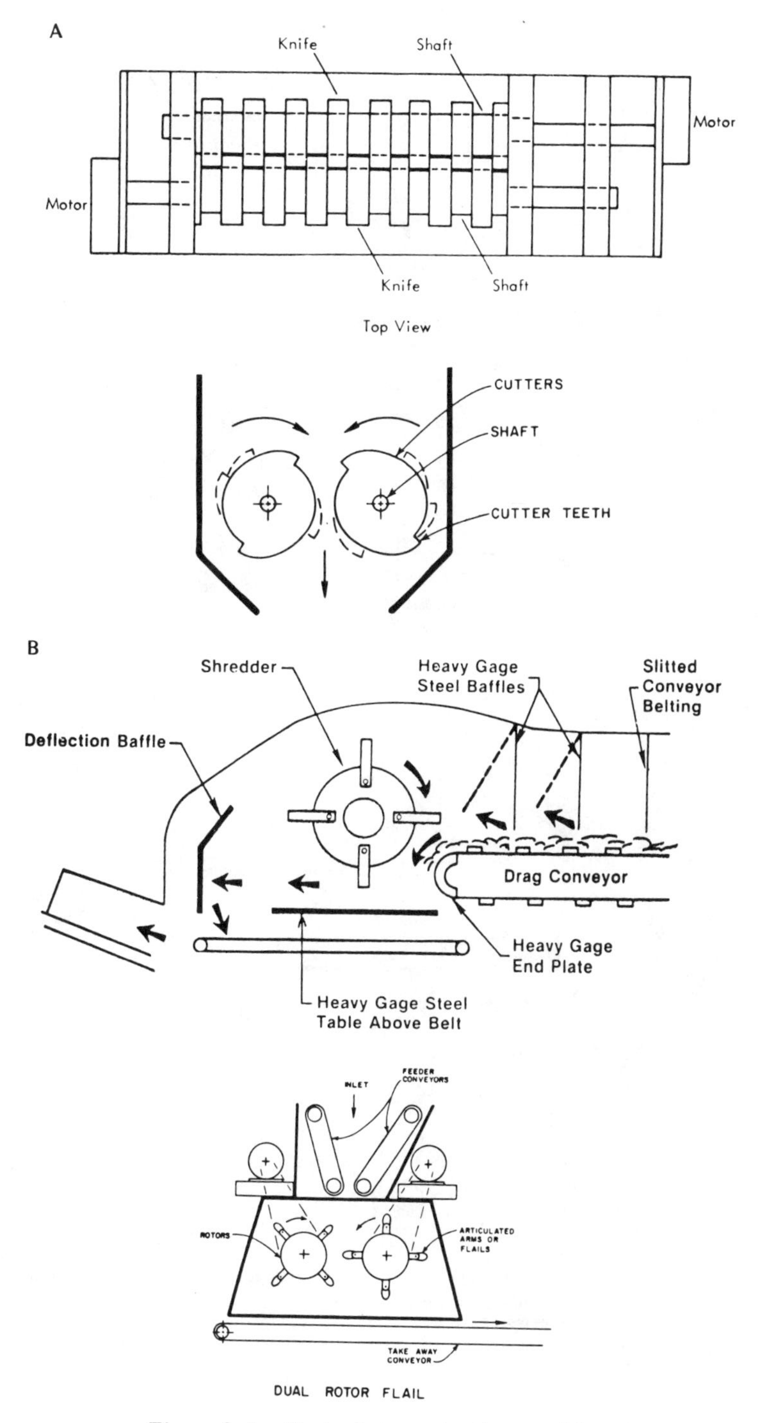

Figure 3-1. Typical size reduction machinery.

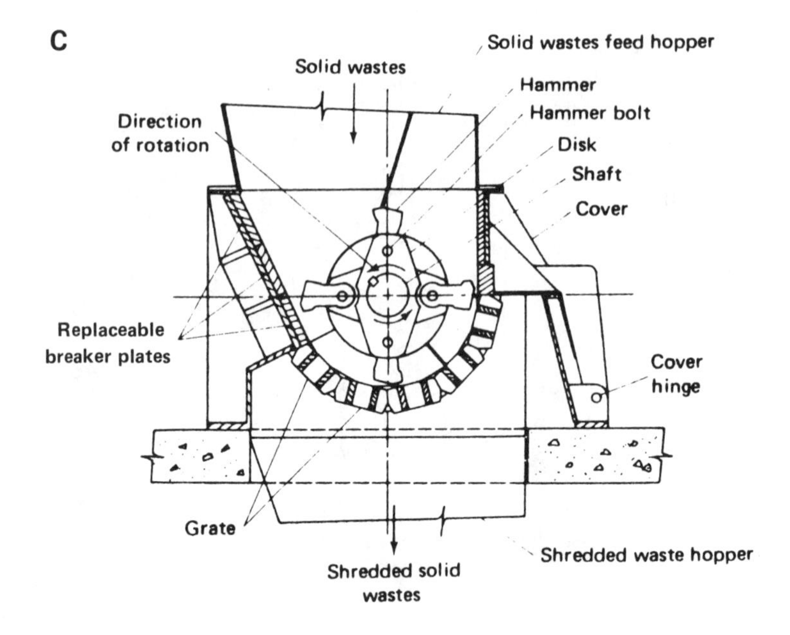

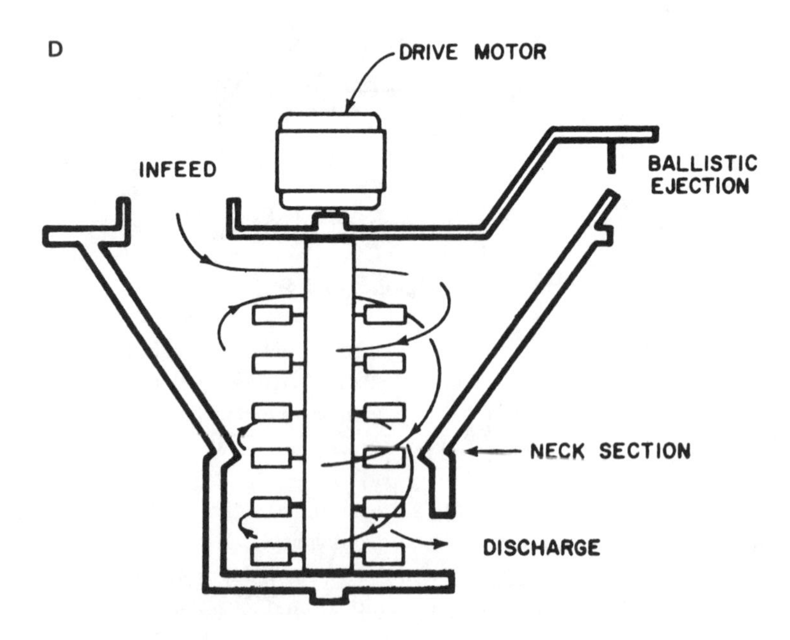

Figure 3-1. (Continued)

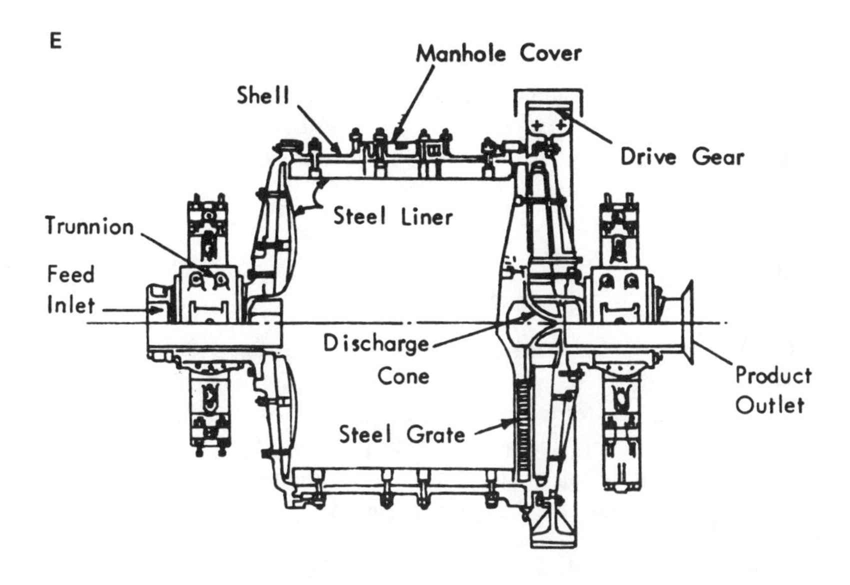

Figure 3-1. (Continued)

Rotary shear cutters are slow-moving devices (20-30 rpm), having one or two rotors. The single rotor type has blades mounted on the rotor that pass by stationary blades, shearing the feed and forcing a portion of it through the gap between the blades. The double-rotor type has crescent-shaped cutters, assembled on the shafts with spacers between them. The rotors move in opposite directions at different speeds, so that the cutters engage the feed material, shear it and carry a portion of it through the space between the rotors. This results in a shearing and tearing action that is effective in cutting tires, metals and other dense materials into slabs of a size determined by the dimensions of the gap. These cutters are particularly suitable for shredding wood, nonferrous extrusions and castings, and aluminum and light steel cans. Municipal refuse without excessive amounts of paper and cardboard can be shredded by these machines. Paper is particularly difficult for this type of machine to handle due to its toughness, which requires great power and sharp knives, and its low density, which reduces capacity. When used for cutting rubber, cutters can cause the rubber to heat up to a point where it gets sticky, ultimately reducing capacity. This heat can be controlled by water cooling or by feeding sufficient amounts of other materials, such as paper and wood.

Rotary shears are usually driven by hydraulic drives, which will relieve when loaded beyond their relief valve setting. Thus, they are not subject to damage when heavy metal objects are encountered or when the piece of material engaged becomes compressed in such a way that it cannot be passed. The Saturn rotary shear is provided with a hydraulic control, which automatically reverses the rotors to discharge the jammed material, then returns to normal rotation.

The capacity of a rotary shear is limited by the length and diameter of the rotors, which represents the size of the feed hopper; the circumferential speed of the rotors and the clearance between them; the density of the feed material; and the ability of the cutters to grab the feed.

The machines are well suited to relatively low-capacity applications. The largest Saturn cutter, equipped with two 300-hp hydraulic drives, is suitable for shredding municipal and industrial refuse, passenger and truck tires and wood pallets.

Flail Mills

Flail mills are relatively high-speed (up to 1800 rpm) horizontal-shaft rotary machines having one to four rotors, to which are attached swinging hammers, either single or double articulated. Single-rotor flail mills are used as bag-breakers. The rotors are spaced so that the hammers nearly touch when running free, to engage and impact material entering from the top or side, cutting or smashing it as it is passed through. The rotor clearance, minus the hammers, determines the dimension of the largest object that can pass through without change of shape, such as a log of wood. The rotors may be operated at different speeds to increase shearing effects.

The hammers fold back or "relieve" when they engage dense objects, passing them through relatively unaffected. This reduces the power requirements of the mill but requires that these objects be separated from the product downstream of the mill. A rotary screen is useful for this purpose.

Flail mills tear or smash the feed without forcing it through a screen. Large or difficult objects are passed through directly, whereas lighter materials may be entrained by the flails and passed through more than one time. Breaker bars, located under the flail discharge, are used to smash objects that otherwise could fly through unaffected.

Hammermills

Hammermills have rotating shafts (800–1200 rpm) onto which are attached stationary or pinned (swing) hammers, which attack the feed material in different ways and drive the material through the discharge, with or without grates. Grates may range in size from a full discharge opening, with only breaker bars to smash the material, to small holes that ensure that all material passing through will be less than the hole size. The thickness of the hammers used ranges from 1/8 inch to 6 inches and that of grate openings from 1/8 inch to 12 inches. The feed handled ranges from automobiles to office wastepaper. There are two basic types:

- **Horizontal-shaft hammermills** have hammers that smash the materials against the breaker bars and the grates; oversize objects are retained until the material is able to pass through the grate, unless it is rejected into a metal trap.
- **Vertical hammermills** control particle size by the clearance between the hammers and the irregular wall, rather than by the size of the grates.

Some rugged, high-horsepower hammermills are designed to shred large and difficult objects, such as appliances, pallets, logs, tires and engine castings. Lighter machines must be protected from such feed materials.

If such objects can arrive in the feed and can fall into the mill, either the mill must be designed to discharge these objects through the grate or another discharge must be provided. Several methods are available to accomplish this.

Horizontal-shaft mills can be equipped with metal traps that catch the metals that do not get cut to grate size. These traps become ineffective if they fill up during the operating period of the machine.

The Heil vertical-shaft mills are designed with conical sides to throw dense objects out of the mill. In this way they are not forced to remain in the machine until they are reduced to a size that can pass out the discharge openings. Along with the undesirable heavy objects, a considerable amount of other material is discharged to a conveyor.

Vertical-shaft mills rely on the clearance between the hammers and the conical cylinder within which they rotate to determine the size of the particles discharged, rather than the size of the discharge openings. Vertical bars are attached to the cone to provide more effective cutting of tough materials, such as textiles. They are sensitive to air movement through the grates and work best with an induced air flow.

Many of the mechanical and design features of hammermills influence their performance, such as feed configuration, grate size and shape, means of adjusting hammer-grate clearance, ability to relieve under jamming situations, and ability to operate in both directions to avoid reversing hammers. The metallurgy of the liners and hammers, methods of building up worn hammers by hard-facing, and optimum methods of replacing hammers have been the subject of extensive investigation, particularly by the user.

Densifying Machinery

Pelletizing machines are used to compress shredded refuse fuels into dense pellets, which can be stored, retrieved and fed to be burned on stoker-fired boilers. The pelletizer feed must have sufficient moisture content and be reduced to a small enough particle size before feeding.

Powdered refuse fuel can be fed to a briquetting machine and densified into briquets suitable for burning in a coal stove or stoker.

The true densities of pellets and briquets must be greater than 65 lb/ft^3 to assure integrity. Bulk density of these densified refuse fuels is about 25-35 lb/ft^3. About 15 KWh/ton of power is required for densification [1].

APPLICATION OF SIZE REDUCTION MACHINERY TO MSW

Hammermills with Grates

The actions that take place in a hammermill with grates are complex and depend on how heavily the mill is loaded relative to the capacity of the grate. As material is fed to the shredder, its particles will be impacted by the hammers and carried past the breaker bars and grate members, thus breaking them into smaller particles. The smaller particles have a high likelihood of passing through the grate, whereas those that don't pass will be carried up by the rotor for another pass.

As loading is increased, more particles remain above the grate and are subjected to repeated impact. At higher loadings, a mass of particles accumulates above the rotor, increasing the density of material within the rotor envelope. This "holdup" of material above the rotor can become unstable, surging up and down and producing a wide range of particle sizes.

The cutting and tearing action that predominates when the particles are isolated changes as loading is increased to a condition in which the particles interact with each other in a wearing fashion, called attrition and autogenous grinding.

As the materials are recirculated more and more, the product of paper and cardboard shifts from flakes toward lint, increasing the amount of "fines" in the product. This may or may not be desirable according to the ultimate use of the product. The power required for this fine grinding is substantially more than for simple cutting, so that the mill draws more power as the loading drives the mill into the grinding mode. Induced air movement is essential to good performance to assist in removing the shredded paper-type materials.

The feed capacity of a hammermill is limited by its drive motor. The motor must have the ability to accelerate the rotor and hammers to the running speed and, under running conditions, must be capable of handling the substantial current surges resulting from the normal ranges of feed variation.

The amount of reserve capacity required in the drive motor depends on the amount and type of material that could drop into the shredder before the overloading can occur and be detected. Most shredder installations are provided with overload controls, which sense the current draw and stop the feed to the mill until the amperage falls below a set level. The range between the average operating point and the motor overload point results in overdesign. Operating motors far below their rating results in a drop in electrical and mechanical efficiency.

Hammermills without Grates

Flail mills, which do not have grates, do not accumulate material above the rotors to the same extent; hence, they have a power draw and a product size almost directly proportional to the rate of feed. As they are essentially "once-through" machines, especially for the difficult materials, they are not subject to as much overload resulting from tough or dense feed material. On the other hand, because they do not have a limiting screen, they will pass oversize material.

Vertical hammermills, which rely on the clearance between the hammers and the walls at the exit, build a holdup in proportion to load as they are loaded and are sensitive to the toughness of the material that must be shredded to pass through the exit clearance.

Rotary Grinding Mills

Ball and rod mills have cylindrical rotary drums containing balls or rods, which cascade as the drum turns, thus grinding the feed material that passes through the drum. They are equipped with internal grates, which allow the ground material to pass out of the grinding area while retaining the grinding media. Ball mills with 15- to 30-mm balls were used in the Eco-Fuel process to grind minus 125-mm shredded municipal waste.

Cascade mills have large-diameter rotary drums containing large (100-mm) balls, which are dropped or cascaded onto the feed, grinding it until it is small enough to be forced through grates that won't pass the balls. They have been demonstrated to be suitable for shredding raw municipal waste to minus 50 mm in size.

Ball mills produce a rolling action as the balls are carried up the drum to a point where they cascade off the wall or the lifters and flow back to repeat the cycle. The grinding action applied to materials such as MSW is a combination of crushing, stretching and autogenous grinding.

In cascade mills, the balls are elevated higher so that their kinetic energy is applied by impact, in addition to the rolling and crushing action caused by the vertical height of the ball bed.

Ball mills and cascade mills advance the feed material slowly down the mill, providing a succession of grinding operations—the equivalent of several stages of shredding. The power required to drive ball mills is nearly constant under all operating conditions because it is determined mainly by the quantity of balls in the mill, the height to which they are lifted and the rotating speed. As the feed is increased, so is the level of the balls slightly, causing a corresponding increase in power draw.

The amount of material that can be fed is limited by the volume between the balls and the rate of grinding. When excessive material is fed, it floats over the ball bed and does not get ground. When too little material is fed, voids will develop and the balls will grind each other.

The amount of grinding on the feed is related to the feedrate because the power is essentially constant. Fluctuations in the rate of feed do not create equal fluctuations in the quality of the product because the mill serves as an accumulator and tends to discharge at a steady rate.

Ball and cascade mills are thus essentially constant-capacity grinders, which should be operated as near as possible to their optimum capacity by careful control of the rate of feed. Material that is not ground must be screened from the product either internally or externally, and rejected or recycled.

Cascade mills produce a steady discharge despite irregular feed and deliver a product similar to that of hammermills with less power input because they are mechanically efficient.

GENERAL NATURE OF THE PRODUCTS OF SIZE REDUCTION MACHINERY

There is a practical limit to the amount of size reduction that can be carried out in a single machine. MSW can be reduced to 2-6 inches in a single machine. Further reduction generally requires a second stage of shredding.

Large materials are generally size reduced by cutting or tearing. Smaller materials are subjected more to wearing or attrition of the surface of the particles. Especially as the particle size approaches its thickness, more and more wear takes place between particles of the material itself. This action is called autogenous grinding, in which the forces are applied to masses of material that grind themselves.

Cut materials are relatively difficult to handle because they are essentially nonflowing in nature, making them well suited to landfill operations in which stability is desired and very unsuited to retrieval from storage for the same reason. On the other hand, ground materials, when dry, are relatively free flowing unless they contain excessive amounts of lint or long fibers.

In between the cut materials and powdery materials are linty materials, resulting from cellulosic fibers, for instance. These are also essentially nonflowing or poor-flowing materials. Lint is generated by most operations applied to paper and cardboard by interparticle action.

When materials are reduced in size, they acquire distributions of particle size that depend on the type of processing. When the machine has a screen or grate through which all the material processed must pass, a maximum particle size is imposed on the product, and most of the particles are much smaller than the screen. Cutting machines may achieve a more uniform size distribution due to the more precise operation of feeding material to knives or cutters.

Each type of material tends to arrive in a characteristic range of sizes and to be reduced to different degrees by mechanical operations. This fact leads to many contradictions in size reduction.

Screening and sizing operations generally can control only two dimensions of the particles. The third dimension, length, can be a serious troublemaker because wires, rods, bolts and long textiles and tapes can pass through relatively small apertures.

While a shredder may cut paper to convenient small pieces, it may pass textiles with relatively little size reduction. While a trommel screen may stop paper from passing, it may unwind a spool of typewriter ribbon and create tangles outside of the trommel drum.

Wine bottles and smaller beverage bottles will be broken into shards or fine splinters and sand, depending on the amount of energy applied to the bottles. To remove and recover beverage cans, they should be opened to release their contents but should not be abused so much that they grab onto combustibles, resulting in contamination of the metals.

Combustibles such as cardboard, paper, plastics, textiles and wood should be cut fine enough that they are easy to convey, store, retrieve, air classify and can be subjected to magnetic separation. When burned, their size should suit the method of combustion, or vice versa. The combustion system must be designed to cope with the fuel product.

Table 3-1 summarizes the relationship of degrees of size reduction, the equipment required for burning and the relative time required for combustion.

HISTORY OF RESEARCH IN SIZE REDUCTION IN THE FIELD OF WASTE PROCESSING

Shredding of MSW developed from the desire to produce a material that could be landfilled without a dirt cover and on which rats and bacteria would not thrive. Hammermills had been developed for wood, mineral and automobile crushing and were rugged and capable of shredding the various materials found in MSW. Some types, particularly the vertical shaft mill, were developed expressly for refuse shredding.

The U.S. Public Health Service (USPHS) and the EPA Municipal Environmental Research Laboratory (MERL) funded research on the application of hammermills to MSW shredding at Madison, Wisconsin under Robert Ham [2]. Subsequent studies addressed resource recovery applications. Data were developed at the University of California, Richmond Field Station, under George Trezek [1] and, later, field test data were important steps in developing an information bank for those considering the use of shredding and wishing to understand shredder performance.

The work done at Richmond included characterization of the properties of MSW, including stress and strain data, density and compression characteristics, the properties of single, double and triple-shredded refuse, as well as analysis of the effect of air classification and screening of the shredded products.

Table 3-1. Relationships Between Combustible Size and Use

Size	*Preparation*				*Burning Method*			*Time to Burn*		
	Shredded					*Suspension*				
(in.)	*Single*	*Double*	*Densify*	*Grind*	*Grate*	*Semi*	*Full*	*Hour*	*Minute*	*Second*
4–6	X				X			X		
2	X				X				X	
1		X			X					X
3/4		X	X		X			X		
3/4		X				X			X	
1/2		X	X		X				X	
1/2		X				X			X	
Lint		X	X		X				X	
Lint		X		X			X			X
Powder		X	X		X			X		
Powder		X		X			X			X

Theoretical principles were investigated to determine to what extent the happenings within the shredder could be simulated by mathematical models, as well as the relationships among shredder speed, shape, hammer configuration, grate opening size, moisture and other factors influencing the power consumption, particle size distribution of the product and the rate of hammer wear.

The major maintenance problem with shredders is hammer wear. Trezek's work developed information on the rate of wear as determined by machine capacity, operating conditions and hammer shape and configuration. The NCRR also did extensive research to further the recovery of resources from MSW.

Many types of size reduction devices have been tested. As the performance of a shredder is determined by the properties of the feed material, comparisons of different shredders are difficult to make in most cases because raw MSW is so highly variable in nature. Only in the case of secondary shredders is information available on both the input and output material. Weinberger [3] reported on tests performed on two shredders in series in an attempt to determine the optimum capacities and grate sizes for a Heil vertical-shaft shredder followed by a Gruendler horizontal-shaft machine.

Even when tests are performed in a laboratory and efforts are made to feed the machine at a constant rate, power consumption varies significantly during the test due to the highly variable material and consequent variations in feedrate. Peak values encountered during the tests, the ratio of average to peak and other pertinent information is difficult to obtain. The high variability of the waste stream tends to obscure the variations we are trying to discern and measure.

The tests reported by Vesilind [4] on a full-scale shredding installation at Pompano Beach, Florida are unusual in that the plant was operated for two consecutive weeks at full capacity, generating data on the degree of variability that is characteristic of shredding MSW. The results show that the shredder was operated, on average, at about 75% of its motor rating, although fed at the highest rate possible without causing it to kick out on overload. The PSD of the product was consistently inconsistent, as indicated by the many samples taken.

A thorough and innovative study by the Cal Recovery group, conducted for the EPA, developed many new concepts of shredding and first introduced the principle of matrix notation and breakage functions to solid waste processing [1].

Much of our present knowledge of size reduction comes from the coal and ore processing industry. The pioneering work of Austin [5], Gaudin [6], Evans and Pomeroy [8] and Broadbent and Calcott [9, 10] has been especially useful.

PARTICLE SIZE DISTRIBUTION OF RAW AND SHREDDED MSW

MSW contains many types of materials, each of which has a characteristic range of particle sizes and is treated in a distinctive way by the shredding process.

In a classic study, Ruf [11] analyzed typical household refuse, raw and shredded. Figure 3-2 shows the "cumulative" PSD of the main species found in the

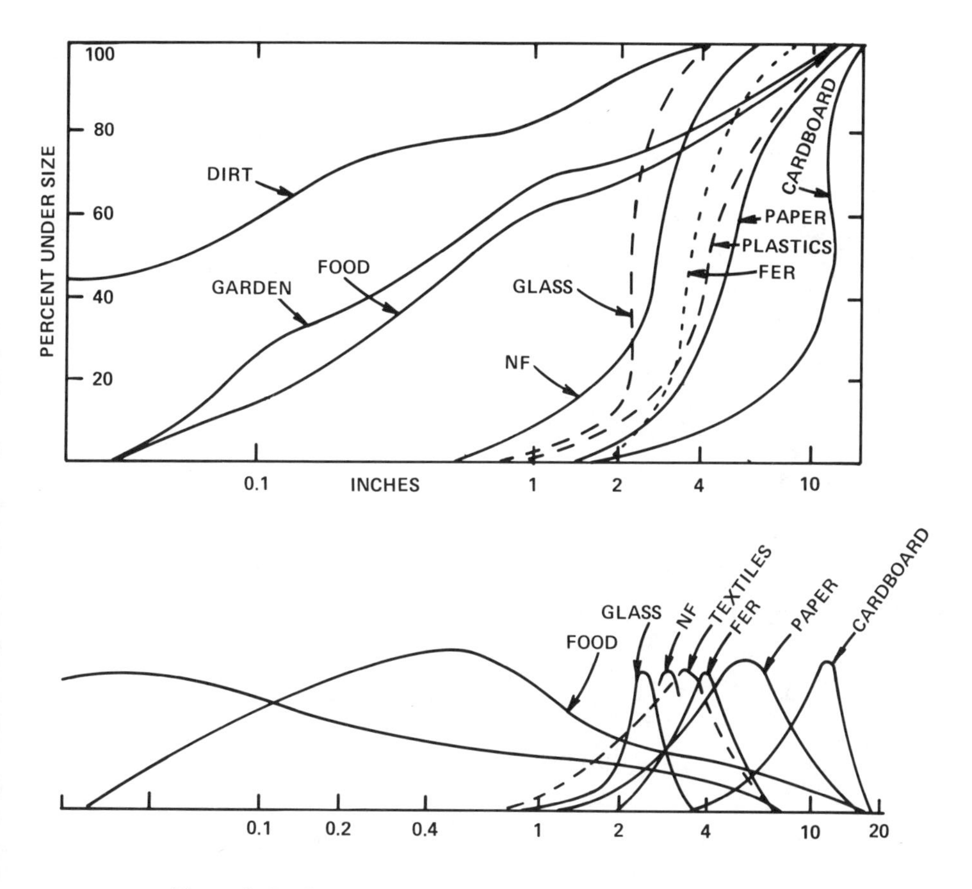

Figure 3-2. Particle size distribution of raw household refuse.

raw refuse: cardboard, paper, plastics, textiles, glass, wood, garden waste and fines (grass, dirt and sand). Figures 3-3 and 3-4 show what happens to the size distribution of the different species as they are processed through one and two stages of shredding.

The "histograms" of these distributions are shown below the PSD graphs to visualize the characteristic peak or mean sizes of the various constituents of the refuse. These are listed in Table 3-2 for comparison.

Note this information about shredding:

- The first stage of shredding reduced the particle size of all species by 3 to 4 times.
- The second stage reduced the size of some materials by 3.5 times, but some smaller-sized materials, such as food, yard waste and sand, were not reduced at all.
- The particle sizes of various ingredients in the product varied widely in size: the shredder did not reduce them to the same size.
- The size screen through which 90% of the material would pass was not reduced as much as the median size: 2.3 times in one stage, 1.3 times in the second stage of shredding.

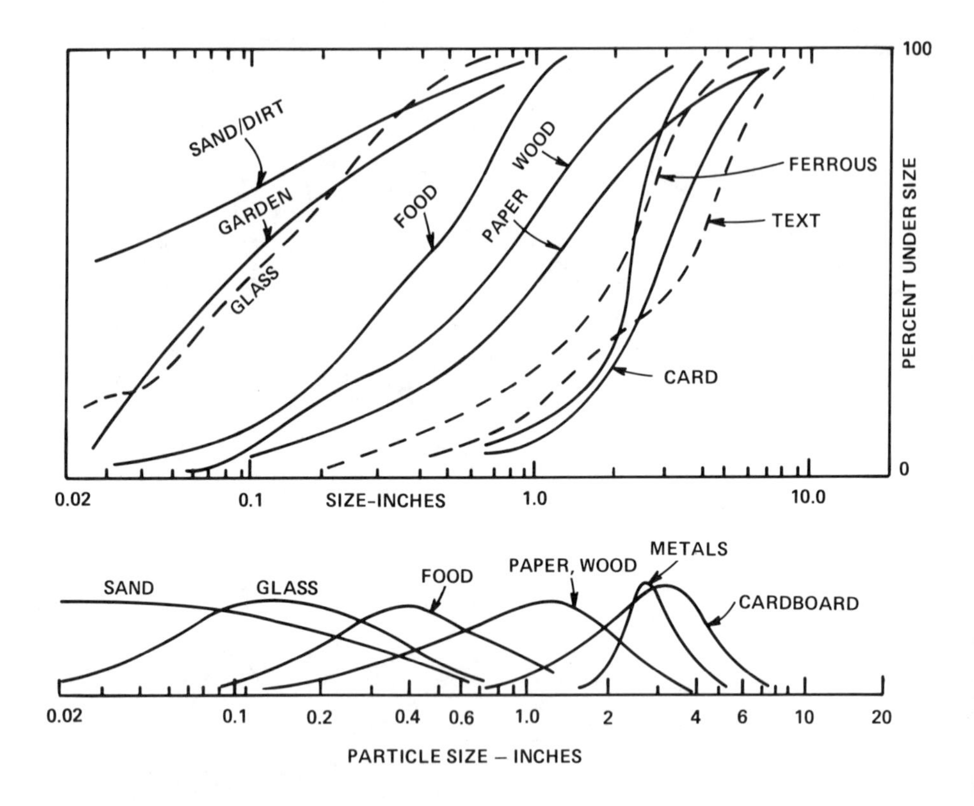

Figure 3-3. Particle size distribution of single-shredded MSW.

- The composite of all materials was reduced 3 to 4 times in each stage of shredding.
- Textiles were not affected much by the first stage, but were reduced three times in the second stage.
- Glass was drastically reduced in both stages of shredding: 23 times in the first, 20 times in the second stage, turning it into sand.

Also note these important principles:

- Sheet materials appeared to be affected in relation to the size at which they entered, being reduced by the same multiple in each shredding operation.
- Some materials were not affected by shredding (food, yard waste and sand).
- Glass was smashed into fine particles.

The semilogarithmic plot used by Ruf conveniently shows the entire span of sizes, but the particle size distributions resulting from the linear vertical axis show as S-curved lines, sharply curved near the axes.

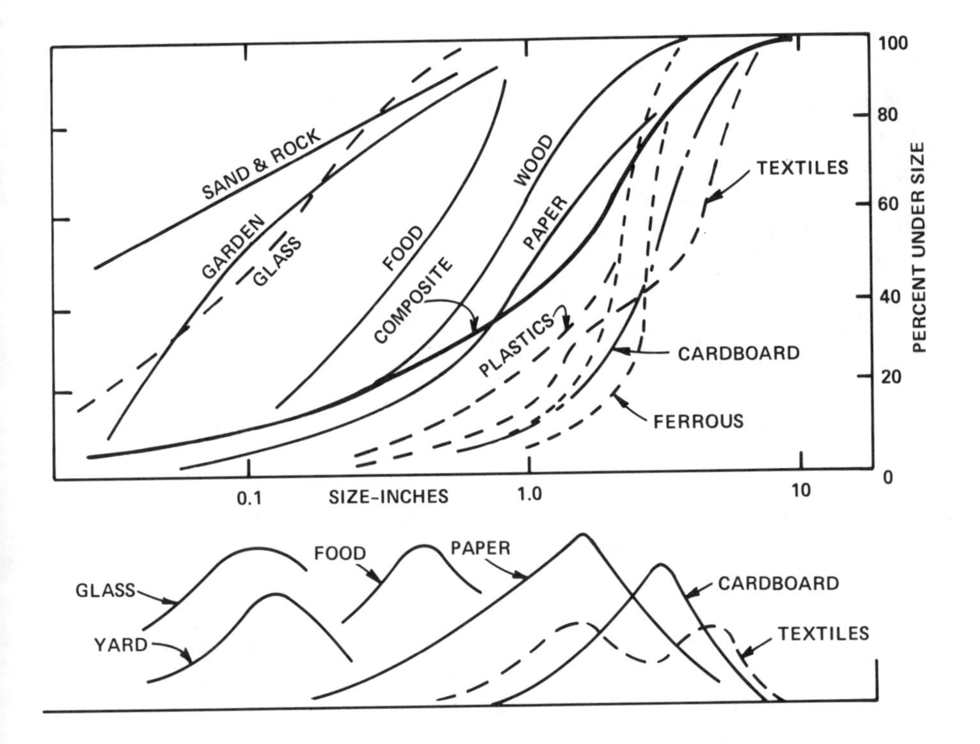

Figure 3-4. Particle size distribution of double-shredded MSW.

Table 3-2. Comparison of Mean Sizes of Species and Degree of Shredding

	Medium particle size (50% by wt), in.		
Species	*Raw MSW*	*Single Shred*	*Double Shred*
Cardboard	11	4	1.2
Paper	5	1.25	0.35
Plastics	4.5	2.0	1.0
Textiles	4.0	4.0	1.4
Ferrous	3.8	2.5	1.5
Wood	2.3	0.8	0.4
Glass	2.3	0.1	0.05
Food	0.6	0.4	0.4
Garden	0.4	0.1	0.1
Sand and Rock	0.07	0.02	0.02
Composite	4.5	1.5	0.4
Size (90% passing)	15.0	6.5	5.0

Rosin and Rammler [12] found that particle size distributions of broken coal, when plotted on double-log versus log paper, generally fall on straight lines with characteristic slopes, depending on the material and the operations performed on it. Basic physical principles relating to breakage of particles account for this mathematical phenomenon.

The following equation was developed to express the relationship between the fraction of the total sample of material that would be retained on a given screen size and that screen size:

$$\text{Fraction retained:}\quad R = -\exp\left(\frac{-D}{X_o}\right)^n$$

$$\text{Fraction passed:}\quad P = 1 - \exp\left(\frac{-D}{X_o}\right)^n$$

where R = the fraction of sample retained on a particular screen size, D, representing all material larger than that size

P = the fraction of sample passing a particular screen size, D, representing all material smaller than that size

D = the screen size, ASTM standard square mesh, expressed in microns, millimeters or inches

X_o = the "characteristic," or peak, particle size, which is the size screen through which 63.21% of the sample passes or on which 37.79% is retained; same dimensions as D

n = the slope of the particle size distribution line as it passes through the characteristic particle size or the slope of the line drawn through all the data points

Other points of interest are X_{80}, used in the Bond work index equation: X_{90} and X_{95}; both are often used to define size of the product of shredding and grinding operations.

The slope of the line, "n," can be determined by measurement of the graph, or from properly selected data points, using the following expression:

$$R = 1 - \exp\left(\frac{-D}{X_o}\right)^n$$

$$\ln\left(\frac{1}{1-R}\right) = \left(\frac{D}{X_o}\right)^n$$

$$\ln\left[\ln\left(\frac{1}{1-R}\right)\right] = n\ln\left(\frac{D}{X_o}\right)^n$$

$$n = \frac{\ln D - \ln X_o}{\ln\left[\ln\left(\frac{1}{1-R}\right)\right]}$$

When the double log, $(\ln(\ln(1/1-R)))$, is plotted vs lnD on log-log paper, a straight line often is obtained for broken or ground materials. This is the basis of Rosin-Rammler and Weibull graph papers. When Rosin-Rammler or Weibull paper is not available, it is convenient to plot $\ln(\ln(1/(1-R)))$ versus (lnD) on linear paper, where D is the particle size in any units. The slope is "n" and the size at $\ln(\ln(1/(1-R))) = 0$ is the characteristic size, X_o, as shown in Figure 3-5.

The value of using Rosin-Rammler (RR) plots for interpreting breakage of brittle materials can be demonstrated by plotting size distributions for broken glass, a well-known component of refuse. Figure 3-6 shows a series of lines representing the PSD of glass that has been broken by a variety of methods. The largest particles were obtained by dropping glass beverage bottles from a height of 9 feet. The data

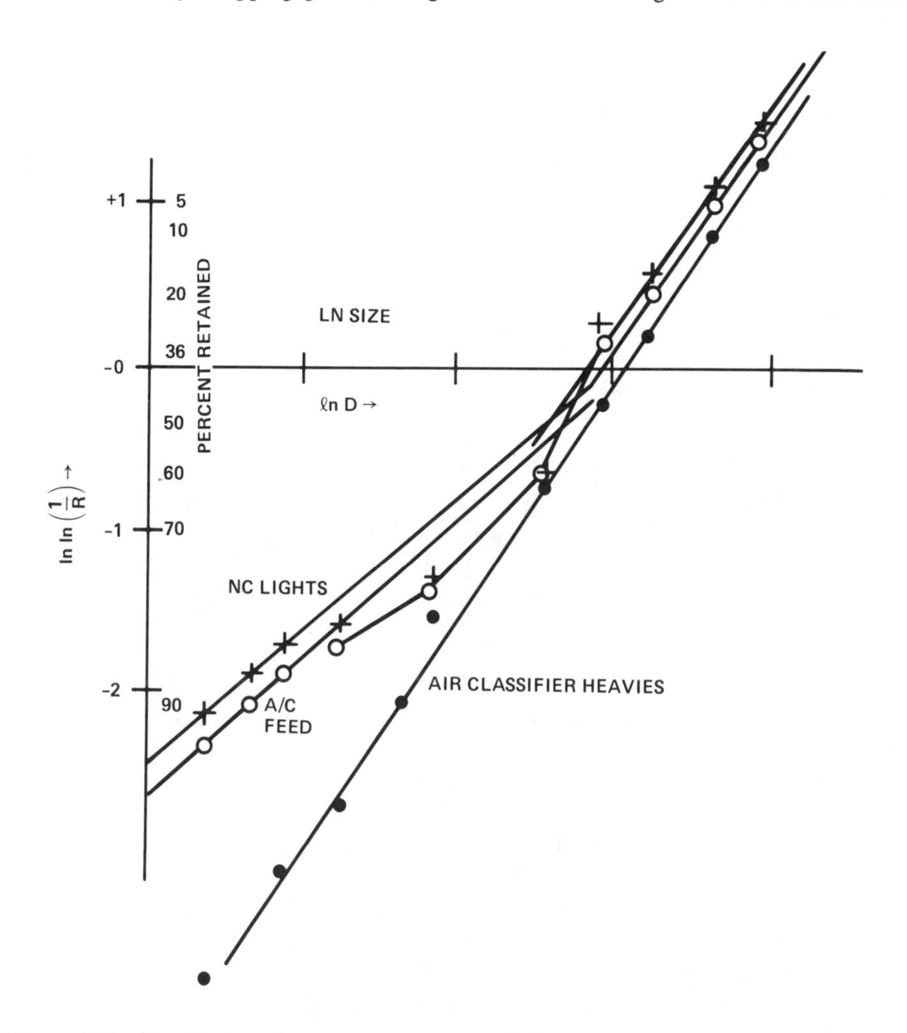

Figure 3-5. Double log of percentage retained (R) versus log of size for air classifier feed, lights and heavies, showing straight line fit of data.

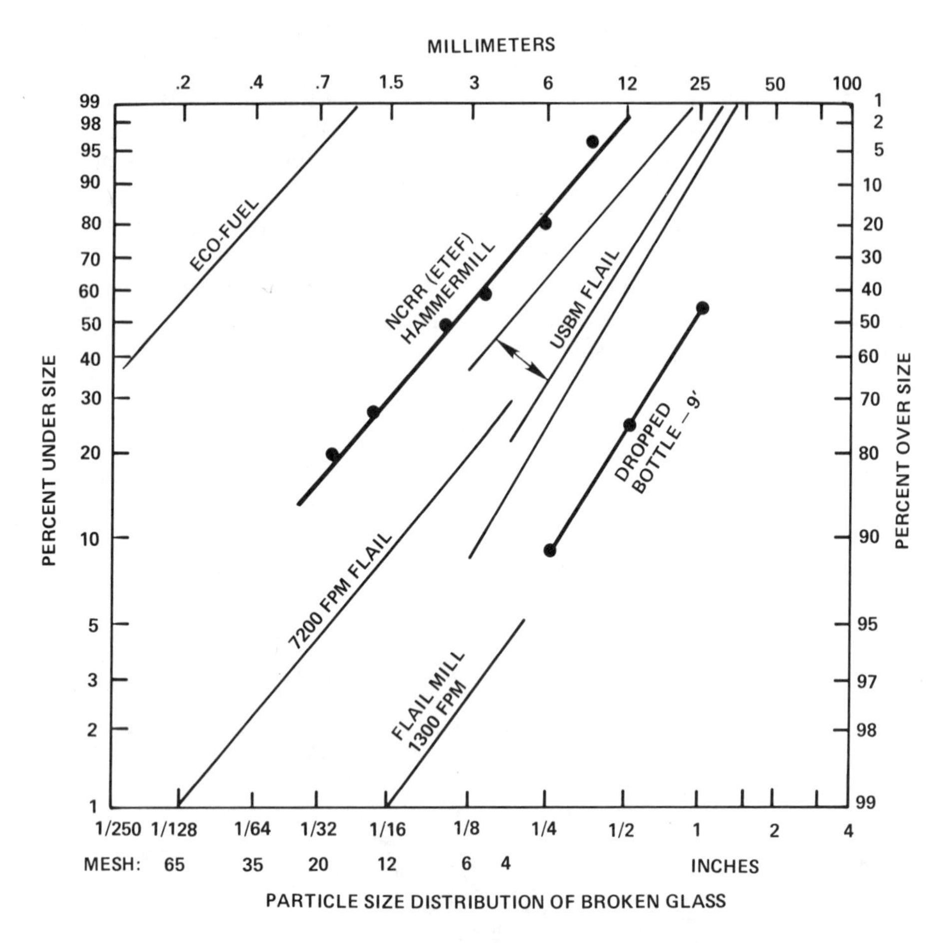

Figure 3-6. Particle size distributions of glass, dropped and shredded, plotted as straight lines on Rosin-Rammler graph paper.

lie on a straight line, except for odd particles such as the neck and the base. A low-speed shredder (flail mill) produced smaller pieces of glass. As higher velocities are used, the glass approaches sand. The data from the BuMines flail mill at College Park show the effect of various shredder speeds on glass size. Data obtained from tests of a Williams hammermill at the NCRR's ETEF show that the glass has been reduced to sandy fines, a size that will fly in an air classifier. Also, these imbed in moist organics and paper and produce a high-ash refuse fuel.

When rags are torn in a shredder, they come out in rectangular strips, due to the fact that the way textiles are woven causes them to be stronger in one direction. Figure 3-7 shows the "three-dimensional" PSD of shredded textiles sampled at Bridgeport before and after changes were made to the American Pulverizer hammermill. With the 8 × 9-inch grates originally provided, 1% of the rags had a long dimension of over 4 feet! By installing thinner hammers, more effective "breaker bars" and

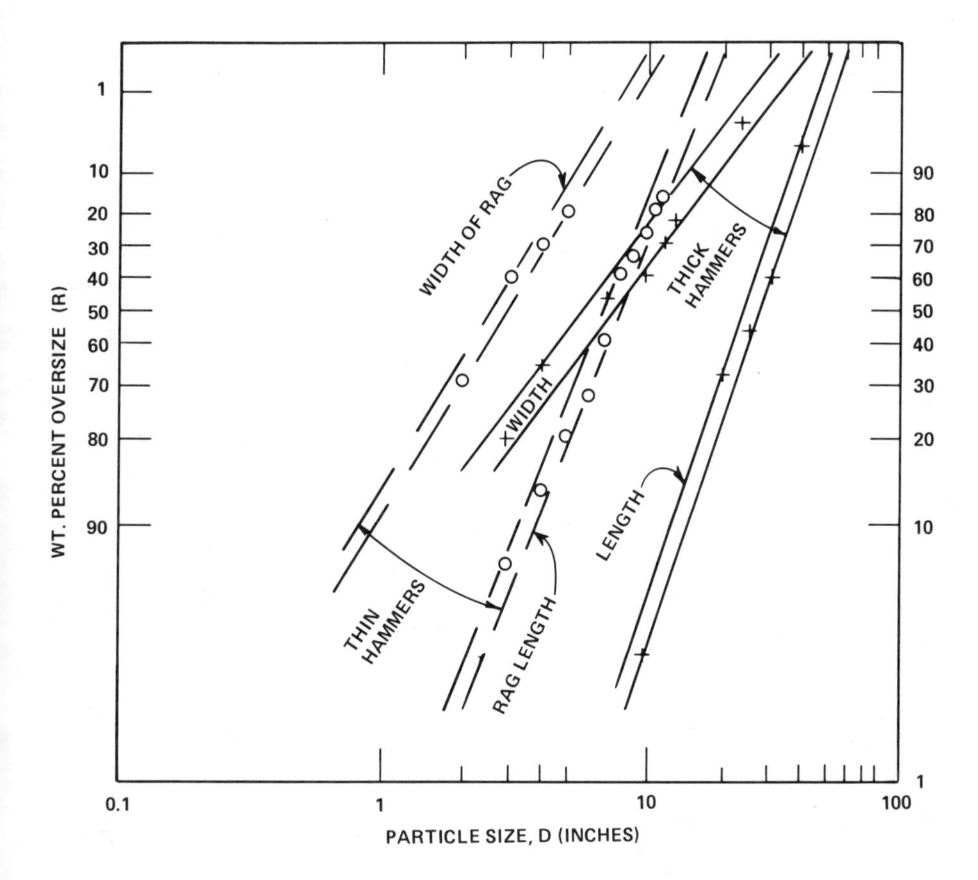

Figure 3-7. Width and length distributions of shredded rags plotted as straight lines on Rosin-Rammler paper and affected by hammer width.

dividing the grate openings into three, 99% of the rags had dimensions less than 15 inches or so, and the short dimensions were less than 4 inches.

This figure attests to the value of the RR plot: with a relatively small number of samples the data showed a straight line that could be extrapolated to the 99% line, giving an indication of the amount of troublesome rags in the product. One percent is a significant amount and can cause a great deal of trouble in the course of a few hours.

Figure 3-8 shows the curves for paper and cardboard, raw, primary and secondary shredded. Here we note the following:

- The particle size data of raw, primary- and secondary-shredded paper and cardboard plots on straight lines for both paper and cardboard.
- The cardboard lines slope toward an apex in both stages, indicating that the larger sizes are not being affected by the shredder as much as are the smaller sizes. This is a sign of grinding rather than cutting, producing proportionally more fines.

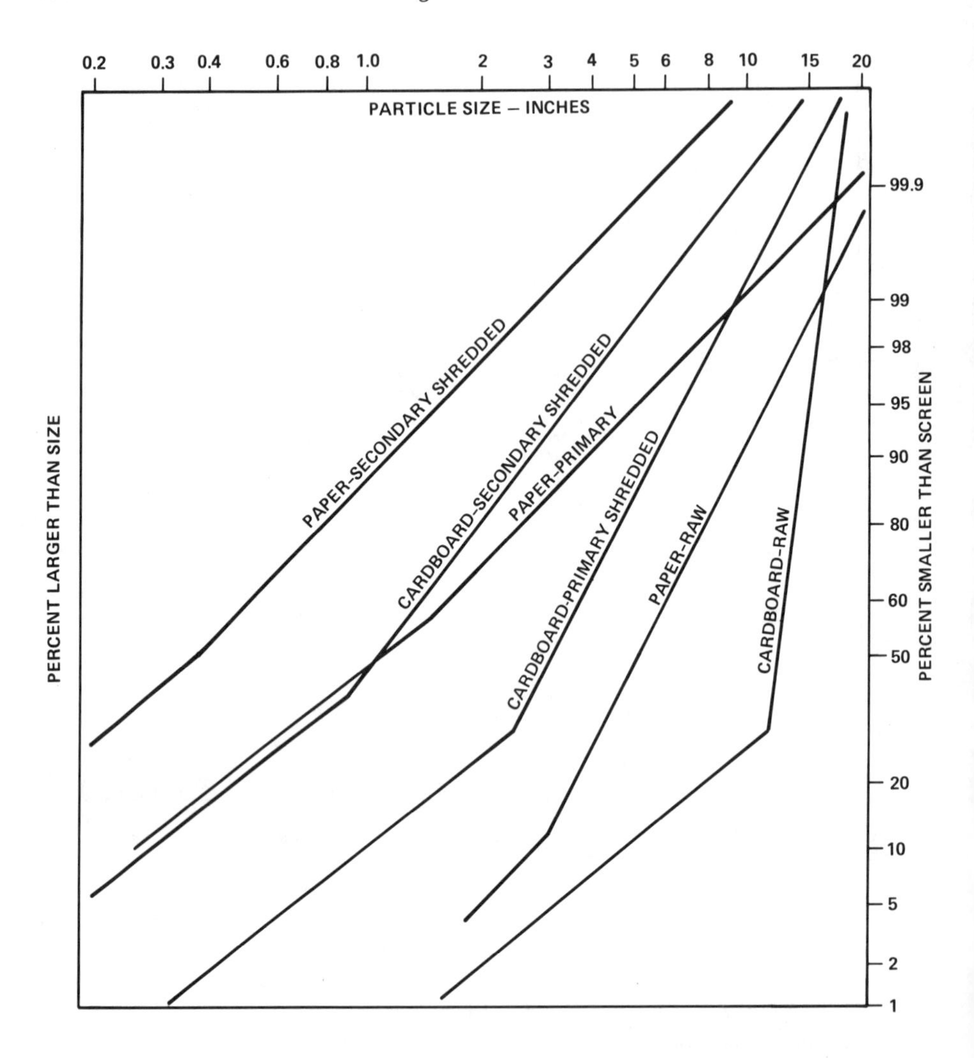

Figure 3-8. Size reduction of paper and cardboard shows changes in slope.

- The paper lines point to an apex in the first stage but move parallel to the left in the second stage, indicating that the large sizes are size reduced by cutting in the same ratio as the smaller ones.

The PSDs of shredded MSW do not necessarily plot as straight lines on RR paper because the material is a mixture. Figure 3-5 shows the size distributions of the feed, accepts and rejects of RDF separated by an air classifier plotted in this fashion. The rejects (heavies) plot on a straight line. The size data of the feed and the accepts (light fraction) fall on two lines, both fairly straight.

Alter and Crawford [13] analyzed samples from ten shredder installations (listed in Table 3-3) to obtain the characteristic size, X_o, and the distribution slope, n.

Table 3-3. Particle Sizes of Shredded Refuse

No.	*Location*	*No. of Samples*	*n Distribution*	X_o *Size*
1	Washington, DC	36	0.6894	2.774
2	Wilmington, DE	55	0.6292	4.561
3	Charleston, VA	42	0.8234	4.033
4	San Antonio, TX	9	0.7685	1.044
5	St. Louis, MO (Gruendler)	11	0.9946	1.610
6	Houston, TX	7	0.6389	2.485
7	Vancouver, WA	8	0.8815	2.2
8	Pompano Beach, FL	7	0.5857	0.666
9	Milford, CT	6	0.9233	1.885
10	St. Louis, MO (Williams)	12	0.9388	3.812

In most cases, the size distributions of the shredded MSW samples deviated significantly from the linear relationship on Rosin-Rammler graphs, showing curves similar to Figure 3-9, in which the data correlated to straight lines extending down to about 20-30% of the feed. The slopes ranged from 0.58 to 1.0, and the characteristic size from 0.7 to 4 inches (20 to 100 mm). Particle size distributions obtained from the ETEF Williams shredder, as well as those found by sampling products of other shredders, were similar to those reported by Trezek.

Mixtures of materials generally display "knees," kinks or s-curves on Rosin-Rammler paper, determined by the relative amounts of the two species and the amount of overlap or separation in particle size. A mixture of two materials, each having a roughly normal distribution with widely separated peaks, will have an RR plot containing an S-curve, both ends of which are asymptotic to the two populations. The crossover gives an indication of the relative quantities of the two types of material and the degree of overlap, if any.

Figure 3-9 is the Rosin-Rammler plot of PSD analyses reported by ten laboratories, which were given samples from a pile of RDF produced by the NCRR at ETEF by double-shredding. These samples were part of a round robin RDF testing program funded by EPA (MERL) and supervised by ASTM E-38-01 (Energy). The results reveal the variation within a single lot: the median (33% retained) size ranged from 3/4 inch to 1-1/4 inch, and the quantity in that size range from 25 to 50% retained.

Raw refuse shredded in a hammermill contains gritty and linty fine material derived from sand, dirt and glass breakage, and from grinding and attrition of paper and cardboard, especially when it has high moisture content. These materials can be discerned on a histogram, as well as by visual observation of the screened sample.

The histogram plotted from the round robin data (Figure 3-10) shows a double population, one peaking in the 28-mm range, the other at about 0.6 mm. A physical examination of the samples in each of the screen fractions revealed that the larger

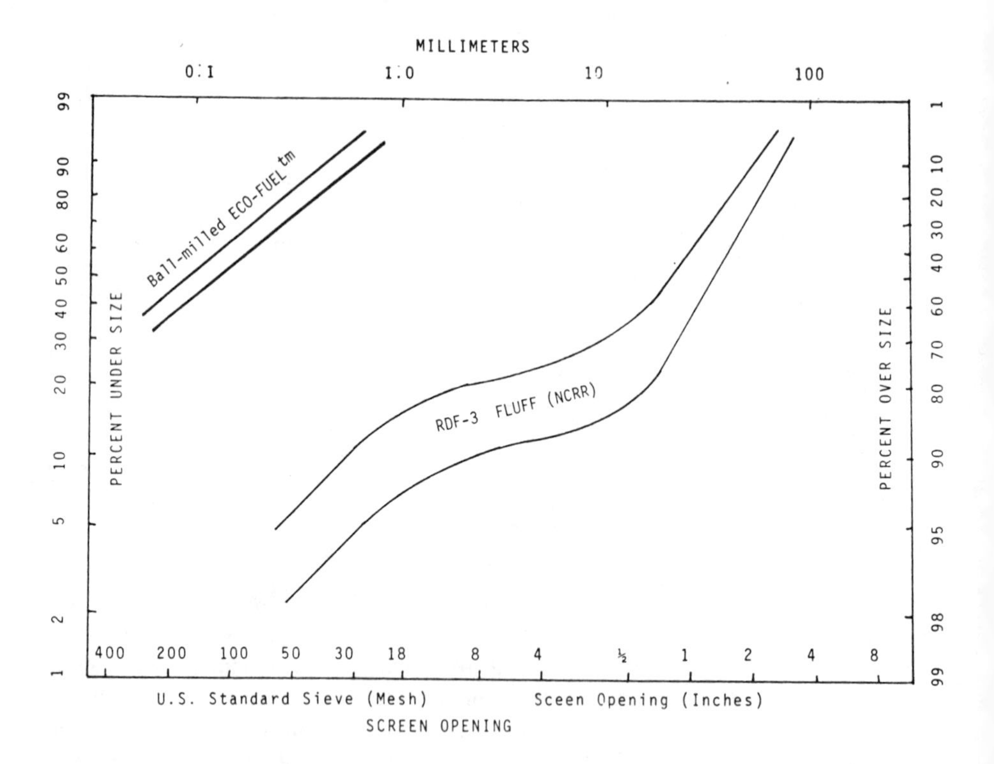

Figure 3-9. Particle size distribution of RDF-3 (ASTM round robin) and RDF-4 (Eco-Fuel) on Rosin-Rammler paper.

population was "light fraction" combustible material, whereas in the smaller screen sizes the samples graded down from mostly lint to mostly sand and grit. The amount of gritty fines ranged from 10 to 25% of the sample, although the samples were drawn from the same "lot." Those who handled this RDF noted that the grit had a great tendency to segregate or settle out and that it was difficult to assure that all samples taken from the lot had the same proportion of grit, thus interfering with the accuracy in analysis of the sample.

PARTICLE SIZE DISTRIBUTIONS OF RAW AND SHREDDED REFUSE

Figure 3-11 shows typical particle size distributions of industrial, commercial and municipal (household waste), first- and second-stage shredded RDF; products from a cascade mill with 4-inch balls and 3-inch grate holes; and powdered RDF (Eco-Fuel) produced by a ball mill with 1-inch balls and passed through a 20-mesh screen. In general, the more shredding and grinding, the steeper the slope of the lines, indicating a larger proportion of fine material produced by grinding and attrition and less by tearing. The powdered RDF can be defined by the slope of the graph.

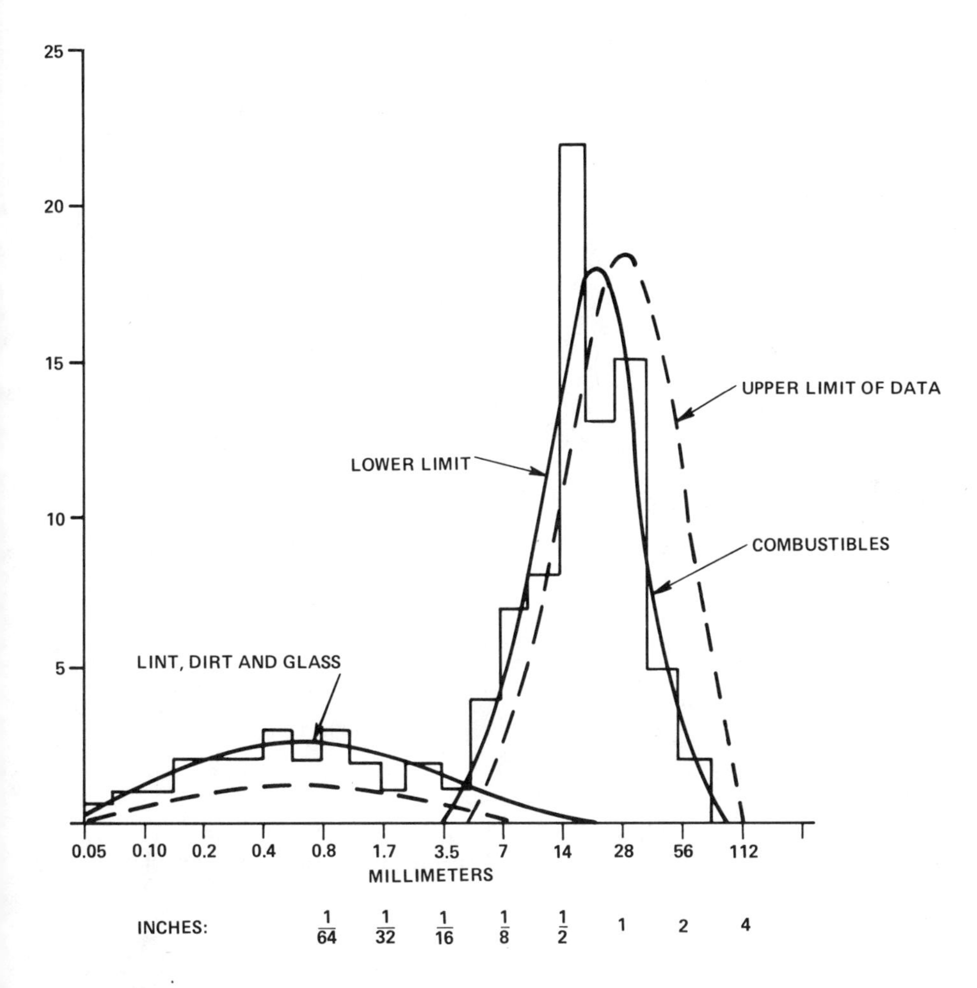

Figure 3-10. Histogram showing combustibles and lint, dirt and glass distributions of RDF-3 (ASTM round robin).

ENERGY REQUIRED FOR SHREDDING OR GRINDING

Work Requirements

Shredding and grinding take place simultaneously to some extent. However, in size reduction of MSW, the first stages exhibit primarily rupture of materials in tension. Under these conditions, it would seem that measuring the work necessary to rupture materials in tension would be directly related to the work needed for shredding.

The tensile force and the work required to rupture materials can be determined by using standard laboratory equipment that measures the force and the distance traveled due to the stretch of the material up to its rupture point.

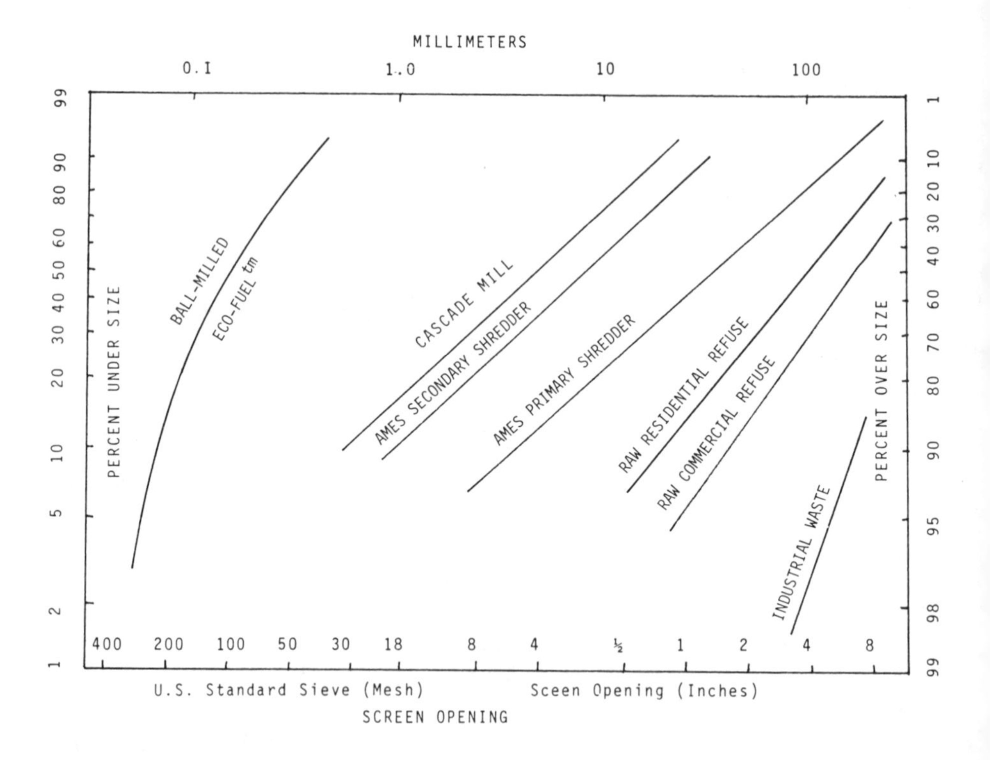

Figure 3-11. Particle size distributions of raw industrial, commercial and residential refuse compared with primary and secondary hammermilled and cascade-milled RDF-3, ball-milled RDF-4 and heavy fractions.

Trezek [1] lists the tensile force per unit cross-sectional area (stress) and the ultimate strain (total stretch per unit length of material at the point of rupture) for a number of materials normally found in refuse. In addition, the rupture energy, which is the work done in rupturing the specimen, was determined in these tests and is summarized in Table 3-4.

The ultimate stress of the paper samples was only twice that of aluminum, but five times that of steel. Despite the 20:1 difference in ultimate strength between steel and paper, the rupture energies were fairly close. Trezek noted that the rate of stretch greatly affected the rupture energy of the plastic sample: at 0.1 in/min it was 111; at 1.0 in/min it was 44; and at 10 in/min it was 19 ft-lb/in^3 due to the flow of material under stress.

When a force is applied to a material, the stress, or force per unit area, causes the material to stretch, so that the forcing or forced object has to move, resulting in work being done on the material (work = force × distance). The amount of stretch resulting from this force depends on the length of the material, so that the work

Table 3-4. Stress, Strain and Rupture Energy of Materials

Material	*Thickness (in.)*	*Ultimate Strength (psi)*	*Ultimate Stress (in./in.)*	*Rupture Energy (ft-lb/in³)*
Steel Cans	0.007	82,000	0.005	9.4
Aluminum Cans	0.006	31,000	0.012	26.5
Cardboard	0.025	6,400	0.025	8.3
Brown Paper	0.009	4,000	0.025	5.1
Newspaper[a]	0.003	4,000	0.025	8.0
Plastic Bottles	0.03-0.2	4,000-5,000	0.06-0.36	19-111

[a]Not determined by Trezek.

done also depends on the length of the material. These principles are essential to understanding rupture breakage of materials that stretch, as distinguished from truly brittle materials, which have little tensile strength.

Trezek tested specimens having a stretching length of 2 inches, so the rupture energy applies to this length only. The rupture energy, which is correctly determined by integrating the force-distance curve plotted by the testing machine, can be approximated by taking the average force at rupture and multiplying it by the stretch. This method is illustrated by this calculation:

$$\begin{aligned}
\text{Force} &= \text{stress} \times \text{area} = \text{stress} \times \text{thickness} \times \text{width} \\
&= 4000 \times 0.009 \times 0.5 = 18 \text{ lb (ultimate)} \\
&= 4000/2 \times 0.009 \times 0.5 = 9 \text{ lb (average)} \\
\text{Work} &= \text{average force} \times \text{distance} = 9 \times 0.025 \text{ in/in} \times (2 \text{ in}) \\
&= 9 \times 0.05 = 0.45 \text{ in-lb} \\
&= 9 \times 0.05/12 = 0.04 \text{ ft-lb}
\end{aligned}$$

The rupture energy may be defined and expressed as the work done per square inch of cross-sectional area of specimen length:

$$\text{Rupture energy} = 0.04/(0.009 \times 0.5 \times 2) = 4.4 \text{ ft-lb/in.}^2$$

This result is close to the laboratory measurement of 5.1 for this material.

The author carried out similar experiments by another method: using 2-inch-long by 0.5-inch-wide newspaper specimens, a weight was dropped from a measured height to determine the foot-pounds of energy required to break the specimen. A number of trials determined that an 18-g weight falling one foot was sufficient to rupture the paper specimen. This information can be calculated as follows:

$$\text{Rupture energy} = \frac{(18 \text{ g})\,(0.025 \text{ in.})\,(12 \text{ in.})}{(454 \text{ g/lb})\,(0.003 \text{ in.} \times 5\text{in.})\,(2 \text{ in.})} = 4.0 \text{ ft-lb/c in.}^3$$

In performing these tests, it was noted that the moisture content of the paper is a major factor in the tensile strength: when very dry it is brittle; at normal room humidity it is strong; but at high moisture contents it becomes soft and loses its strength.

From rupture energy it is possible to calculate the work that would have to be done to tear a sheet of paper into successively smaller pieces and, therefore, to estimate the work needed to shred all the pieces in a given size distribution to a smaller size distribution. This estimate would be valid for particle sizes that greatly exceed the thickness of the paper. For smaller sizes, a different set of rules would be expected to apply.

Bond Work Index and Rosin-Rammler Size Distribution

Much research has been devoted to determining the amount of work necessary to break materials, as a function of the initial and final particle sizes and size distributions. Bond [14] derived and exhaustively confirmed a relationship expressing work and product size based on the concept of crack length in friable ores, as follows:

$$W = 10\ Wi\left(\left(\frac{1}{P}\right)^n - \left(\frac{1}{F}\right)^n\right)$$

where W = work input, kWh/ton of feed
Wi = work index, kWh/ton
F = screen dimension, through which 80% of the feed passes, μ
P = screen dimension through which 80% of the product passes, μ
n = a dimensionless constant

Bond and others have found that the power requirements of various materials could be correlated by using the screen dimensions through which 80% of the feed material would pass, F_{80}, and 80% of the product would pass, P_{80}. Bond considered "n" to be 0.5, whereas Holmes [15] proposed that this exponent would vary with the type of material being reduced in size. Austin and Luckie [16] believed that the feed and product size exponents affect the slopes of the respective Rosin-Rammler feed and product size distributions. The author has found that the exponent shifts to about 0.25 when the material is ground by surface wearing or attrition, as in the case of ball-milling RDF into powder, partially accounting for the surprisingly low power required to make powdered Eco-Fuel from shredded MSW (about 40 kWh/ton is required).

Using the Bond equation to define and determine them, work indices have been measured and found to be constant for many brittle materials over a wide range of input feed sizes, proving the utility of this approach.

Stratton and Alter [17] wondered whether the Bond relationship would apply to the components of municipal waste and to the mixture because the breakage of brittle materials is basically by shear or by tension, which are also the actions that are effective in shredding paper products and textiles. They analyzed samples of shredded refuse from full-scale commercial shredding installations believed to be representative of the types of refuse shredders currently operating in the United States (Table 3-5) and found that the size distributions obtained by using standard laboratory sieves could be correlated by straight lines, giving the characteristic size, X_o, and a slope, "n." Eight of the ten shredders normally received municipal solid waste: one was an automobile shredder and one was designed for shredding OBW. All were receiving MSW or packer truck refuse.

The authors found that the size distributions obtained by using standard laboratory sieves could be correlated by straight lines, giving the characteristic size, X_o, and a slope "n." They were unable to make direct measurements of the power required to make these samples. However, assuming the power consumption of these shredders was the same as that reported by Trezek, they calculated the work index based on X_{80} sizes determined by graphic analysis and found that the work index of all the shredders was about the same.

Encouraged by this result, they ran tests to obtain direct information on the work index of various materials contained in MSW. Replicate batches of mixed office wastepaper, glass bottles, aluminum cans and crushed steel cans were shredded by a Heil vertical hammermill at NCRR's ETEF. The power was measured on a recording ammeter, idling power was subtracted to find the net power required for shredding and the product size distribution determined. The results of these tests are given in Table 3-6.

To calculate the work index from test data, the size distribution of the feed and the product are needed, from which P and F and the kWh/ton of feed are determined. To compare the feedstock independent of the machine, it is necessary to subtract from the measured power input the mechanical friction of the shredder, which is normally taken as the no-load power. It is convenient to use 10 inches for the value of F because it does not contribute much to the calculation.

The work index for paper only was roughly 200, whereas the values they had calculated for their samples ranged from 387 to 481, reflecting the relative toughness of the other materials, such as cardboard and textiles, and the significant effect of moisture in MSW. The work index for glass was only 8. Steel was 50% higher than paper, whereas that for aluminum cans was two to three times higher than paper, according to these data.

Power Required for Size Reduction

The Bond equation can be used to calculate not only the work index but inversely to predict the power required for shredding, kWh/ton, given the value of P_{80} (and assuming F_{80}). When values of X_o are available, they can be converted to P_{80} by either looking at a PSD plot or from a knowledge of the slope, n. The Bond equation can be used to construct a plot, such as Figure 3-12, which shows the "specific power," kWh/ton, in terms of the "characteristic" particle size, X_o, the work index, Wi, and "n," the distribution coefficient that measures the width of the histogram.

Table 3–5. Shredder Installations Producing NCRR Samples [13]

Shredder Number:	*1*	*2*	*3*	*4*	*5*	*6*	*7*	*8*	*9*	*10*
Manufacturer:	*WMS*	*GRN*	*H*	*NEW*	*GRN*	*WMS*	*EID*	*TOL*	*EID*	*WMS*
Type	H	H	V	H	H	H	V	V	V	H
Horsepower	1,000	900	500	2,000	1,250	500	800	200	1,000	800
Rotor Diameter	80	42	100	82	62	44	62	38	62	52
Rotor Length	84	60	83	104	80	75	70	83	70	60
Rpm	620	900	600	600	900	900	369	1,350	369	900
Grates	6	–	–	–	2 X 3	6 X 11	–	–	–	–
Hammer weight, lb	120	–	–	–	200	100	NA	14	NA	135

Table 3-6. Work Index for Various Materials

	Feed		Product		Wi
Material	*X*	*n*	*X*	*n*	*Kwh/ton*
Glass Bottles			0.0927	0.979	8
			0.0480	0.622	7
			0.0759	0.697	8
Paper (office)	13.7	4.468	2.247	2.516	198
			2.060	2.389	188
			2.227	2.715	195
Steel Cans	23.07	2.833	4.742	2.361	281
			4.247	2.178	250
			4.382	2.358	254
Aluminum Cans			3.578	2.723	679
	3.578	2.723	2.585	2.524	721
	2.585	2.524	1.889	1.941	763
	1.889	1.941	1.280	1.801	452
MSW, Washington			2.774	0.6894	463

The latter has a relatively minor effect on the kWh/ton, whereas the particle size, X_o, has the major effect. The work indices of paper, MSW and cans are also shown on this log-log plot, which covers a wide range of conditions.

The relationship of X_o, X_{80} and other definitions of particle size can be determined by using the slope, n, and the Rosin-Rammler equation:

$$R = -e^{(-(D/X_o)^n}$$

Table 3-7 gives the factor K by which X_o must be multiplied to obtain X_{80}, X_{90} (nominal size) and X_{95} (top nominal size defined by ASTM E-828-81).

Specific Energy and Specific Capacity

When the "specific energy" determined by the Bond equation (kWh/ton) is plotted against increasing particle size for a typical value of n, such as n = 0.8, we get the hyperbolic curve shown in Figure 3-13. Such hyperbolic curves are difficult to interpret. Also shown are data points from Trezek.

Ton/kWh, the reciprocal of kWh/ton, may be called the "specific capacity." This plots as a straight line, with a zero intercept, the expected complement of a hyperbola. It is a much more useful and fundamental term for the measurement and evaluation of shredding operations, as we shall see in subsequent analyses, due to the better ease in understanding linear relationships, especially when trying to compare the effects of other variables and making correlations.

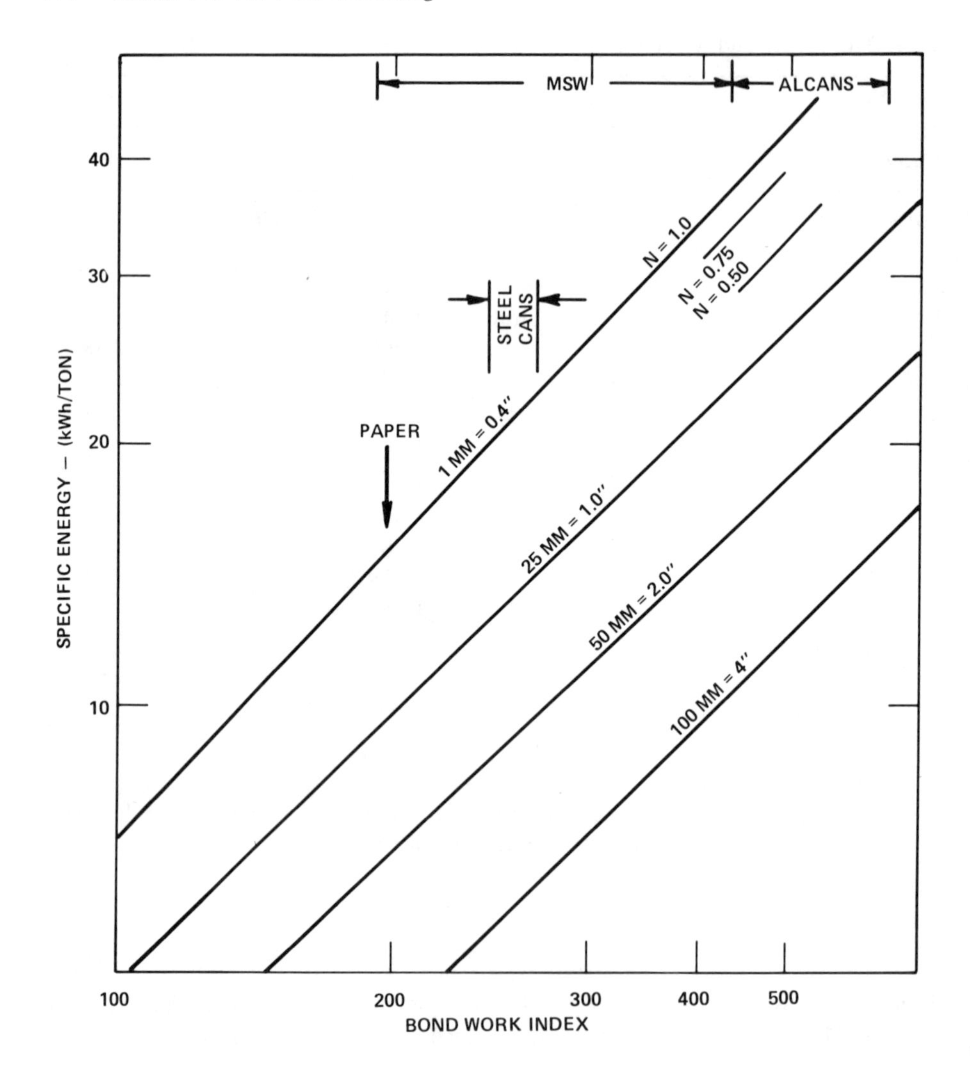

Figure 3-12. Bond work index and specific energy versus sizes of product, X_o, for various values of slope, n, showing ranges for typical species in MSW.

How does actual shredder test data compare with the Bond theory? This can be answered by plotting data obtained by measuring the net power consumption of shredders under measured feedrates and determining the characteristic particle size of the product.

When we plot ton/kWh versus the characteristic size, X_o, on linear coordinates, we find that the data obtained from many different shredders fall close to a straight line, as shown in Figure 3-14. Included are data reported by Savage et al. [18], Trezek [1], Weinberger [3] and Vesilind [4]. A correlation of the data of Savage for six shredders (one omitted as an outlier) follows a line passing through solid

Table 3-7. Multiplying Factor, K, to Get Other Sizes from X_0

Slope, n	K_{80}	K_{90}	K_{95}
1.0	1.61	2.30	3.00
0.9	1.53	2.10	2.68
0.8	1.46	1.95	2.41
0.7	1.40	1.80	2.16
0.6	1.33	1.65	1.93

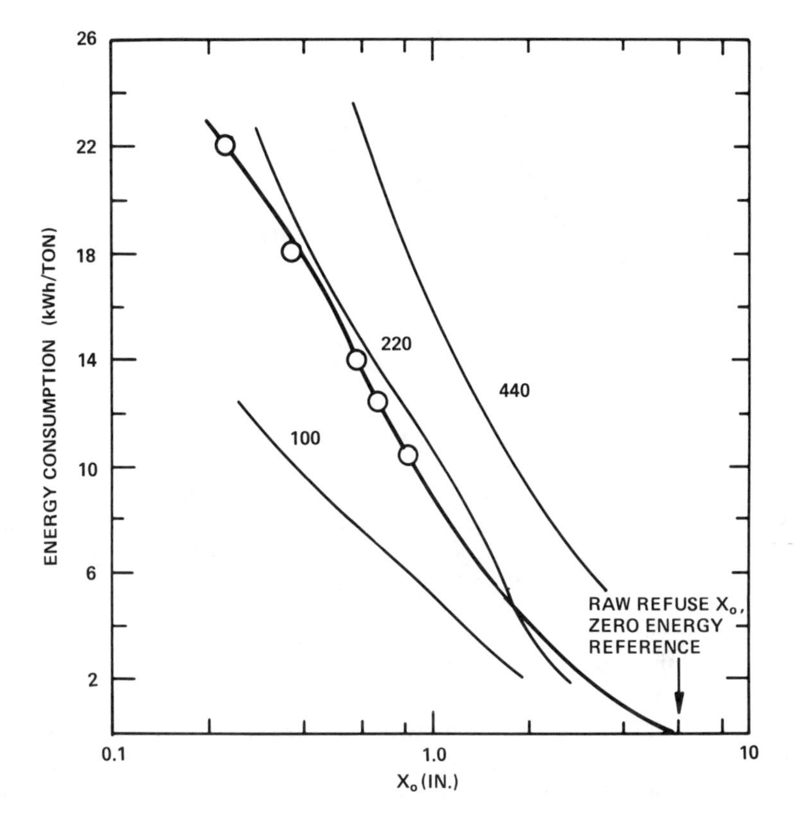

Figure 3-13. Specific energy consumption to achieve a given characteristic particle size, X_0, showing lines of constant Bond work index.

data from the test of a Heil shredder at Pompano Beach and is close to data generated by the flail mill at East Bridgewater. On the other hand, the range of variation in individual shredder tests is disturbing, probably resulting from different MSW characteristics.

A comparison between commercial and residential wastes, previously reported by Trezek, is also shown, indicating a substantial difference in the power requirements.

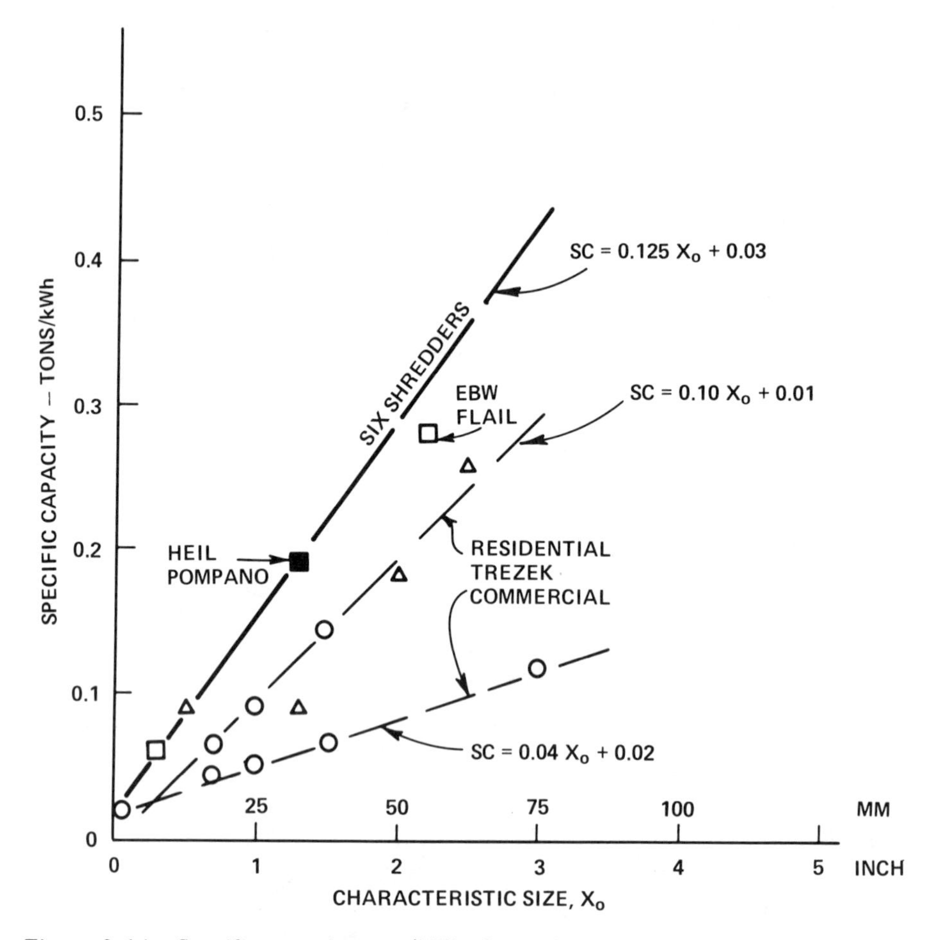

Figure 3-14. Specific capacity, ton/kWh, for various shredders versus characteristic particle size, X_o, showing correlation lines and equations.

The various shredders appear to have processed wastes ranging from residential to commercial. Using the data explicitly presented in this figure, we can obtain the equations of the lines, as follows, where X_o is the characteristic size in inches. (These equations can be converted to X_{90} or X_{95} by use of Table 3-6, where n is known):

Correlation of 6 shredders: ton/kWh $= 0.03 + 0.125\ X_o$
Residential waste (Trezek): $= 0.02 + 0.10\ X_o$
(work index = 230)
Commercial waste (Trezek): $= 0.04 + 0.04\ X_o$
(work index = 411)

The work indices shown here were calculated from the data from which the correlation was obtained. The method of calculating the work index from test data is illustrated below, using the Pompano Beach data published by Vesilind, which is averaged over one week of operation:

$$\text{Given: } X_o = 1.3 \text{ in} \times 24{,}000 = 32{,}500\ \mu \text{ (average)}$$
$$\text{kWh/ton} = 5.25 \text{ (average over one week)}$$
$$n = 0.62$$

$$\text{Work index} = \frac{W}{10} \Big/ \left[\frac{1}{\sqrt{P_{80}}} - \frac{1}{\sqrt{F_{80}}}\right]$$

The relationship between X_o and $X_{80} = P_{80}$ can be determined mathematically when the slope n is known, from the Rosin-Rammler relationship:

$$R = 1 - \exp\left(\frac{-D}{X_o}\right)^n$$

Rearranging

$$-\left(\frac{D}{X_o}\right)^n = \ln(R)$$

At 80% passing, R = 20% retained on screen, so

$$\left(\frac{D}{X_o}\right)^n = \ln\left(\frac{1}{0.20}\right) = 1.6094$$

Then,

$$P_{80} = D = X_o(1.61)^{(1/n)}$$

For n = 0.62,

$$\left(\frac{1}{0.62}\right)$$

$$P_{80} = X_o\,(1.61) = 2.16 \times 24{,}000 = 51{,}840$$
$$F_{80} = \quad = 10 \times 24{,}000 = 240{,}000$$

The work index is

$$Wi = \frac{(5.25/10)}{(0.0044 - 0.002)} = 223$$

The calculation of the Trezek data is as follows:

For commercial waste,

$$\text{At 1 in.} = X_o, P = 1.0\,(1.61)\,(25{,}000) = 40{,}250\ \mu$$

$$W = 20.5 \text{ kWh/ton}$$

$$Wi = \frac{(20.5)\,(200.6)}{10} = 411.2$$

For residential waste,

$$\text{At 1 in.} = X_o, P = 1.0\,(1.61)\,(25{,}000) = 40{,}250\ \mu$$

$$W = 11.5 \text{ kWh/ton}$$

$$Wi = \frac{(11.5)\,(200.6)}{10} = 230.7$$

The substantial difference in work index between residential and commercial refuse is consistent with the difference in shredder power that has been reported.

Comparison of Bond Work Index and Rupture Energy Models

The lines calculated for work indices of 100 to 400 are plotted on Figure 3-15 for comparison with points representing performance of full-sized shredders. The curves published by Trezek show that commercial refuse, which is drier and tougher, follows the Bond line with a work index of 400, whereas household refuse follows a work index closer to 200.

Using the rupture energy of the material as the mathematical model, we can calculate the amount of work that would have to be expended to tear paper with a given PSD into a smaller size distribution. This is done on the reasonable assumption that when paper is shredded the particles are cut the same number of times into the same fractions, so that the shape of the PSD remains about the same while the mean particle size is reduced. The work that the shredder must do cannot be less than the energy calculated to tear the pieces individually.

The estimated power requirements for shredding paper, based on rupture tests and calculated for various efficiencies, are shown in Figure 3-16 for comparison with the Bond curves and actual shredder performance. The trends of the Bond curves do not appear to follow the data in the small particle sizes. Likewise, the curves for rupture energy show that the actual power used to produce small particles approaches a higher efficiency of rupture energy. In summary, both models overestimate the work needed to grind to small particles. This has been borne out in practice, as shown by the plotted point for 40 kWh/ton used to produce Eco-Fuel.

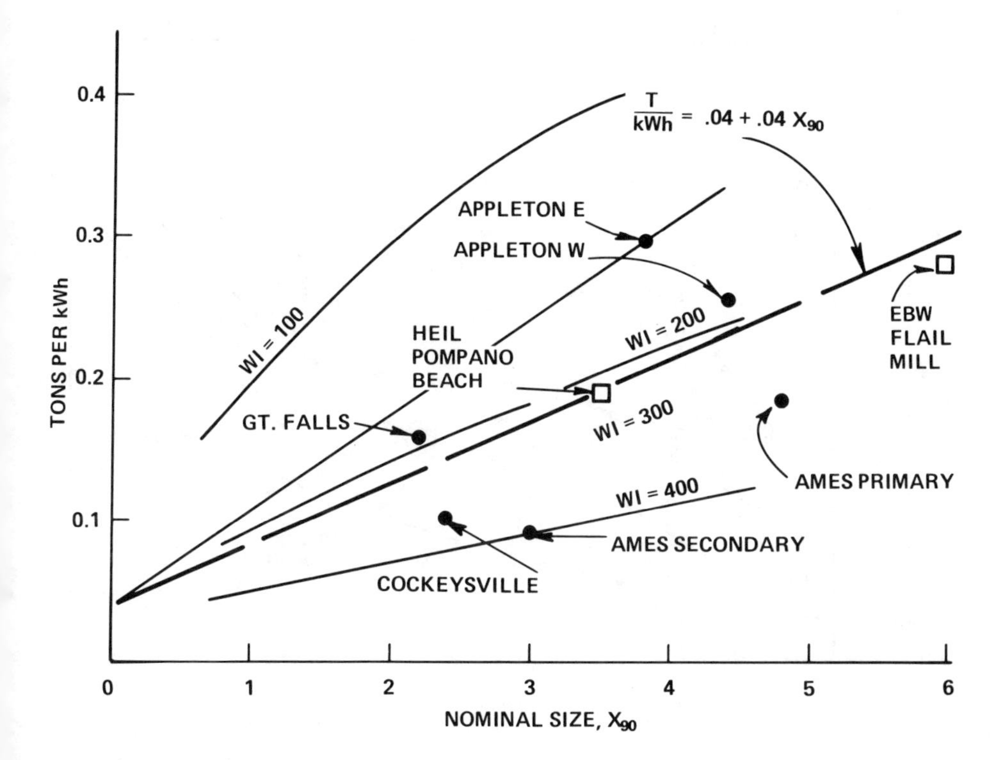

Figure 3-15. Specific capacity versus nominal size, X_{90}, for various shredders, showing lines of constant work index.

Various researchers have stated that the exponent of the Bond expression (0.5) is not correct for small particles and is closer to 0.3.

The grinding energy required to produce Eco-Fuel (powdered RDF) is indicated on Figure 3-16. This point is consistent with an exponent of about 0.25, which would cause the Bond curves to fall closer to the rupture curves.

Summary

So far it appears that the nature of the material and the degree of size reduction determine the work that must be applied to the feed material to produce a given degree of size reduction and that this work is largely independent of the type of shredder that produced the product.

The next question is, given a particular shredder, how do shredder configuration and the nature of the feed material affect the product size distribution and power requirements? What other factors are influential?

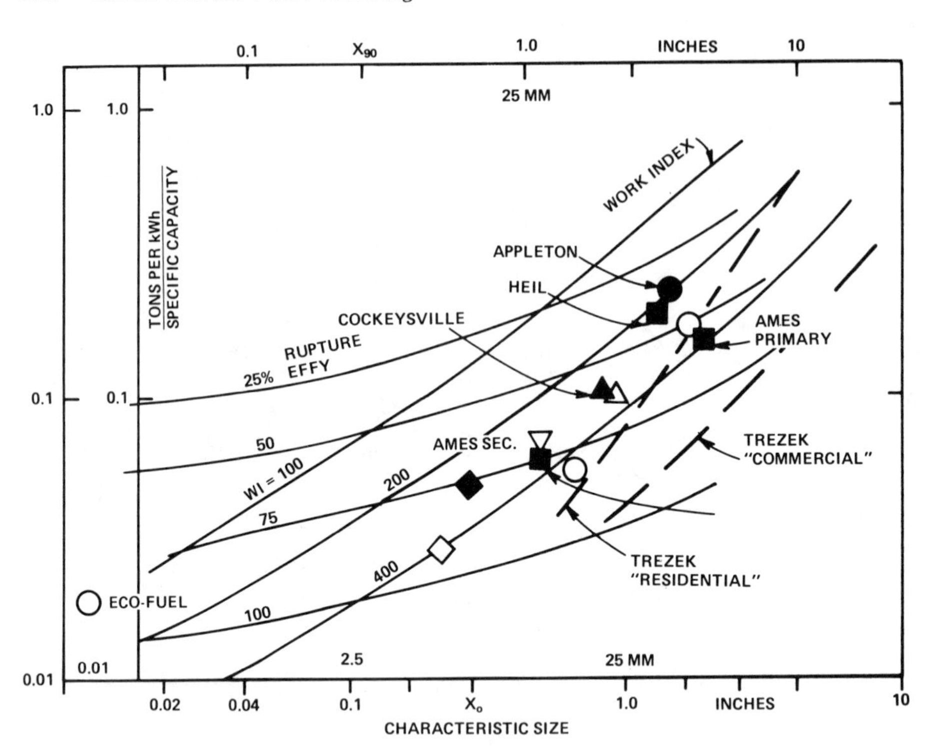

Figure 3-16. Specific capacity versus characteristic size, X_0, showing lines of constant rupture efficiency and Bond work index and shredder operating points.

FACTORS AFFECTING SHREDDER PERFORMANCE

Up to now we have examined how much energy is required to achieve a given reduction in size for a given material. Now let us consider the performance of the machine that produces this size reduction. What can be achieved by a given machine and how is a machine selected or adjusted to produce the desired result? Given the machine, how does it perform and what product will it deliver from a given feed?

A given comminution device can be characterized by the size distribution it can produce with a given material feed and the consequent specific energy consumption or specific capacity per unit of power consumption (its reciprocal).

Some of the major factors that influence the size distribution and the actual gross specific energy consumption are as follows:

- Feed size distribution (source and nature of feed)
- Flowrate of material (how near to limiting capacity)
- Amount of "holdup" above the rotor
- Work index of the materials in the feed
- Moisture content of feed

- Grate or extraction spacing
- Hammer shape and number
- Rate of hammer wear
- Relative velocities (rpm or tip speed)
- Airflow through machine

Feed Size

The size of the largest particles in the feed will determine the size of the feed hopper of the machine, hence, its frame dimensions. It has only a small effect on the power required. The size distribution of the feed material is difficult to measure. Its range has been indicated by published studies such as those shown in Figure 3-11 for several localities. Plotted on RR paper, it generally shows the characteristic straight line, which can be extrapolated to the maximum particle size at "99% less than" and, thus, approximates the size distribution of the MSW. For instance, referring to Figure 3-11, this intercept ranges from 30 in. to 50 in. for residential and commercial refuse.

The maximum particle size is sometimes known by the method of containing and transporting the MSW. For instance, household garbage is usually contained within standard garbage cans, and the largest piece contained is represented by the height of the container. Municipal refuse trucks can pick up pieces having dimensions up to almost 8 feet, and industrial waste often contains large cardboard cartons and shipping pallets.

Flowrate

The flowrate of the material entering the machine is especially critical in grate machines when the flow approaches the capacity limit of the grate.

The shredder has a fixed exit configuration, containing the orifices through which the material must pass. Some force must be exerted on the material to pass it through these orifices, and the pressure must increase with the flowrate. As the flowrate increases, the cross section through which the material flows is reduced by the material itself until the point is reached at which the material is flowing at its characteristic density. MSW becomes more dense as resistance to flow increases, ultimately stalling the machine. If the person responsible for overloading the machine has to dig it out, overloading will be less frequent. He will tend to operate at excessively light capacity.

Holdup

As a horizontal grate shredder can "hold up" a significant amount of material, the feed and discharge rates are not always the same at any given moment, causing power surges. Grateless shredders respond to the rate of loading. The particle size produced is mainly influenced by feedrate.

The average flowrate is the major factor in the specific energy consumption and particle size distribution and, at the same time, elusive to measure because flowrates tend to vary constantly, even when carefully maintained.

Moisture Content of the Feed

Moisture content of the feed has a significant effect on the power draw of a shredder. Paper products are tough when at equilibrium with atmospheric moisture levels, and much easier to tear when they have absorbed more moisture.

The effect of moisture content on specific energy is illustrated in Figure 3-17. The minimum appears to be in the range of 35 to 40%. After shredding, high-moisture material contains more larger particles and fewer fines.

Figure 3-18 represents data taken at two shredded speeds, with measurements of the kWh/ton of MSW fed to the Gruendler hammermill at different levels of moisture content. The points are scattered to such an extent that their trend is uncertain, and regression analysis must be used to try to make sense out of the data points.

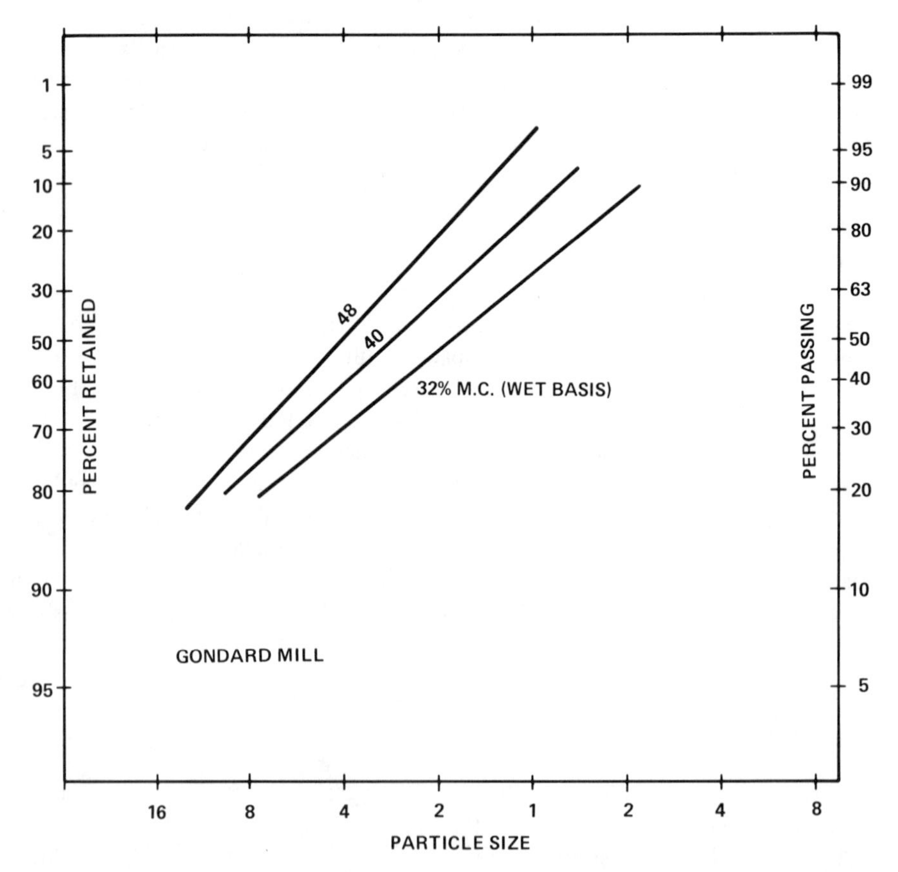

Figure 3-17. Effect of moisture content on particle size distribution [1].

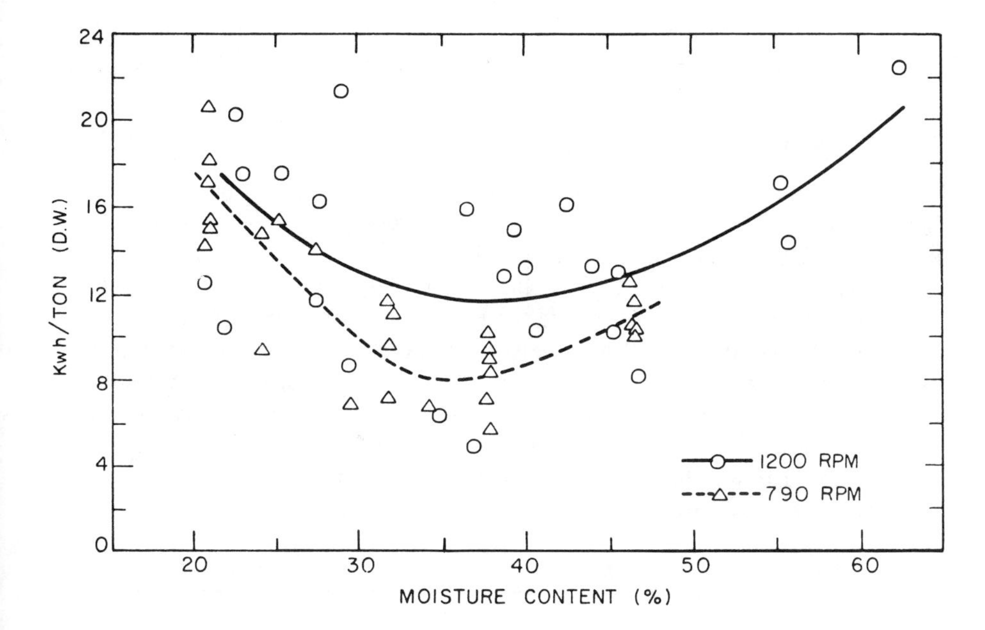

Figure 3-18. Effect of moisture content on specific energy [1].

As the curve goes down, then up, the data on each side of the minimum are taken for analysis. The data up to 32% moisture give these regression equations: Up to 32% moisture,

$$790 \text{ rpm: } \frac{\text{kWh}}{\text{ton}} = 32.97 - 0.790\ (\text{MC})\ (R = 0.80)$$

$$1200 \text{ rpm: } \frac{\text{kWh}}{\text{ton}} = 33.19 - 0.763\ (\text{MC})\ (R = 0.80)$$

Above 32% moisture,

$$1200 \text{ rpm: } \frac{\text{kWh}}{\text{ton}} = 0.38\ (\text{MC}) - 4.5\ (R = 0.68)$$

The regression coefficient, R, is fairly high, indicating a good (not excellent) correlation. There is a remarkable agreement between the sets of data taken at the two shredder speeds. The increase in power above 32% is reported to be due to the result of wadding the material, which interferes with smooth flow through the shredder.

Other data indicate that in this and other similar hammermills the particle size does not substantially change due to change of moisture in this range. The change in power can be attributed mainly to the change in ultimate or tear strength of the paper as a function of moisture content.

Shredder Grate Size

The grate size and hammer clearance directly affect the product particle size. The Madison research produced extensive data on PSD versus grate size, in testing a horizontal-shaft 150-hp, 1200-rpm Gondard mill [19].

Figure 3-19 shows the PSD obtained from these tests of the Gondard. These data show that the grate size intersects the PSD curves at the 98-99% undersize lines on this graph and that the "characteristic size" of the material ranged from 0.75 to 1.20 in. as the grate varied from 3.5 to 6.25 in. The PSD of product collected by Alter and Crawford [13] from a Williams horizontal-shaft hammermill and from tests of the vertical-shaft Heil mill at Pompano Beach, reported by Vesilind [4], as well as the product of a 10-foot-diameter cascade mill, are shown for comparison.

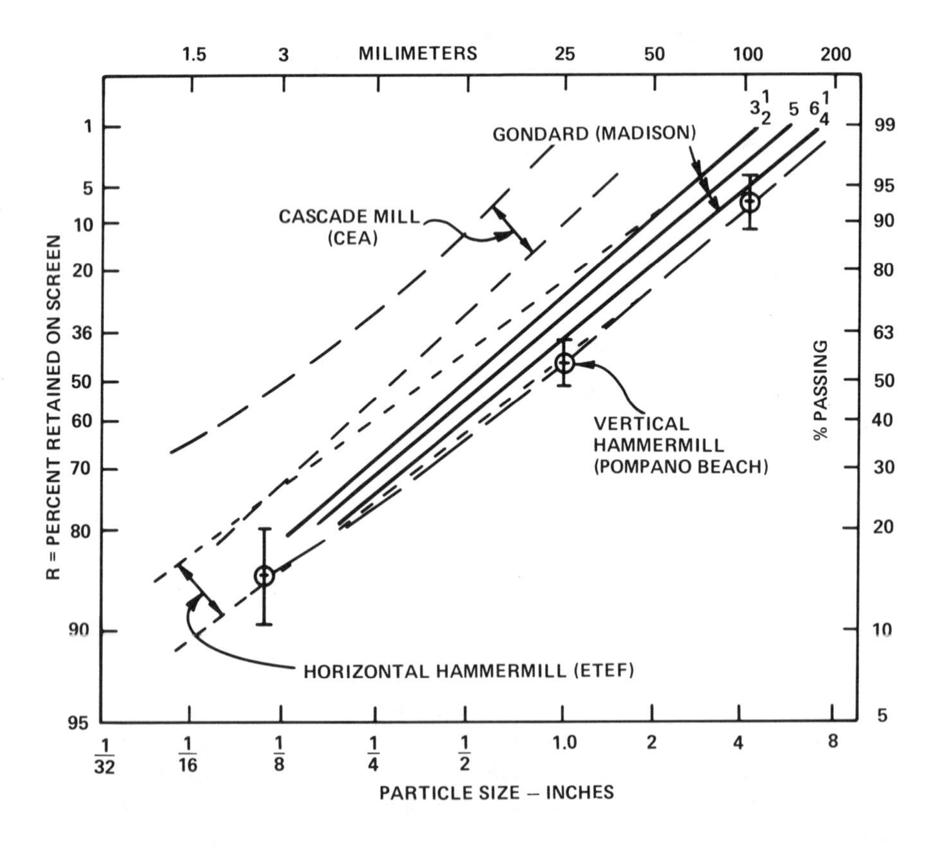

Figure 3-19. Particle size distributions for various shredder grate openings, compared with horizontal hammermill, grateless hammermill and cascade mill products. One standard deviation is marked.

Product samples taken from a given mill will vary constantly within a range, as shown on the graphs, requiring many samples to be taken to obtain both a good estimate of the mean and the standard deviation. The Pompano Beach data show the means and standard deviation of daily samples taken on two successive weeks.

Air Velocity through Shredder

Shredder manufacturers generally recommend that an air flow be provided through the shredder grate when light materials are to be shredded—on the order of 5000-10,000 ft^3/min for full-sized shredders. This helps assure steady flow through the shredder and more consistent particle size.

Hammer Wear versus Particle Size

Hammer wear changes not only the shape of the cutting surfaces, but also the exit clearance and, thereby, the particle size. In addition, the wear affects the way the particles are hit and the angle of impact. The effect of hammer wear is illustrated by Figure 3-20.

There is a rather drastic effect of hammer wear on the mean particle size over a period of time. The relationship between the amount of material shredded and the characteristic size has been described in terms of an exponential decay, as determined by the Madison tests:

$$Y = A + B\,e^{(-CT)}$$

Here, T is the cumulative tons of refuse shredded since the installation of new hammers, and A, B and C are constants.

The effect of hard facing of hammers has been studied by Savage et al. [20]. At Appleton, Wisconsin and Cockeysville, Maryland, different hard facing materials were applied to hammers and their rates of wear determined by logging the refuse shredded and the hammer weights. The tested Rockwell hardness ranged from 20 (no hard facing) to 54, as shown in Figure 3-21. A clear trend is indicated between the tonnage processed per unit of hammer weight loss and hardness. The explanation for the lower wear at Cockeysville is probably the fact that more glass was removed before shredding.

The optimum hardness range is shown, based on this analysis: above about 35% the hammers lost weight due to chipping, which negates the benefits of hardness. Figure 3-22 shows that the wear data can be correlated against the hardness and the relative size reduction, $Z = (F - P)/F$, where F is the feed size and P is the product size.

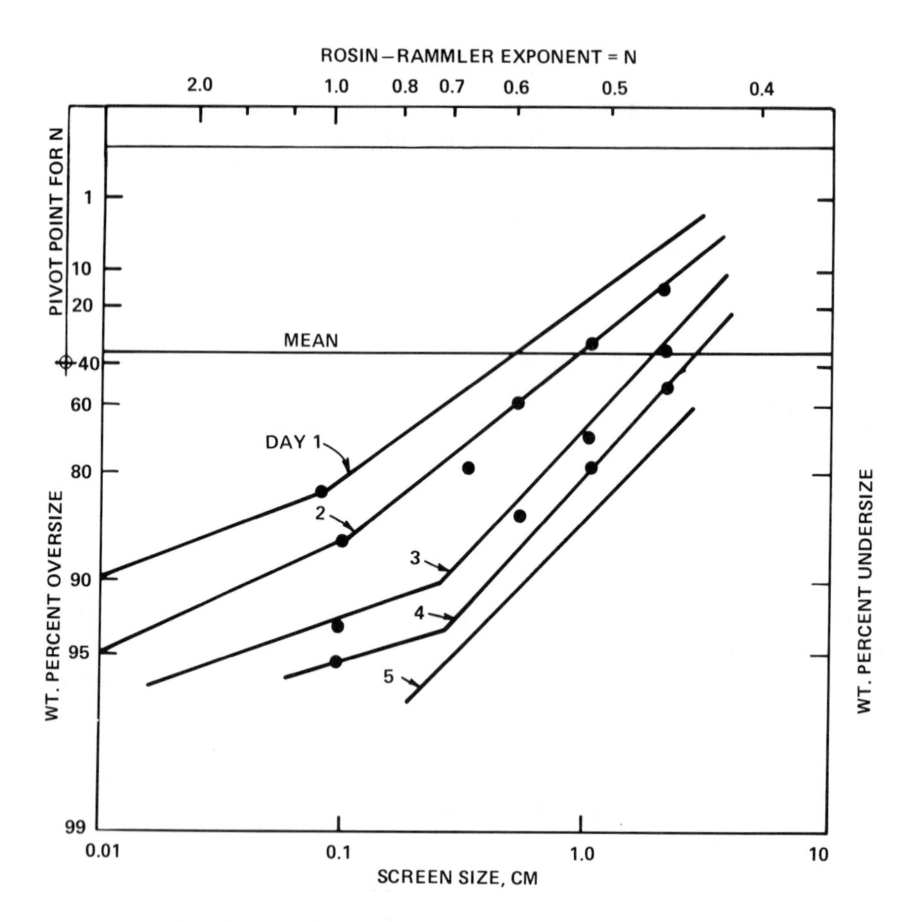

Figure 3-20. Effect of hammer wear on particle size distribution on four successive days [1].

Relative Velocity (Rotor Speed)

The influence of shredder rotor speed on the size of product is illustrated in Figure 3-23. The higher rotor speed would be expected to generate finer material at the expense of more power consumption, everything else being the same.

ANALYSIS OF SHREDDER PERFORMANCE AND EFFICIENCY

So far we have found that MSW and its shredded products can be characterized as to size distribution, moisture content, density and work index. We have seen that their mechanical configuration influences the PSD of the product, and that wear, exit clearance and rotor speed are important influences on particle size. Finally, the specific energy, kWh/ton, and specific capacity, ton/kWh, are related to the characteristic particle size.

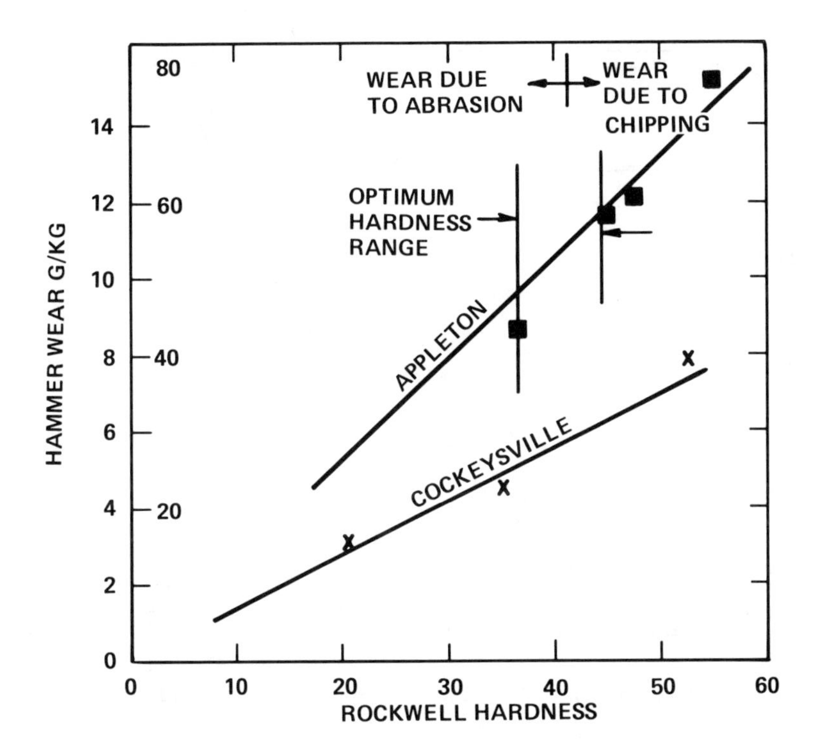

Figure 3-21. Harder hammers reduce wear in shredding.

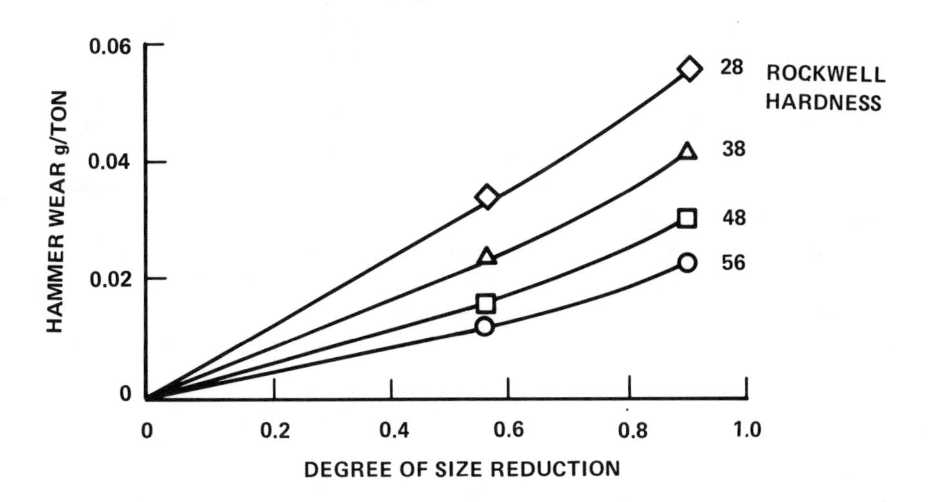

Figure 3-22. Degree that size reduction and hardness affect hammer wear.

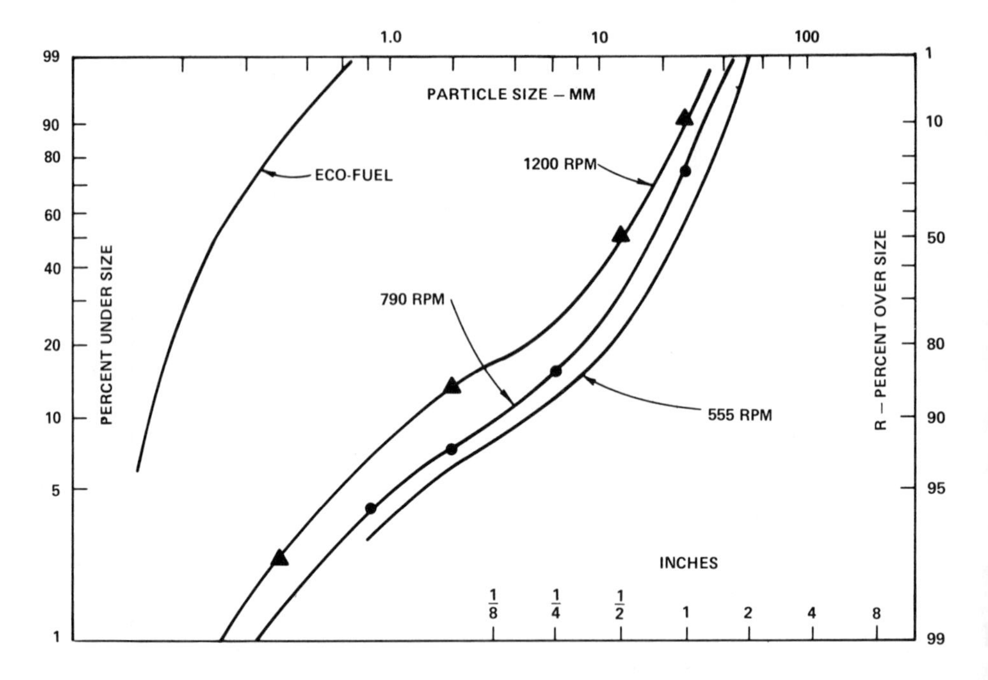

Figure 3-23. Effect of rotor speed on product [1].

Mechanical Efficiency

How well does a given machine perform? In addition to the power needed for shredding, the drive motor must overcome bearing and windage losses. Also, if a machine produces more fines than another, it will require more power.

Just to rotate a machine at normal operating speeds without feeding material requires considerable power. For instance, a 250-hp shredder may use 50 hp just to overcome motor friction and windage, bearing friction and windage of the rotor. The rotor windage does some useful work in moving air through the shredder, which helps drive the material through.

When the 250-hp shredder is operating at its practical feedrate, to avoid overload on surges it may be drawing an average of only 150 gross horsepower because at any time a slug of dense feed or a tough material may enter and demand more power. Subtracting the idling friction of 50 hp, we get 2000 net hp, which is called the useful work:

$$\text{Mechanical efficiency} = \frac{\text{useful work}}{\text{input work}} = \frac{100}{150} = 67\%$$

The specific energy and work index data referred to above were determined by subtracting the windage to get the net power used because it would be impossible to correlate data that relate to the feed material when the friction of the machinery varies from machine to machine and with the loading of the machine. In actual use, however, the horsepower is what has to be paid for.

Performance Curves

It is illuminating to compare the performance of three basic hammer-type shredders: horizontal- and vertical-shaft hammermills, and a flail mill.

The relationship between specific energy consumption and the rate of feed to the shredder is given by Trezek in Figure 3-24 for two different shredder speeds. There is a considerable amount of scatter in the points, indicating the difficulty in maintaining constant feedrates and relating the measured feedrates to the simultaneous power consumption.

By plotting the reciprocal of the specific energy consumption and the specific capacity, the mean values take a parabolic form, as shown in Figure 3-25. The envelopes contain all the data points. These envelopes define parabolic curves clearly. The specific capacity at 790 rpm rises to a higher mean peak (0.13 ton/kWh, or 8 kWh/ton) than that at 1200 rpm (0.09 ton/kWh or 11 kWh/ton). This represents a 26% reduction in power.

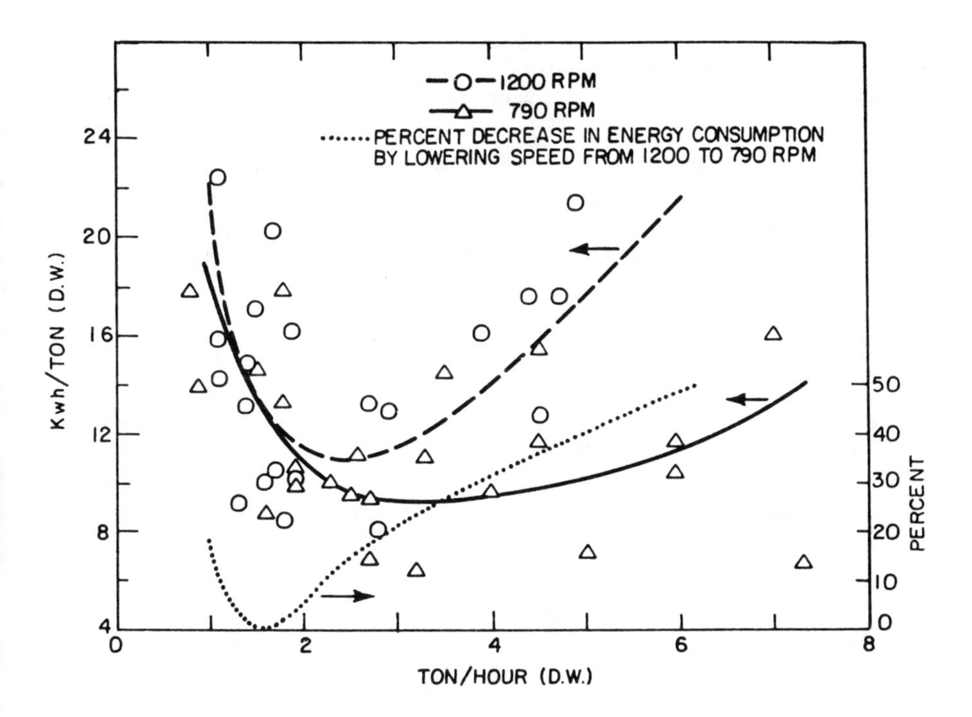

Figure 3-24. Specific energy required at 790 and 1200 rpm [1].

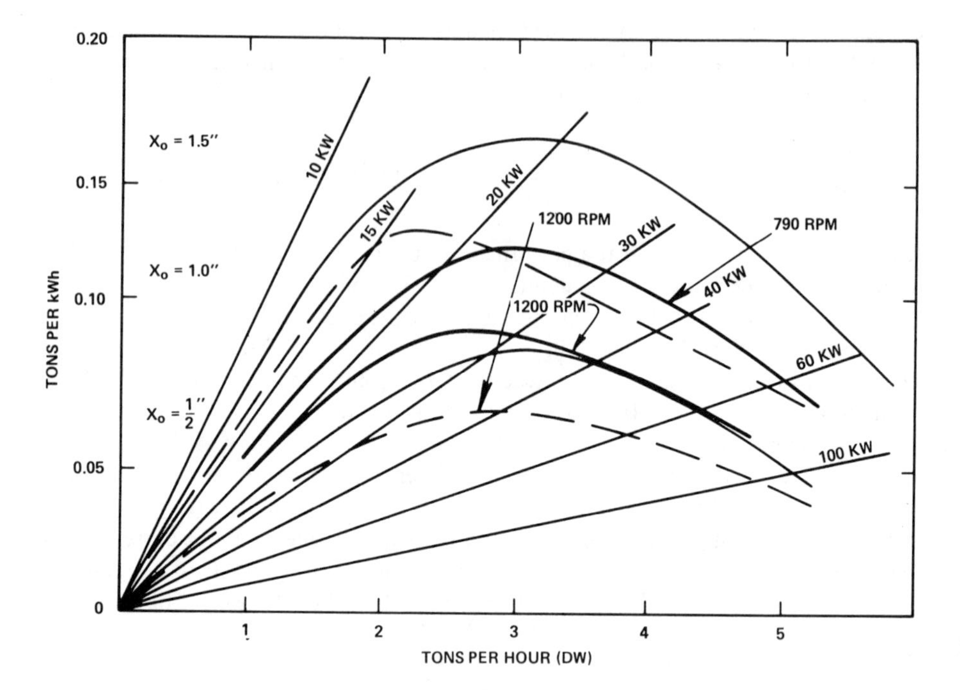

Figure 3-25. Specific capacity of the horizontal shredder of Figure 3-24 showing power consumption lines and performance characteristics.

This power saving also results in a change in PSD. As the kWh/ton is a direct function of mean particle size, then the reciprocal ton/kWh is also a direct function and the particle size, S_o, can be indicated on the ton/kWh axis, as shown. Now we can see that the mean particle size at the peak tonnage of 3 TPH at 790 rpm is 1.1 inches. On the other hand, at 1200 rpm, the peak occurring at 2 TPH yields a smaller particle size of about 0.9 inches.

The plot of ton/kWh versus TPH lends itself to plotting the power consumption lines. These help us to understand how the shredder will actually behave when operating. As the shredder is loaded, the mean particle size increases, with no appreciable change in power consumption, following the 20 kW line until about 2 TPH, at which point the power consumption increases rapidly without a corresponding increase in the particle size. What is happening is that the material is not flowing out of the shredder as readily. And, as feed is increased, it is recirculated within the shredder, maintaining a fairly constant particle size between 2 and 4 TPH, then falling off under an overloading condition as the feed increases to 5 TPH.

When a shredder is operated near the peak of this curve, it is likely to surge, alternately holding back material and discharging it, while holding a large mass suspended above the rotor. The shredder that produced these data was a Model 48-4 Gruendler hammermill with 44 hammers, driven by a motor rated at 250 hp, but seldom loaded anywhere near that limit except on unexpected surges.

The wide range between the upper and lower limits of the data points represents about a 2:1 range in power draw at any given feedrate, more likely reflecting variations in load above and below the average rather than changes in work index because the characteristic particle size also fluctuates accordingly. Difficulties in feeding the machine steadily will influence particle size. Also, surging of material inside the shredder causes a cycle: the shredder alternately holds back material, grinds it down and then discharges it, accepts a new charge of material that was floating above the rotor and builds up weight until it is heavy enough to force material out of the shredder. It should be mentioned that the graphs in Figure 3-25 were plotted on a dry basis in an attempt to eliminate the variable of moisture.

Figure 3-26, from Weinberger's tests in Baltimore County, shows total horsepower drawn by a vertical hammermill versus the load. In this case, the machine was lightly loaded because it was associated with a smaller secondary shredder. For this reason, the idling or no-load power is relatively large: unless a machine can be loaded heavily, this idling penalty can be costly. Note the difference in power reported for different degrees of opening of the shredder discharge, corresponding to different product particle size.

Figure 3-26 shows the ton/hphr minus the idling power to show the work actually done on the material, in which case the loading curves pass through zero at no capacity. This plot shows the effect of high loads on the efficiency of the machine: up to 5 TPH. The unit capacity in ton/hphr is linear, gaining rapidly as loading is increased. Above this it falls off, so that we get more tons without a corresponding increase in power: the machine is becoming more efficient. It is to be expected that there will be a loss of efficiency at higher capacities.

The particle size of the vertical mill product can be adjusted significantly by changing the amount of opening of the shredder outlet, thus increasing the material retention time in the machine, as shown in Figure 3-26.

The double-rotor articulated-hammer flail mill at East Bridgewater was tested over a period of three weeks to determine the power consumption and the rate of hammer wear, while processing Brockton MSW. This machine has 72-inch-long by 24-inch-diameter rotors, which were belt-driven at 1200 rpm by 125 hp motors. The top particle size was reported to range from 6 to 8 inches. The size distribution has been found to parallel that of 6- to 8-inch grate shredders, except that the oversize objects can be much larger because this mill has no grate.

Short tests were run confirming that the machine could process 50 TPH without overloading the 125-hp rotor drives. Conveyor limitations would not permit steady operation at more than 38 TPH of feed. Figure 3-27 shows the specific capacity of this shredder versus TPH feedrate.

The power consumption increased linearly from an idling power of 70 kW to 160 kW at 50 TPH. In contrast to grate-type mills, no fall-off of specific capacity was found within the rated range. The net power requirement was 1.5 kWh/ton over the range of capacity, indicating that the machine did about the same work on the MSW at all loadings. This is understandable because the machine has no grate that would hold back material and force the shredder to work harder as loading is increased.

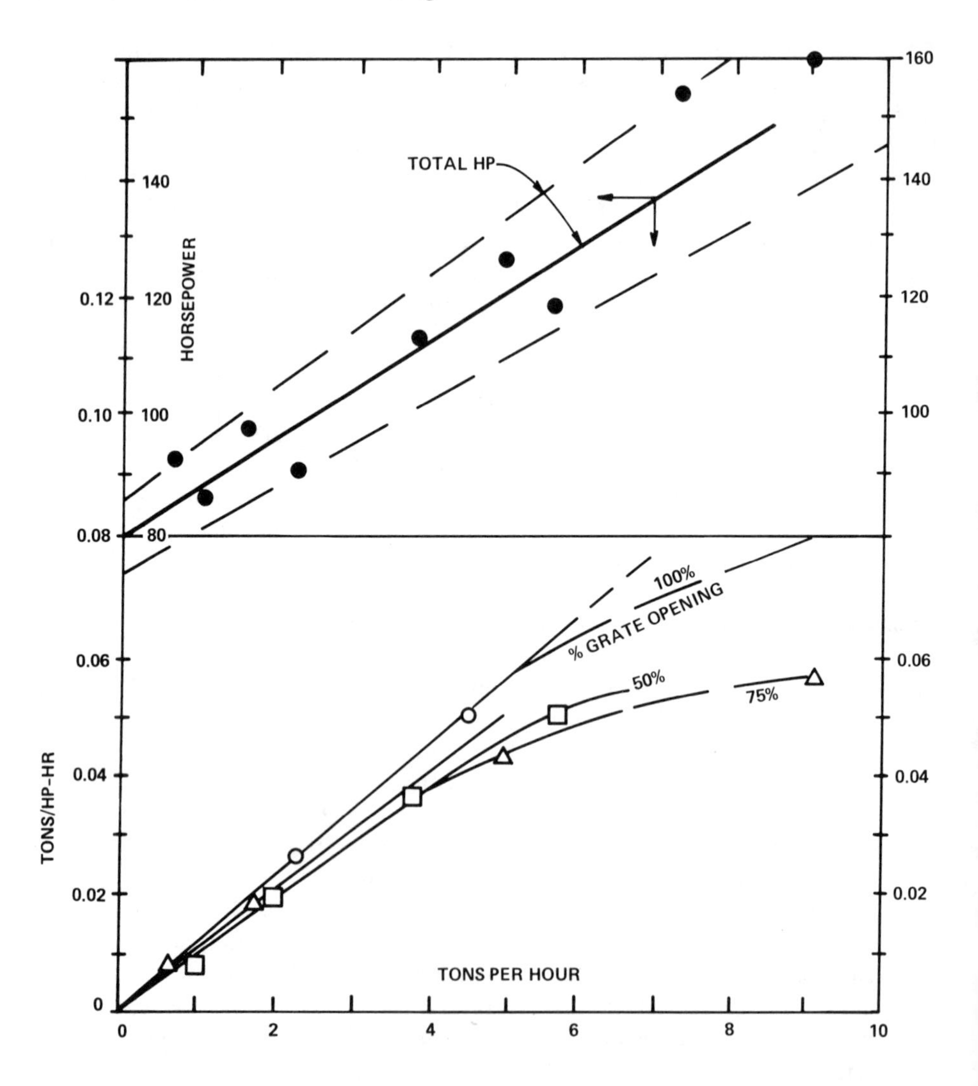

Figure 3-26. Performance of a vertical shredder showing effect of various grate openings on specific capacity versus load [3].

While the flail mill has the advantage of a low power requirement, it produces a large amount of oversize material, which must be screened out downstream. The East Bridgewater mill discharged to a magnetic separator and, thence, to a trommel, which screened the minus 5-in. product as unders and recirculated the oversize back to the mill. The capacity of the system was limited by the surface area of the trommel (about 200 ft^2) to about 12 TPH. The tests described above were run without recirculation.

Curves of the type shown above can be the basis for selecting shredders that will process a specified amount of material per hour and produce the desired product size, based on the type of refuse anticipated.

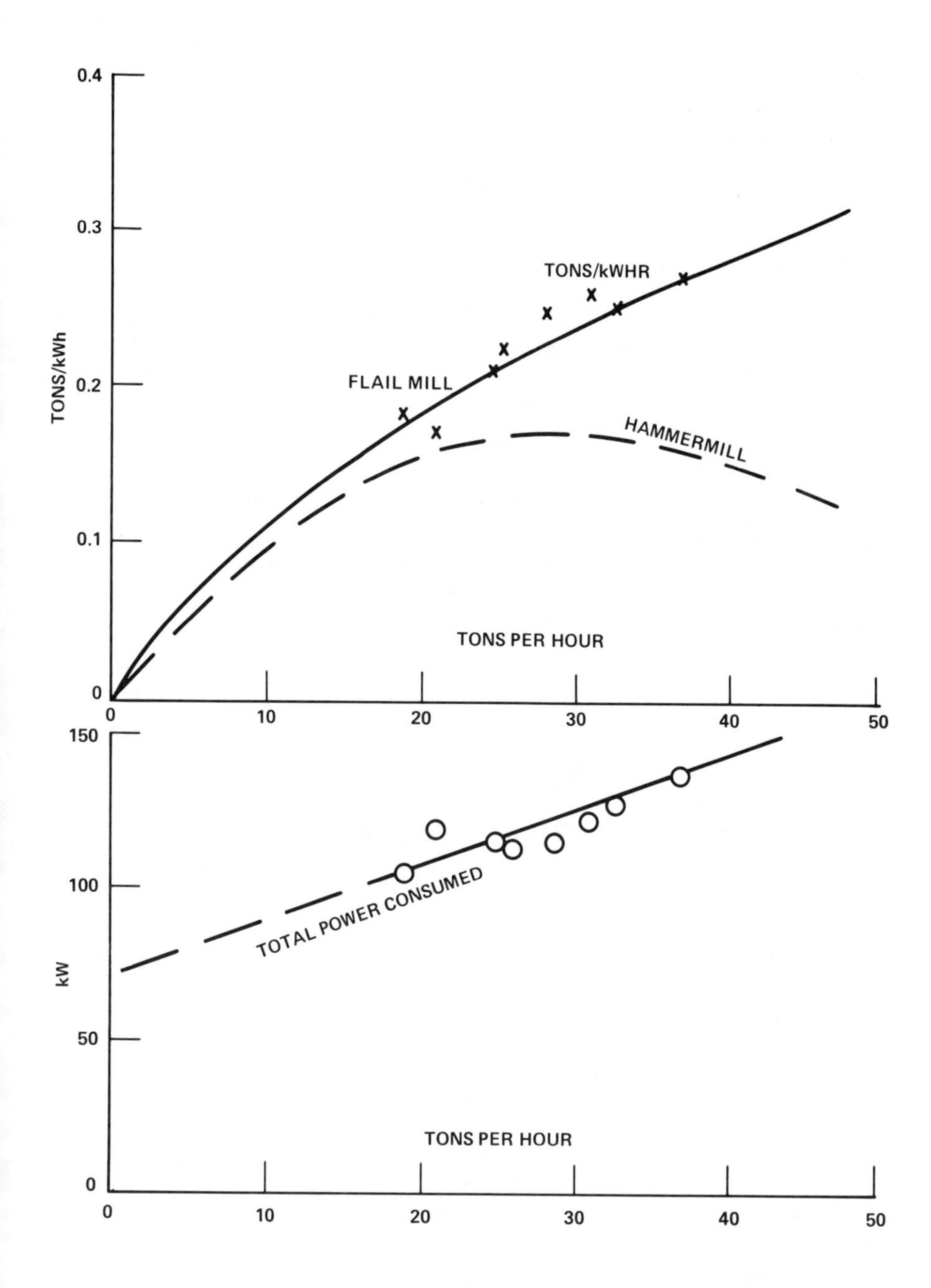

Figure 3-27. Performance of a flail mill at East Bridgewater (double rotor, 1200 rpm, 90% passing 6 inches), showing rising specific capacity characteristic of grateless flail mill.

For a given shredder equipped with a given motor, the operator can load the shredder to the motor limit: this capacity may or may not correspond with the estimates of the system designers, requiring adjustments. More or less time will be required each day and on different days to process the incoming waste stream. The size of the product also may not correspond with original estimates. Once the shredder is in operation, grate sizes and hammers can be changed to match the product size, power and capacity to suit the refuse as it comes. Unless belt driven, speed normally cannot be changed.

Performance of Rotary Shear-Type Shredders

Rotary shear shredders are made in single- and double-rotor configurations. In either case, moving knife cutters pass by stationary bars, creating a shearing action as they pass.

The "displacement" of these shredders can be calculated by determining the volume of the cavities per revolution and multiplying this volume by the number of revolutions per minute. As an example, the displacement of the single-rotor Garbolizer can be calculated as follows:

Model S40, with a feed opening of 42-inch × 42-inch, 26-inch rotor and 200-hp drive has 12 blades that pass through 3-inch × 5-inch openings while sweeping through a 10-inch cutting area at 27 rpm. It is suitable for shredding passenger tires.

$$\begin{aligned} \text{Volume swept} &= \text{length} \times \text{width} \times \text{depth} \times \text{blades} \times \text{rpm} \\ &= 5 \times 3 \times 10 \times 12 \times 27 \\ &= 48{,}600 \text{ in.}^3 \\ &= 27 \text{ ft}^3/\text{min} \end{aligned}$$

Model S520 has a 5-foot × 4-foot feed opening, a 400-hp drive and a displacement of 42 ft^3/min or 93 yd^3/hr and is suitable for truck tires and wooden pallets.

The actual weight capacity depends on the density of the feed material and how effectively the cutters "bite" the feed. Compression feeders may be needed to assure a good bite.

Double-rotor cutters have two cutter rotors rotating at different speeds to facilitate grabbing the feed. The displacement can be calculated in the same manner as above. The weight capacity, which depends on the density and nature of the material, has been estimated for various materials as follows:

Saturn Model 96-50 (feed opening 95 × 50 inches), 600-hp drive:

Type of Feed	Approximate Capacity (TPH)
Garbage	35-60
Steel Cans	10-15
Paper and Cardboard	7.5-10
Rubber Tires	6-8
Aluminum Scrap and Cans	5-7
Light Ferrous Metal	4-5

The specific power required is estimated to be about 7.5 kWh/ton, based on a minus 4-inch product. Due to the highly variable nature of refuse, the capacity and power requirements of shear cutters must be determined by actual test on typical materials, rather than by theoretical considerations.

Performance of Shredders in Refuse-Fuel Production

Long-term operating data are available for a number of facilities producing refuse fuel, which show the relationship among installed horsepower, product size and actual power consumed.

St. Louis Demonstration

This facility was tested in great detail from September 1974 to September 1975. This plant has a Gruendler hammermill driven at 874 rpm by a 933-kW (1250 hp) drive motor. The following data represent average conditions:

Hammermill product: 95% minus 38 mm (1.5 in.)
63% minus 17 mm (3/4 in.)

Capacity, TPH:	42 (rated)	50 (maximum daily average)
Horsepower:	1250 (rated)	610 (average for year)
Amperage:	140 (rated)	66 (average for year)
	50 (idling)	50-230 (range)

kWh/ton of feed:

daily range	– 12-27
monthly range	– 15-20
Average for year	– 17.5

These data show the wide range of performance (12-27 kWh/ton) and the low utilization of installed horsepower (610/1250 = 49%) experienced in this demonstration test. After identifying the reasons for this variation, it should be possible to reduce the width of range. The shredder feed was limited by a motor amperage control, which could stop the feed but otherwise was at the discretion of the operator. A better method for controlling the feed to the shredder should make a significant improvement.

Ames, Iowa

The City of Ames RDF facility showed roughly the same power requirements for a similar RDF product. Extensive data were taken at this plant prior to installation of disc screens, giving these results:

	Installed kW	Monthly Average kW	kWh/Mg
Primary Shredder	750	200	7
Secondary Shredder	750	340	12
TOTALS	1500	540	19

After the installation of disc screens, the power draw of the two shredders was balanced better and performance improved at the expense of increased rejects.

Pompano Beach, FL

The performance of a Heil 92B hammermill was monitored for two weeks at Pompano Beach. The following data were obtained:

Horsepower	1000 (installed) 570 (average)
Feed Capacity, TPH	60 (rated) 80 (average)
Specific Energy, kWh/ton	12 (nominal) 5.3 (average)
Product	95% minus 100 mm
	90% minus 75 mm
	63% minus 30 mm (characteristic size)

SUMMARY AND CONCLUSIONS

The relationship between shredder power consumption and the degree of size reduction for the paper and cardboard fractions in municipal refuse has been found to be predictable and highly dependent on moisture content. Although correlation is difficult, it appears that the energy required to rupture materials and the work index are closely related for many materials.

The separate or separated components of refuse generally have size distributions that can be defined mathematically and follow Rosin-Rammler or probability distributions closely. Mixtures, however, show nonlinear distributions. As MSW varies continuously in composition and size of constituents, the product of size reduction will vary accordingly.

The degree of size reduction that will be accomplished by a given machine or grate size varies with the tensile strength of the material; therefore, textiles are not reduced as much as paper and cardboard when shredded at the same time. As a result of this, to reduce textiles to a size that will not cause problems with handling machinery, the paper and cardboard may have to be reduced more than otherwise necessary.

The grate size and configuration of a grate-type shredder can be adjusted to achieve the desired product size. However, the capacity and power consumption will be changed correspondingly. It is important to make these matches in the design phase to avoid mismatches in the operating plant.

Residential and commercial wastes are significantly different in power requirements: the shredder must be able to cope with the variation in composition to which it will be subjected. Grateless shredders can be provided with external sizing devices to control the maximum particle size, the oversize being recycled or rejected.

Cutter-type shredders can be expected to produce products with more consistent top sizes due to the restrictive nature of the apertures through which the material must pass.

Grinding, as compared with tearing operations, has a different exponential relationship in relating work to particle size, probably closer to 0.25, as mentioned above. Table 3-8 lists some characteristics of actual operating shredders.

Table 3-8. Dimensions, Drive Power and Approximate Rated Capacity of Size Reduction Machines

Flail Mills				
Single-rotor bag breaker	Width:	60 in.		
	Drive:	40 hp		
	Capacity:	50 TPH		
Double-rotor flail mill	Feed opening:	48 in. high × 72 in. wide		
	Drive:	2 @ 100 hp		
	Capacity:	45 TPH		
Horizontal Hammermills				
Ames primary mill	1000 hp	Capacity:	50 TPH	
Secondary mill	1000 hp		50 TPH	
St. Louis primary mill	1250 hp	Capacity:	80 TPH	
Vertical Hammermill				
Pompano Beach, FL	1000 hp	Capacity:	80 TPH	
Rotary Shears				
Garbalizer	300 hp	Capacity:	20 TPH	
Saturn	600 hp	Capacity:	45 TPH	
Ball Mill Bridgeport, CT	14 feet wide × 30 feet long		1500 hp	40 TPH
Cascade Mill Newark, NJ	28 feet wide × 20 feet long		4000 hp	150 TPH
Pellet Mill	150 hp	Capacity:	10 TPH	

REFERENCES

1. Trezek, G.J., et al. "Significance of Size Reduction in Solid Waste Management," EPA-600/2-77-131 (July 1977).
2. Ham, R.K. "The Role of Shredding Refuse in Landfilling," *Waste Age* 6(12): 22 (1975).
3. Weinberger, C.S. "Evaluation of Secondary Shredding to Enhance RDF Production as Fuel for Cement Kilns," 1980 ASME NSWP Conference, Washington DC (1980), p. 373.
4. Vesilind, P.A., A.E. Rimer and W.A. Worrell. "Performance Characteristics of a Vertical Hammermill Shredder," 1980 ASME NSWP Conference, Washington, DC (1980), p. 199.
5. Austin, L.G. "Introduction to the Mathematical Description of Grinding as a Rate Process," *Power Technol,* 5:1 (1971–72).
6. Gaudin, A.M. "An Investigation of Crushing Phenomena," *Trans. AIME* 73: 253 (1926).
7. Gaudin, A.M., and T.P. Melox. "Model and Comminution Distribution Equation for Single Fracture," *Trans. ASME* 223:40–43 (1962).
8. Evans, I., and C.D. Pomeroy. *The Strength, Fracture and Workability of Coal* (Elmsford, NY: Pergamon Press, Inc., 1966).
9. Broadbent, S.R., and T.C. Callcott. "Coal Breakage Processes," *J. Inst. Fuel* 29:524 (1956).
10. Broadbent, S.R., and T.C. Callcott. "Coal Breakage Processes," *J. Inst. Fuel* 30:13 (1957).
11. Ruf, J.A. "Particle Size Spectrum and Compressibility of Raw and Shredded MSW," Ph.D. Thesis, University of Florida (1974).
12. Austin, L.G., D.T. Luckie and R.R. Klimpel. "Solutions of the Batch Grinding Equation Leading to Rosin-Rammler Distributions," *Trans. AIME* 252:87 (1972).
13. Alter, H., and B. Crawford. "Materials Recovery Processing Research," NCRR for USEPA/OSWM, Washington, DC (October 1976).
14. Bond, F.C. "The Third Theory of Comminution," *Trans. AIME* 193:484–494 (1952).
15. Holmes, J.A. "Contributions to the Study of Comminution," *Trans. Inst. Chem. Engr.* 35:124–141 (1957).
16. Austin, L.G., and P.T. Luckie. "Grinding Equations and the Bond Work Index," *Trans. AIME* 252:259–266 (1972).
17. Stratton, F.E., and H. Alter. "Applications of Bond Theory to Solid Waste Shredding," *J. Environ. Eng. Div., ASCE* (1978).
18. Savage, G.M., G.R. Shiflett, L.F. Diaz and G.J. Trezek. "Evaluation of Performance of Hammermill Shredders Used in Refuse Processing," Paper presented at the Fifth Annual Research Symposium, Orlando, FL, USEPA/MERL, 1979.
19. Gawalpanchi, et al. "Particle Size Distribution of Milled Refuse," *Waste Age* 4(5):34–74 (1973).
20. Trezek, G.J., G.M. Savage and L.F. Diaz. "Highlights of Shredder Research in Resource Recovery Processing," in *Proc. Seventh Annual Res. Symp.*, EPA MERL, Philadelphia, PA (1981), pp. 10–17.

CHAPTER 4

Metals Recovery and Wet Processes

The purpose of processing MSW is to dispose of the refuse in a consistently safe, reliable and environmentally acceptable manner, while creating the optimum quantity and quality of revenue-producing products commensurate with the capital, labor and operating expenses. The residue that cannot be made into profitable streams must be disposed of in a suitable form at a suitable disposal site.

To produce valuable products it is necessary to refine them by eliminating undesirable ingredients, such as metals, moisture and glass in the case of fuel products, and combustibles in the case of metals and glass. Reduction in product particle size enhances the effectiveness of processing operations, increasing the marketability for further use or refining of the fuel, metals and glass.

This chapter will trace through the sequence of separate operations that changes the nature of MSW to create the product and waste streams that achieve our objectives.

It is important to emphasize the differences among what is technically possible to do, what can practically be done and what is economically feasible. The nature of the input material determines what can possibly be done. Paper products can be made out of wastepaper, fuels from the combustibles, ferrous products out of "tin" cans and aluminum from nonferrous scrap. Some of the glass present in the MSW can be converted to useful form, but not the dirt and fine silica sand that is also present. A major fraction of what enters the plant can be recovered. What we cannot or do not recover must be disposed of at a cost that must be included in overall economics.

The operations performed must be possible specifically with the actual materials. This means that the operations that specific machines or devices do or can perform must be suited to the properties of the materials themselves. This coordination is easier said than done, but it generally helps to define what is wanted and what can be done and get some agreement between these statements before expecting the machine or the material to behave in the way we would like it.

This chapter will review the basic systems used to process MSW into products and to minimize residues. Flowsheets and machinery configurations will be described and material balances reviewed to show the performance of these processes, as well as the component operations. The following processes will be investigated:

- pilot plants, to determine the components of MSW and the separation potentials of various operations;

- testing and evaluation of separation operations: (1) dry processes for recovering refuse-derived fuels and products; and (2) wet processes for recovering pulp, fuel and products; and
- processes for recovery of paper and plastics from MSW.

The total process consists of a matrix of operations, all interdependent. The performance of these operations singly, and as a part of the whole, will be investigated and evaluated.

Methods and definitions used in evaluating processes and systems will be presented that can be used to quantify the effectiveness of the processes in terms of recovery, purity, yield and the quality of the products.

The various species in the MSW feedstream are diverted by processing operations in various ratios, called "splits," which can be measured during testing or operation and used as machine characteristics in computer programs.

The performance of individual operations or processes can be analyzed as a part of operating configurations of the system and the total performance of the process section or the entire plant being evaluated. The factors that affect the performance of the individual operations must be determined by testing to understand the effect of parameters such as loading, configuration and the nature of the feed material. The performance of the entire system, or the individual operations, can be optimized by process analysis. Possible configurations can be tried and the consequences of these arrangements traced along the separating and refining processes from input to output. In this manner, the quality of fuel products can be improved by optimizing removal of inert materials. Processing of residues can recover combustibles and remove contaminants from metal and glass products.

The theory and practice of such operations as magnetic and wet separation are investigated to show the operations that can be used and results that have been reported. The effect of loading and configuration on the efficiency of recovery and the purity of products will be investigated.

We will investigate how data on processing MSW is obtained and analyzed; how processing operations can be assembled and the entire operation analyzed; how we can determine what results could be achieved; and how close to perfection the actual process is.

UNIT OPERATIONS AS PART OF THE TOTAL PROCESS

There are many options available for processing refuse into energy, ranging from minimal processing and burning in boilers to those that produce liquid and gaseous fuels. Processing can:

- recover valuable materials, such as metals and glass,
- refine products to increase their economic value,
- remove contaminating materials, such as dirt and glass,
- mix materials for greater uniformity,
- reduce particle size to permit classification,

- reduce particle size to obtain better storage and feeding properties,
- reduce moisture to improve storage and combustion properties,
- reduce ash and inert contents of fuel products; and
- remove organic contaminants from metals and glass.

The following operations can be used to recover resources:

- **Size reduction**: cutting, shredding and grinding
- **Size separation**: screening
- **Metals recovery**: magnetic and wet processing
- **Density separation**: fluid drag forces versus gravity
- **Inertial separation**: drag forces versus inertia
- **Drying**: direct contact, radiation, convection
- **Densification** and other forms of beneficiation
- **Combustion**: dedicated boilers, remote boilers
- **Conversion**: digestion, pyrolysis, gasification

The degree of processing that should be done under a given set of conditions is a matter of practicality and economics. This book emphasizes evaluation of the processing operations and systems and their effectiveness in producing valuable products at full-scale production rates. With this information, alternate processes can be assembled and evaluated in terms of their capabilities, the value of their products, estimates of plant capital, labor and operating costs. Then the complete economic, institutional, environmental and social evaluations can be made to decide which is the optimum solution.

ANALYSIS AND EVALUATION OF THE PERFORMANCE OF UNIT OPERATIONS

Properties of the Feed

A first step in resource recovery plant design is to identify and quantify carefully the properties of the various materials in MSW. In Chapter One, the chemical and other physical properties were explored, and the results of standard methods for testing MSW and RDF for chemical and thermal properties were given. The relative amounts of the basic ingredients fall within certain ranges, which, despite their variations, are predictable. "Consistently inconsistent" expresses this exact inexactness.

Now we are interested more in the physical properties that are related to processing of the materials, such as size, shape, density, the combustible content, moisture content, metals and glass, and miscellaneous contents, and also in objectionable materials, such as putrescibles, tires, mattresses, appliances, and especially hazardous materials such as combustible liquids.

These properties are related to the operations we would like to perform: separation, classification, size reduction, drying, conveying, storage and transportation.

Sampling

To determine these properties, samples must be collected from landfills, packer trucks or from refuse put out on the street. These can be weighed, sorted and analyzed for moisture, the various species included in them and for the properties of each species such as moisture, ash and heating value. Size analyses can be performed on the entire lot and the individual species. This type of data has been reported widely. The same methods of sampling can be carried out on the various streams within a processing plant.

Many of the properties we must measure or understand have special application to specific machines or operations, such as the lifting velocity as applied to air classifiers, toughness as applied to shredders, and particle size and shape as applied to screens. These properties generally have ranges or distributions.

Capacity

A major property of a machine or operation is capacity, in relation to the size of the machine. Capacity and the effectiveness corresponding to it are closely related, especially at economic capacities. Some machines will give excellent performance at light loads; however, as we increase the load to get more production out of the machine, the effectiveness of the operation and the quality of the product usually are reduced.

Energy

Energy use must be considered from the point of view of cost of energy, availability and form of energy. Sometimes the use of energy is a very efficient way of accomplishing an operation; sometimes it is spent in useless friction, and sometimes friction is useful and necessary.

Operation and Maintenance

Labor is a major cost in RR plants. Both operation and maintenance (O&M) costs will play a major role in determining optimum operations. Wear and tear on machinery, reliability, downtime and many other factors also will enter the picture.

Before looking at individual unit operations, some definitions should be reviewed so that these measures of performance can be used consistently.

MEASURES OF EFFICIENCY AND EFFECTIVENESS

Thermal Efficiency

When evaluating thermal equipment, where heat is converted to work we use the term "thermal efficiency" to evaluate the performance of the equipment or process, and define it as the work output divided by the heat input, expressed in the same units, such as kJ, Btu, kW or hp, so that the ratio is dimensionless (the dimensions cancel).

The efficiency of conversion of heat to work is limited to less than 50% by the maximum temperatures we can produce in practical equipment and the temperatures at which the heat must be thrown away. The amount of work that can be obtained from a given amount of heat depends on the temperature at which the heat is generated and the temperature at which the heat is removed from the engine. The amount of work that can be obtained from a heat engine is reduced as the "exhaust" temperature is raised so that at atmospheric temperatures it is only 10-35%.

Mechanical Efficiency

All mechanical operations entail some frictional losses, which convert work into normally useless heat. A frictionless gyroscope can spin forever in a vacuum but only for a limited time in air. If one pound of water is lifted 100 feet, it cannot deliver more than 100 ft-lb of work as it returns to the level from which it was lifted.

Energy Efficiency

The energy used to perform operations must be measured properly and related to the results of these operations. One should always bear in mind that work energy is a much more valuable form than heat energy, that high-temperature energy is more valuable because it can be converted to more work than lower-temperature energy and that friction converts valuable work to low-temperature heat. The dollar cost of energy and the dollar value created by the use of this energy generally are more important. However, theoretical principles are useful in determining the ultimate and practical limits of conversion and how well the process is operating.

Effectiveness

The term "effectiveness" should be reserved for operations involving other than mechanical and thermal efficiency, such as the degree of removal or separation of materials. Complete separation of a substance from a mixture can never be achieved in practice because it would take too much time or effort. We usually compromise and accept more reasonable levels: 50% of the ultimate is easy to achieve; 75% is not too hard; 96% is hard; 98% is twice as hard, and 99% twice again as hard.

Effectiveness may be defined, simply, as "what is achieved versus what could be achieved." To make this comparison, we must determine what should be achieved, and measure, in the same terms, what we can or did achieve, thus permitting comparison of various devices and operations on the same scale, from the same reference, so we can make fair comparisons.

An important benefit of this definition of effectiveness is that it tells us how well we are doing, that is, how near we are to perfection. If we are near to perfection, we know we have to make a great effort to make a small improvement, whereas when far from it, a small effort will be very rewarding.

From an economic point of view it is extremely useful to know the incremental value of an increased expenditure, as related to the benefit that results. If we do not know this, we will tend to spend money in the wrong places and not achieve our objectives efficiently.

Measurements of effectiveness of processes dealing with MSW streams are difficult to obtain due to the continuous variations in flowrate and stream properties. It is essential that they be made in relation to the capacity of the equipment because the heavier the load on most of these operations, the lower the effectiveness. This fact requires us to optimize rather than blindly select equipment.

Quality

We commonly use the word quality to describe the level of some desirable feature or characteristic of a material, or even of an operation. Each situation requires a definition of the characteristic we desire, a way to measure it and a way to compare this measurement with others or with standards of quality. ASTM standards have this purpose: to measure by standard methods, express the results in standard ways and then to relate these results to other standards or measurements. These standards have been developed by the consensus of the producer and the user of the products, with the laboratory as the common ground for determining the properties and quality and with other agencies often involved for various reasons. ASTM Committee E-38 has developed many standards for refuse-derived products. (See Appendix B.)

GENERAL SEPARATION

Separation operations require the existence of a difference, on which basis the separation can be made, such as size, shape, density, magnetic properties, wettability, friability and so forth.

Separating devices are evaluated in accordance with their effectiveness or completeness of function: how much they did compared with how much they could have done under ideal conditions. Effectiveness can be expressed in terms of recovery of the desired materials, elimination of contaminants or purity of the product.

Recovery and Purity

Consider a separating device having a feedstream containing two kinds of materials, x and y, entering at (o), with flowrates X_o and Y_o, and having two exit points, (1) and (2). The device is intended to separate material x from material y on the basis of some property of x and y.

Consider a magnetic separator. A stream of nonmagnetic material, x, mixed with magnetic material, y, passes under a magnet, which picks up the magnetic, but not the nonmagnetic, particles.

A perfect magnetic device discharges all x particles to exit (1) and all y particles to exit (2), giving a recovery fraction of 1.0:

$$RF = \frac{X_1}{X_o} = 1.0 \qquad X_o \rightarrow X_1 = X_o$$

$$Y_o \rightarrow Y_2 = Y_o$$

If our purpose is to remove all magnetic materials from the feedstream, then we are interested in the "purity" of the product stream. In this case, perfect purity is

$$P = \frac{Y_2}{Y_o} = 1.0$$

As the device is less than perfect in both recovery and purity, some X_o reports to Y_2 and some Y_o reports to X_1, resulting in less than 100% recovery and purity:

$$\text{Recovery to (1)} = R_1 = \frac{(X_o - X_2)}{X_o} = \frac{\text{(input-loss)}}{\text{input}}$$

and

$$\text{Recovery to (2)} = R_2 = \frac{(Y_o - Y_1)}{Y_o} = \frac{\text{(input-loss)}}{\text{input}}$$

$$\text{Purity of (1)} = P_1 = \frac{X_1}{(X_1 + Y_1)} = \frac{\text{pure}}{\text{contaminated}}$$

and

$$\text{Purity of (2)} = P_2 = \frac{Y_2}{(X_2 + Y_2)}$$

As an example, consider a mixture of 10% ferrous (F) and 90% paper (P) passing under a magnet, with the ferrous exit containing 80% of the entering ferrous and 10% of the paper. The remaining 20% ferrous contaminates the paper and only 90% of the paper is recovered:

$$\begin{array}{lcll} F_o & = & 0.1 \times 0.8 \rightarrow 0.08 & = F_1 \\ P_o & = & 0.9 \times 0.1 \rightarrow 0.09 & = P_1 \\ \hline \text{Feed } W_o & = & 1.0 \qquad\quad \rightarrow 0.17 & = P_1 + F_1 = \text{product} \end{array}$$

$$\text{Underflow} = \begin{array}{l} 0.1 \times 0.2 \rightarrow 0.02 = F_2 \\ 0.9 \times 0.9 \rightarrow 0.81 = P_2 \\ \hline \qquad\qquad 0.83 = F_2 + P_2 = \text{underflow} \end{array}$$

$$\text{Recovery (1)} = R_1 = \frac{(F_o - F_2)}{F_o} = \frac{(0.1 - 0.02)}{0.1} = 80\%$$

$$\text{Recovery (2)} = R_2 = \frac{(P_o - P_2)}{P_o} = \frac{(0.9 - 0.09)}{0.9} = 90\%$$

$$\text{Purity (1)} = \frac{F_1}{(P_1 + F_1)} = \frac{0.08}{0.17} = 47\%$$

$$\text{Purity (2)} = \frac{P_2}{(P_2 + F_2)} = \frac{0.81}{0.83} = 97.6\%$$

In the case of magnetic recovery, the difference is physical—the magnetic property of the ferrous material. In the case of screening, the difference is the size of the particles, as compared with the screen size. Recovery is then defined as the fraction of particles (that are less than screen size) that actually passes through the screen compared with the fraction in the feed that could have passed. Then,

$$\text{Recovery of fines} = \frac{(\text{recovered fines})}{(\text{fines in feed})}$$

Purity of product is the ratio of true oversize material to the total material passing over the screen:

$$\text{Purity of oversize} = \frac{(\text{true oversize})}{(\text{actual overs})}$$

Efficiency

It is appropriate to use the term efficiency when relating actual performance to possible performance. The use of efficiency is an attempt to develop a single-valued function for describing the effectiveness of a separation device because both purity and recovery are necessary measures (one cannot adequately describe the process). Two different expressions for efficiency have been developed. Rietema [1] defined separation efficiency in terms of the fractions of the input streams:

$$\text{Separation efficiency} = \frac{X_1}{X_o} - \frac{Y_2}{Y_o}$$

Worrell [2] developed another form:

$$\text{SE (Worrell)} = \frac{(X_1 \times Y_2)}{(X_o \times Y_o)}$$

The significance of these can be illustrated with this example:

A magnet receives 80 TPH of paper and 20 TPH ferrous, of which 15 TPH is collected along with 5 TPH of paper:

$$\text{Recovery of ferrous} = \frac{15}{20} = 75\%$$

$$\text{Purity of ferrous} = \frac{15}{(15+5)} = 75\%$$

$$\text{SE(Rietema)} = \frac{15}{20} - \frac{5}{80} = 68.75\%$$

$$\text{SE(Worrell)} = \frac{(15 \times 75)}{(20 \times 80)} = 70.3\%$$

The two expressions for efficiency usually yield similar values, and there is no overriding reason to choose one in preference to the other.

FERROUS MATERIALS RECOVERY

Magnetics

The force of magnetic attraction or repulsion can be used to remove ferromagnetic materials (called ferrous) mixed with nonferrous materials. Nonferrous metals can be separated from nonmetallic materials, which can hold a charge after passing through a fluctuating field in an eddy current separator, after which these charged particles will be repelled by a magnetic field.

Magnetic fields can be generated in a magnetic core by winding a coil of wire around the cores and passing an electric current through the wire. The intensity of the field so generated is

$$H = \frac{(N\ I)}{1}$$

where N = number of turns
1 = length of core (cm)
I = current (amperes)

The field intensity creates a magnetic induction, B, defined by Faraday's law:

$$\frac{-dB}{dt} = \frac{V}{A}$$

where B = flux density
V = voltage
A = area normal to flux

When a conductor is "cut" by lines of magnetic flux, a voltage, V, is generated. The magnetic flux, Φ, is defined in Webers (W), in which 1 W = 1 volt second.

Then

$$B = \frac{\Phi}{A} = \frac{W}{m^2}$$

Magnetic materials have a property called magnetic permeability, μ, defined as

$$\mu = \frac{B}{H}$$

Ferromagnetic materials have large positive values of μ, which cause a high concentration of flux lines. Work has to be done to magnetize a volume of material:

$$\text{Work} = V\, I\, H\, dB = V\, I\, \mu\, H\, dH$$

$$\text{Work} = \text{force} \times \text{distance}$$

or

$$\text{Work} = Fdl$$

The force with which a magnet attracts a ferromagnetic material, in one dimension, is

$$F = \frac{dW}{dl} = VH\left(\frac{dB}{dl}\right) = V\, HGH\, \mu\left(\frac{dH}{dl}\right)$$

Force is a function of field intensity, field gradient and magnetic permeability. Ferromagnetic materials will move in the direction of increasing field intensity, hence, toward the magnet.

Ferromagnets may be either permanent or electrically maintained, in which case they generate heat, which must be removed and dissipated externally.

Figure 4-1 shows the relationship between the working distance of a magnet from a ferrous object and the diameter of the magnet. Curves for permanent and electropermanent magnets are shown, permanent being used mainly for the smaller sizes. The horsepower of the magnet drives is shown for reference. The wattage drawn by these magnets is about the same as the horsepower indicated.

The performance of magnets depends on the distance between the material and the magnet and the thickness of the bed of material from which the magnet draws ferrous material. When the bed is heavy, the weight of material over the magnetic object has to be overcome to extract the object. Sometimes the twisting action of the object under the action of the magnetic field helps to loosen the object from the burden. Other times the object cannot be lifted, reducing separation efficiency.

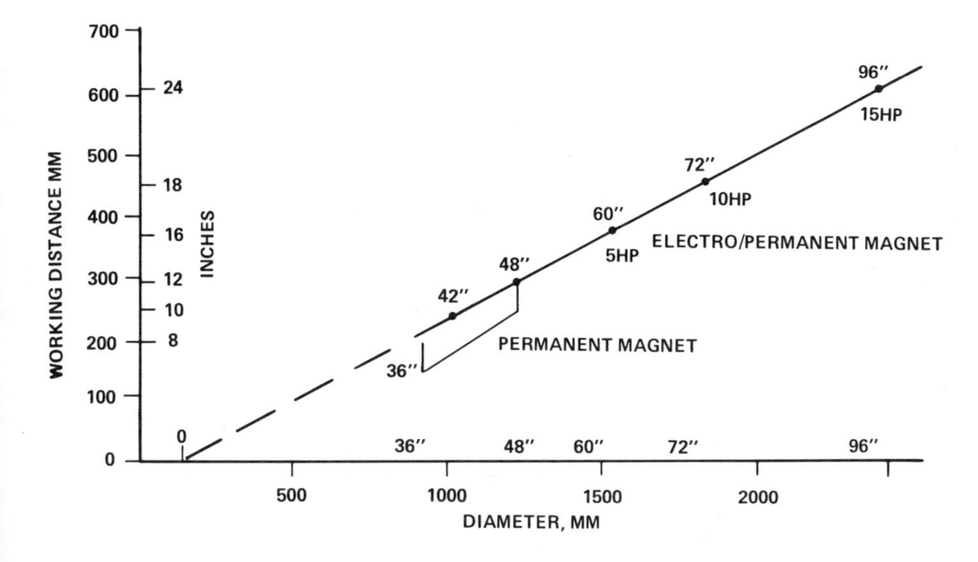

Figure 4–1. Working distance versus diameter of magnets (Data from Eriez Magnets).

Figure 4–2 shows the recovery efficiency measured as a function of the burden of the material above the belt. It is apparent that a severe drop-off in efficiency occurs at a certain critical point. It is important to maintain bed thicknesses below this critical dimension. Bed thickness is a function of the width of the belt, belt speed and feed density. As an example, a belt 4 feet wide and 3 inches deep, traveling at 300 ft/min, will handle $4 \times 300 \times 3/12 = 300$ ft^3/min. Material at 10 lb/ft^3 will then be flowing at $300 \times 10 \times 60/2000 = 90$ TPH. If the maximum particle size is 3 inches (a good number for cans), the bed will have little, if any, particle interference if it is distributed evenly.

Ferrous Metals Recovery Systems

Recovery I

The ferrous metals system at Recovery I has been described in detail, as has been its performance [3]. The original system at Recovery I was as shown in Figure 4–3. MSW was fed to the 45-foot (13.7-m)-long by 10-foot (3-m)-diameter rotary trommel screen with 4-3/4-inch (121-mm) holes. The underflow fraction was conveyed to a 1-meter-diameter electromagnetic drum, while the trommel overflow was size reduced in a Heil vertical-shaft shredder and conveyed beneath a 1.2-meter-diameter electromagnetic drum. The two magnetic scalpings were then conveyed to a vibrating air separator and air classifier, called a ferrous concentrator, to clean and separate the ferrous metals into two fractions, heavy and light. The dense metals were conveyed to a roll-off container. The light can stock was fed to a hydraulic can compactor

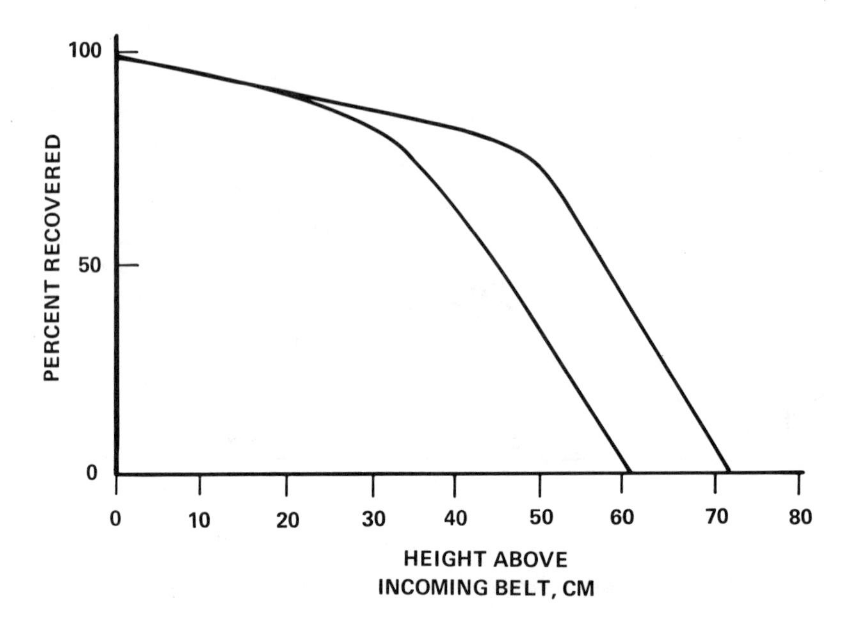

Figure 4-2. Shows how height of magnet above refuse affects efficiency.

(densifier), which compacted the cans into discs, after which they were conveyed to rail cars.

The system as designed and installed was developed at a time when there was a scarcity of data on the characteristics of full-scale systems for the recovery of ferrous from shredded refuse, when no specific product standards existed and no contracts existed for sale of the product.

When the system was started up, many problems were found and corrected, including the following:

- Burden depths of MSW on conveyors were greater than expected. The conveyor speed was increased to 450 ft/min to reduce burden.
- The magnet air gap was increased from 14.5 in. (0.37 m) to 22.5 in. (0.89 m), gaining reliability at the expense of recovery efficiency.
- Light ferrous openings were too small, resulting in jams. Matching contours were changed to reduce misreporting of light fraction ferrous to the heavy fraction.
- Cyclone air flows required balance changes to minimize lifting light metals.
- The compactor power pressure had to be reduced to avoid deformation.
- Lack of storage of light scrap created difficulties in coordinating empty and full rail cars.
- Rail cars could be loaded to only 24 tons. As much as 15% contaminants were found. The result was high penalties imposed by the customer.

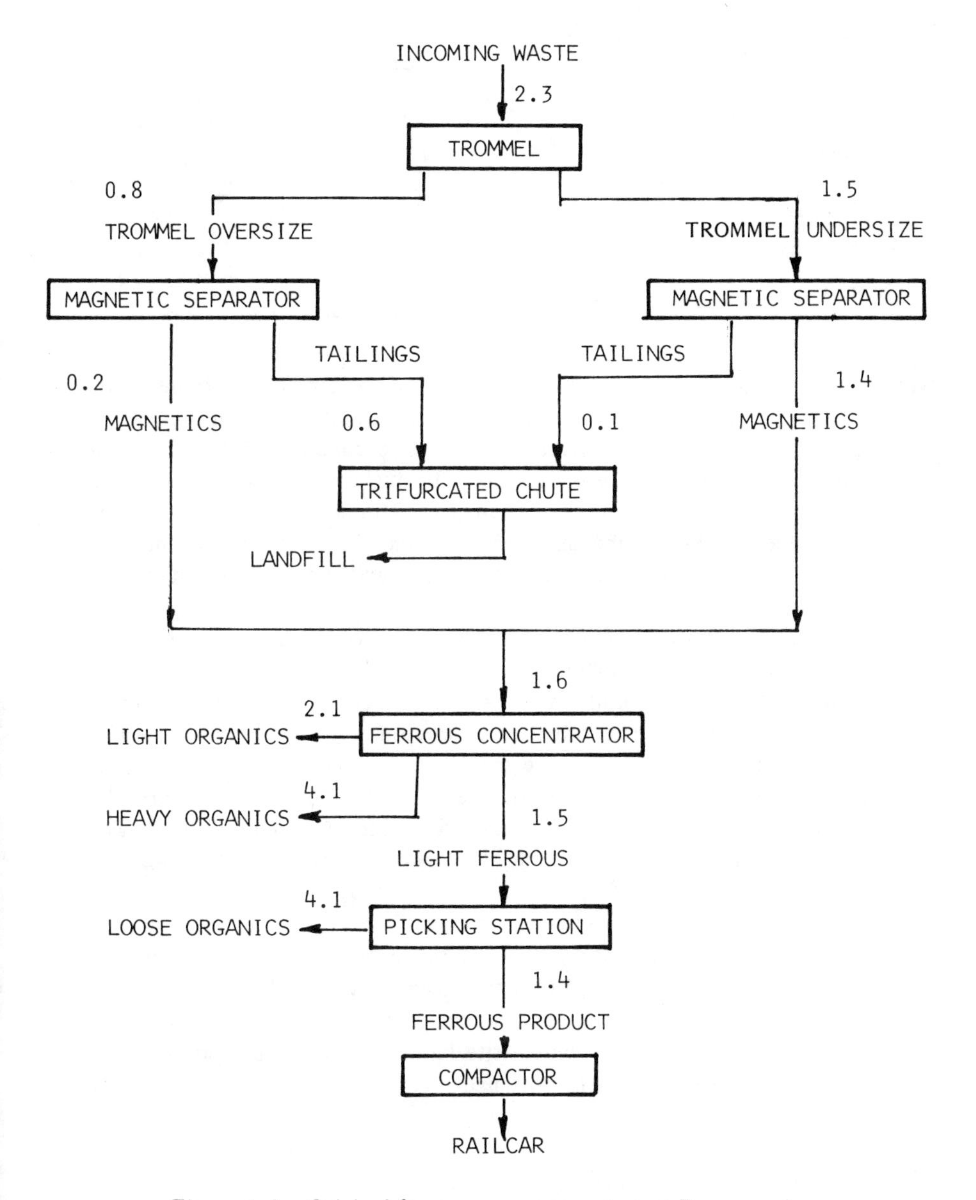

Figure 4-3. Original ferrous recovery system at Recovery I.

An extensive testing program was initiated to identify the sources of contamination and to find ways to produce a less contaminated product.

Tests of the trommel screening efficiencies and magnetic separator recoveries were made. The results are summarized in Table 4-1. The performance of the Recovery I Ferrous Recovery Plant (FRP) has been reported only for the original installation,

Table 4-1. Trommel Efficiency and Magnetic Recovery Results [3]

Component	*Trommel Screen Efficiency (%)*	*Trommel Unders Magnet Recovery (%)*	*Shredded Material Magnet Recovery (%)*
Light Ferrous	79.2	92.6	24.4
Heavy Ferrous	90.0	83.4	38.1
All Ferrous	82.8	90.3	27.2

which has subsequently been revised as described below. The purpose of the test program was to evaluate the performance of each unit operation and that of the entire plant to determine the reasons for excessive contamination and low density of the metal product. Samples were taken throughout the process and analyzed for metal content and degree of contamination. Samples of the product were taken to test the effectiveness of a metals shredder in releasing contaminants and increasing product density. The test results were favorable and resulted in installation of a shredder in the process.

Although the test data reflect the performance of the original, not the revised, plant, it is nonetheless a valuable demonstration of the performance of the magnets, ferrous concentrator and air knife. Table 4-2 shows these performance data in simplified format, which directs attention to the fraction of material recovered. The mass splits of the metals concentrator were measured and reported at a flowrate of 106 TPH, and at a total unprocessed ferrous stream flowrate of 1.84 TPH, or only 1.7% of the input to the plant during this test. A six-month test in 1979 showed that the average amount of clean metal in the raw waste was about 3.74%, half the anticipated 7.5%.

The light ferrous product from the original ferrous metal recovery system contained a significant percentage of organic contamination, mostly attached to the metal. From this information it was concluded that secondary shredding of the light ferrous product was needed to liberate the contaminants and increase density, while reducing particle size. Tests were run on two offsite hammermills and the product evaluated with the customer. The criteria for the light ferrous product were then developed, as follows:

- Maximum contamination, 4%
- Bulk density 21.5-26 lb/ft^3 (344-417 kg/m^3)
- Shred size: maximum 5% less than 1 in (25 mm)

A 37-kW shredder was then tested to optimize the position of the grate, breaker, plates and hammers. After optimization, tests showed that the contamination had been reduced to 3.5%, or, with fiber cans removed, to 2.4% and 2.7% in two tests. The shredder delivered 3.76 TPH at 25.2 kW, or 6.7 kWh/ton processed. Surges were applied at 5 TPH rates: it was found that surges over 15 seconds overloaded the

Table 4-2. Performance Test Results – Ferrous Recovery

Underflow Magnet							
	Feed	–	*Tails*	=	*Product*	*Recovery, %*
Light ferrous metal	311	–	23	=	288	92.6
Attached contaminants	61	–	21	=	40	65.4
Heavy ferrous metal	72	–	12	=	60	83.3
Attached contaminants	19	–	16	=	3	16.3
Loose contaminants	8,500	–	8,492	=	8	0.1

Shredded of Magnet						
	Feed	–	*Tails*	=	*Product*	*Recovery, %*
Light ferrous metal	117	–	89	=	28	24.4
Attached contaminants	10	–	8	=	2	15.0
Heavy ferrous metal	88	–	55	=	33	38.0
Attached contaminants	14	–	13.6	=	0.3	2.4
Loose contaminants	6,559	–	6,558	=	0.6	0.01

Ferrous Concentrator				
	Lights	*Light Ferrous*	*Heavy Ferrous*	*Recovery, %*
Light ferrous	414.9	3.4	403.2	97.2
Heavy ferrous	22.9	14.1	8.4	36.7
Loose contaminants	26.6	0.9	0.5	1.9

Ferrous Product		
	Metal, %	*Contaminants, %*
Light ferrous metal	76.2	10.3
Attached contaminants		10.3
Heavy ferrous metal	11.4	
Attached contaminants		0.9
Loose contaminants		1.2
TOTALS	87.6	12.4

shredder. Dense items were added to evaluate shredder sensitivity and found to contribute substantially to current draw. Clearly, long surges in feed and excessive amounts of heavy items are to be avoided to prevent overload.

The revised ferrous recovery system at Recovery I is shown in Figure 4-4; however, a number of changes should be noted:

- The drum magnet was pivoted to permit height adjustment to obtain maximum efficiency.
- The light ferrous from the concentrator is passed through an air knife, which functions as a safety device to prevent dense ferrous items from entering the shredder and, thence, into the shredder.

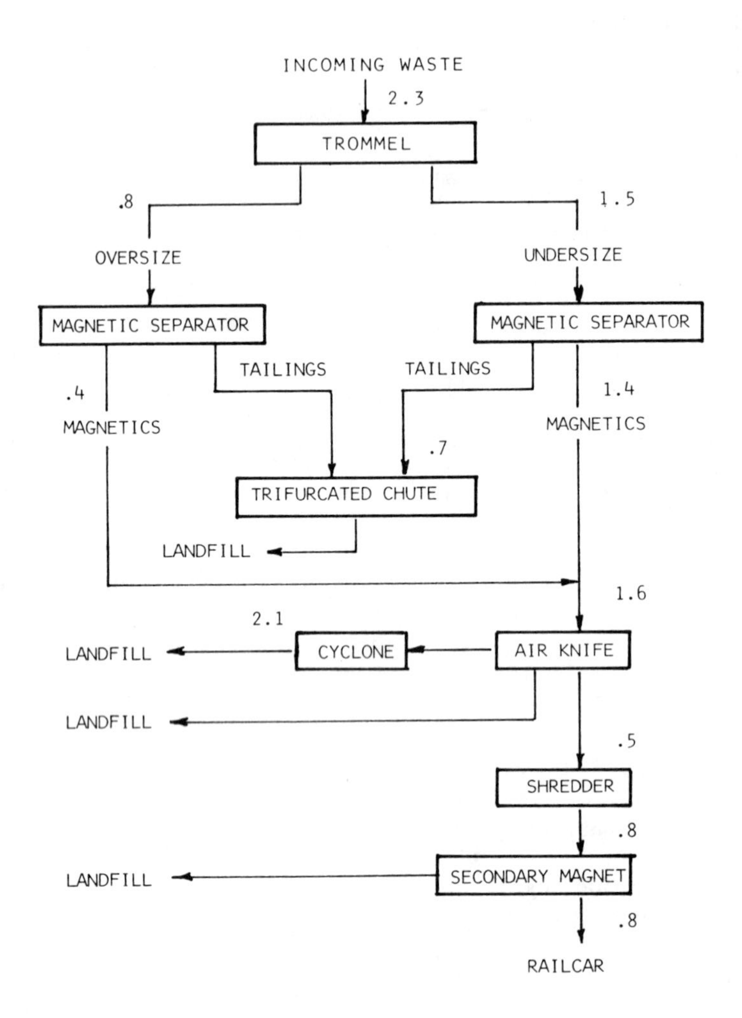

Figure 4-4. Revised ferrous recovery system at Recovery I.

- The outfeed is fed by a vibrating pan conveyor to a second magnetic separator, which discharges product with minimum contamination to a storage bin awaiting rail cars for loadout. After compaction, this product is a nominal 2.5 in. (63 mm), with a density of 21-26 lb/ft^3 (344-417 kg/m^3) and a 4% maximum contamination.
- The pneumatic system removes the light waste material, mostly putrescibles, through a "closed loop" air system, which returns the air to the air knife blower while discharging the waste from a cyclone through a rotary valve. The wastes go to landfill.

The revised ferrous metals system is presumably capable of handling over 4% ferrous metal from an 85-TPH feed from the tipping floor, as shown in Figure 4-5.

Staffordshire County, England

A ferrous metal recovery plant built at the Staffordshire County Council incinerator plant at Hanford, U.K., has been very effective in removing tin cans from domestic refuse before incineration, thus reclaiming high-grade steel scrap and virtually pure tin. A private company, Material Recovery Limited, developed and tested the process in a pilot plant in Benwell in 1976 and proved that an economical process was possible.

The basis of this process is the size reduction of household refuse by a Saturn rotary shredder, which chews the refuse into minus 8-inch material, which is suitable for efficient magnetic separation. After the first stage of shredding, the material is passed through a second rotary shredder, which opens the seams of the cans to free some of the debris trapped in them. A second magnet separates most of the remaining contaminants. At this stage, the magnetic fraction contains less than 10% by weight of the contraries (paper, plastics etc.), after which it is shipped to Batchelor Robinson detinning works, where it is washed and processed. With U.K. refuse it has been estimated that 560 tons of scrap will yield about 500 tons of high-grade steel scrap and 2.5 tons of virtually pure tin. The aluminum ends of the tinplate cans cannot be removed with the process described, but successful methods may yet be found.

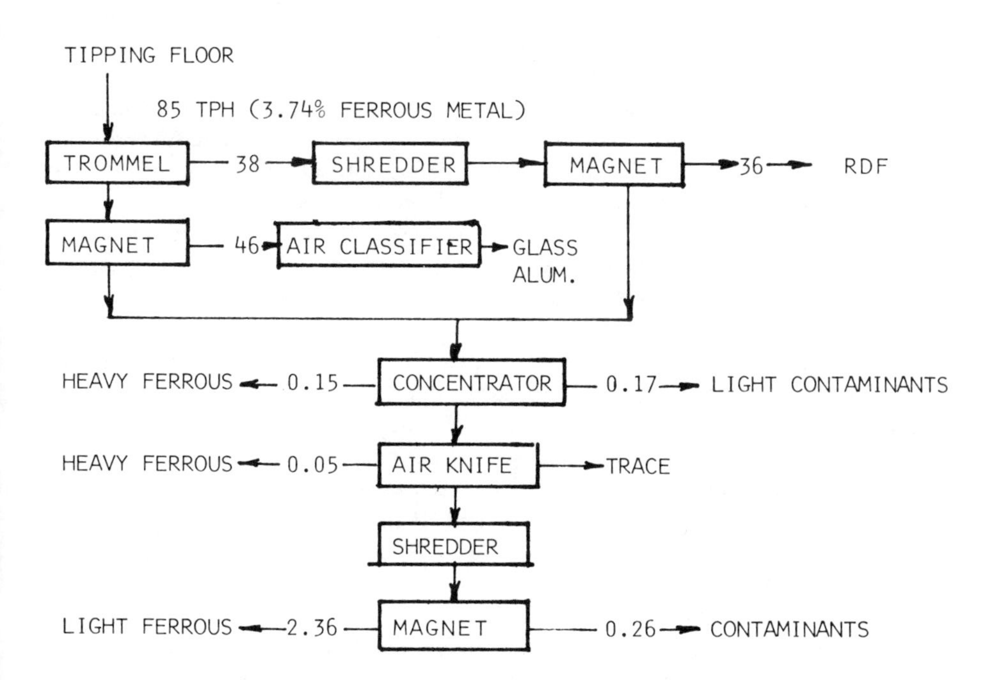

Figure 4-5. Process flowchart of expected materials balance for Recovery I.

Incinerator performance is improved by the removal of metals and size reduction of the incinerator feed.

The Saturn shear-type shredders installed at Hanford are Model 7246 units with 300-hp hydraulic drives. The primary unit is rated at 50 TPH of household refuse, but 30 TPH average rate is sufficient to handle the plant's design capacity. At this rate, the primary shear size reduces the refuse at 7.5 kWh/ton.

Saturn shear shredders have been used in the U.S. to shear aluminum. In an installation at Robnea Metals in Detroit, Michigan, a 7246 shredder with 300 hp reduces aluminum scrap to minus 4-inch size, and a second-stage 52-36 machine with 150 hp reduces the material to minus 1 inch, at a reported rate of 10 TPH. This indicates that 33 kWh/ton are required for this operation.

Shear shredders must be tested on the materials that are anticipated to determine their actual capacity and power requirements because they are sensitive to the properties, size and shape of the feed material. They have an especially difficult time with pure paper, but mixed refuse appears to ameliorate the problem of cutting paper.

ALUMINUM RECOVERY

Although aluminum has been estimated to comprise 0.7% by weight of the municipal refuse stream, resource recovery plants have generally reported much smaller quantities, such as 0.15-0.25% available for recovery. Most of this aluminum comes from all-aluminum beverage cans, but some is associated with steel-bodied cans, flexible aluminum foil, rigid foil containers such as frozen food trays, extrusions such as storm window frames, and castings such as auto parts.

It is reported that it takes only 5% as much energy to produce aluminum from scrap as from virgin raw materials. This results in high prices being paid for recovered aluminum that meets specifications—about $500/ton delivered, compared, for instance, with $40/ton of steel. About 25% of aluminum cans are recovered by recycling without serious contamination. The remainder gets mixed with refuse and is difficult to recover due to contamination with organic materials, mixing of various alloys of aluminum and the presence of other nonmagnetic and magnetic contaminants in the recovered stream.

The organic contaminants include lacquer coatings (3-4%) and "loose combustibles," which become entrapped or snagged with the aluminum during processing, despite efforts to remove them. The organic contaminants are reported to cause 5-10% losses of aluminum when melting clean aluminum scrap and more when the feed is not clean. Current specifications for recovered aluminum limit the amount of loose organic contaminants to 2% by weight. If the recovery plant cannot accomplish this, the metal reclaimer will have to do this cleaning, reducing the revenue to the recovery plant.

MSW contains several types of aluminum alloys, as well as alloys of copper, zinc, iron and other metals. The aluminum alloys found in MSW can be classed either as wrought alloys or casting alloys. Wrought alloys are much higher in magnesium and manganese, but lower in iron, silicon, copper and zinc, than are casting alloys. Some types of aluminum can be mixed together and remelted to make new products, such as can stocks, but other mixtures, such as cast and wrought alloys, have more limited uses or require more processing steps. There are few applications for an aluminum alloy containing as much manganese as a wrought alloy and as much silicon as a casting alloy. Specifications for aluminum recovered from MSW are under development within ASTM Committee E-38 on Resource Recovery; it is on this basis that the quality of recovered aluminum can be judged.

Aluminum Recovery Systems

Recovery I, New Orleans

Aluminum recovery equipment has been installed and tested to varying extents at such recovery facilities as Americology, Ames, Monroe County, Recovery I and Bridgeport. Extensive testing has also been done by the NCRR and by Reynolds Metals in Houston, Texas. For many reasons the results have been disappointing. With the institution of "bottle bills," which encouraged more direct recycling of beverage containers, the already low concentration of aluminum in refuse was further reduced.

The EPA-funded metals recovery facility at Recovery I has generated extensive data, which will serve to illustrate the complexity and low recovery efficiencies that have been experienced. The long-term tests at Recovery I netted about 16% of the aluminum in the feed in a form acceptable for recycling to beverage can manufacture under the normal conditions of plant operation (about 650 TPD).

Figure 4-6 shows the flow diagram of the aluminum recovery plant, with the average rates of flow expressed in pounds per hour. The aluminum in the feed (650 lb/hr) represented only 0.26% of the refuse stream, although 0.9% had been anticipated. Only 107 lb/hr was recovered.

The plant was operating at normal feedrates of 125 TPH, or 250,000 lb/hr during the testing period. The aluminum content of this stream was only 650 lb/hr, or about 2%. The MSW was fed to a primary trommel, which dropped only 35% of the aluminum at these production rates—200% of design MSW feed. The total trommel split at these rates was about 40% to unders, indicating that there was no increase in concentration of the aluminum in the trommel unders. The net result of trommeling was a loss of aluminum to the overs.

As the trommel unders were processed through the line, a series of devices removed the light combustibles and glass. Then an eddy current separator pulled the aluminum from the residue. Another classifier removed heavy nonferrous material, and a final screen removed most of the grit to make the final product. Table 4-3 shows the devices in the processing line, the flowrate, aluminum removal efficiency, loss of aluminum and the concentration of aluminum in the stream at each point.

Beyond the primary trommel, the combined efficiencies of aluminum recovery result in 74% recovery. The largest loss is the result of aluminum passing over the trommel, along with the heavy burden of MSW. The recovery of aluminum through the trommel holes is roughly proportional to the rate of feed. Normally the trommel

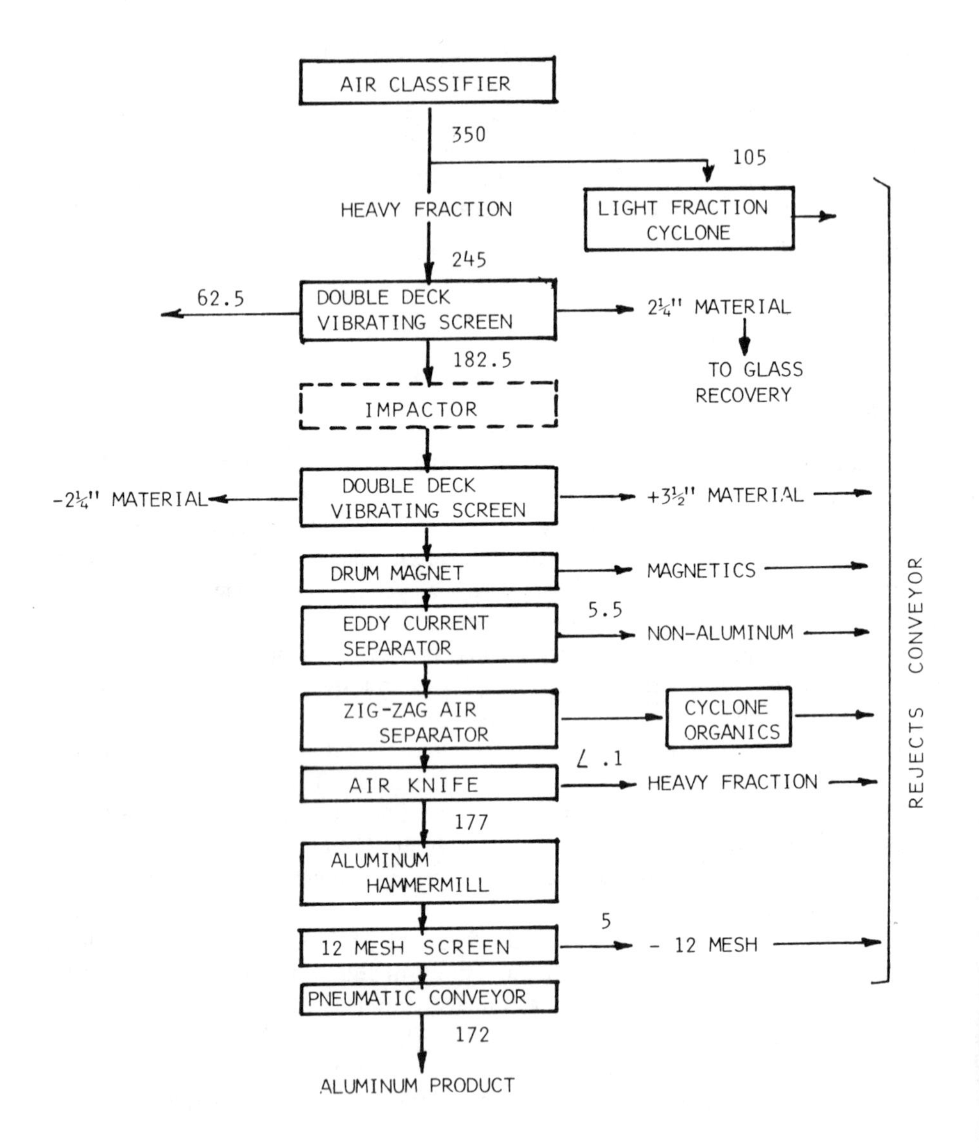

Figure 4-6. Aluminum recovery at Recovery I.

is operated at 200% of design, resulting in reduction of screening capacity from about 66% to 34%. Thus, we might expect the aluminum recovery to increase from 16% to about 35% if the trommel were operated at the design capacity of 66 TPH. To operate the trommel at this capacity would have required two working shifts, for which there was no economic justification.

The major difficulty in recovering aluminum from MSW is the similarity of density of aluminum cans to that of heavy paper and cardboard, and the resultant heavy burden of these materials on the conveyor belts from which the aluminum is to be

Table 4-3. Recovery of Aluminum from Trommel Unders [5]

Device	*Aluminum Feed (lb/hr)*	*Efficiency (%)*	*Loss (%)*	*Aluminum Concentration (%)*
Primary Trommel	650	34.3	65.7	0.26
Air Classifier	223	90.5	9.5	–
Aluminum/Glass Concentrator	185	94.3	5.7	–
Aluminum Concentrator	162	97.9	2.1	–
Magnetic Scavenger	153	100.0	0.0	66
Eddy Current Separator	153	90.2	9.8	70
Air Knife	114	98.2	1.8	94
Fines Screen	107	100.0	trace	94

recovered by eddy current and other types of "aluminum magnets." Some of these devices are designed and operated to maximize recovery of aluminum cans, others to recover mixtures of nonferrous metals, which are further processed for recovery of aluminum, such as dense media separation followed by jigging and electrostatic separation as used in the wet process of Black Clawson Co.

WET PROCESSES

Wet processes for separation of heavy, inert materials from organics are effective and efficient due to the substantial differences in densities between inerts and organics. These processes have been used in resource recovery plants and by glass and metals reclaiming operations.

Hydrasposal® Process

The Hydrasposal wet process developed by Black Clawson (a division of Parsons and Whittemore) provides a comprehensive demonstration of this approach to processes for separation of valuable materials from municipal refuse. It has been operated at full scale at the Franklin, Ohio, Hempstead, New York and Dade County, Florida plants, incorporating operations long used both in the manufacture of paper and in the mining and metals recovery industries. When applied to MSW, three additional elements have to be contended with: putrescible food wastes, highly abrasive glass and dirt, and oddly shaped metals.

The original Hydrasposal process installed at Hempstead is shown in Figure 4-7. The MSW is fed into a water-filled Hydra-pulper®, similar to a kitchen blender, where rotating blades beat most of the combustibles to pulp. Holes in the side of the tank, 3/4 inch in diameter, allow pulp, glass, ceramic and metal particles to escape from the pulper. These are carried in a stream of water to a wet cyclone, which

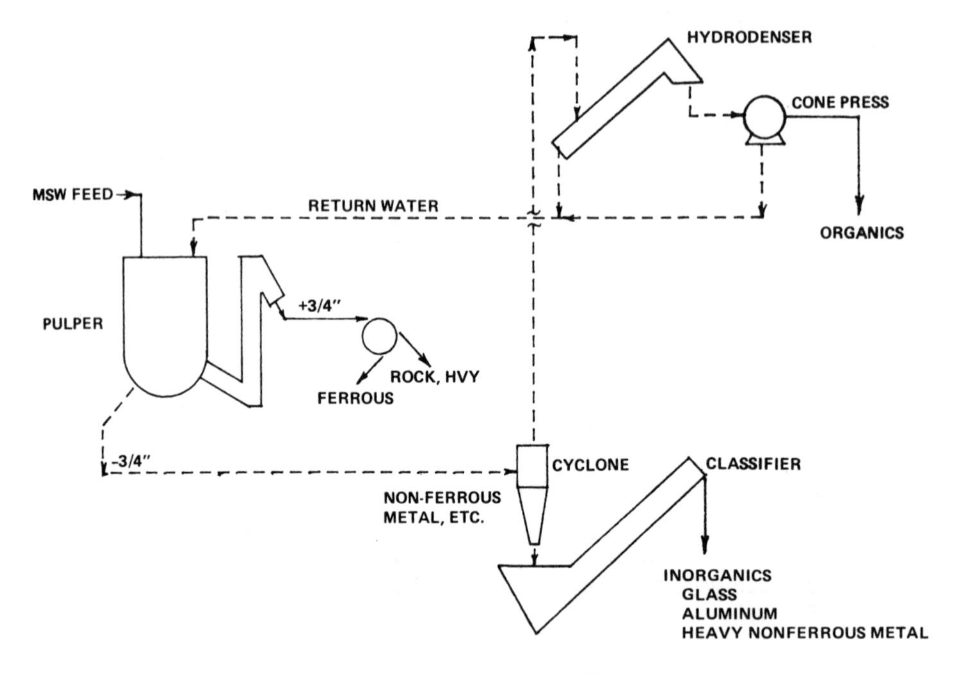

Figure 4–7. Schematic of the original Hydrasposal process.

separates the dense materials, mostly glass, aluminum and heavy nonferrous metals, from the organic pulp. The glass and metals are removed by a spiral classifier, where they are backwashed before further processing. The organics are taken to a storage vessel, after which they are dewatered by a Hydradenser® and a cone press to about 50% moisture content.

The nonpulp materials, such as ferrous and heavy nonferrous metals and rocks, which do not pass through the 3/4-inch holes, are removed from the Hydropulper by a bucket elevator called a "Junker," after backflushing with water.

The dewatered organics become a fuel similar to the bagasse produced from sugar cane and can be burned in boilers designed for high-moisture fuels. The fuel is blown into the furnace, high above the grate, so that a major portion is burned in suspension and the remainder burns out on the moving grate below.

The Hydrasposal process produces clean metal and glass products, which can be separated into ferrous lights (cans), ferrous heavies, aluminum lights (cans) and heavies (extrusions and castings), and glass. The plus-1/4-inch glass particles can be passed through an optical sorter to produce color-sorted glass, which can be sold to glass manufacturers. The sand and glass fines make a clean landfill.

The Hydrasposal process adds water to the refuse as it is processed, raising the moisture content from about 25% as-received to about 50% in the fuel product. This water must be vaporized in the boiler, resulting in a reduction in boiler efficiency relative to that from lower-moisture fuels. Compensating for this loss of thermal efficiency are the revenues produced from the clean metal and glass by-products of the process, as well as from the clean landfill. The overall process does not have any liquid effluent. All moisture is passed through the furnace of the boiler and vented through the emission controls and stack to the atmosphere.

The water within the system is biologically active, so must be carefully contained to avoid odors. Vapors that are emitted during processing can be used as combustion air for the boilers, thus destroying odors. Chemical scrubbers also can be used to destroy odors.

The wet process is especially suited for plants having dedicated boilers coupled with steam turbines, which generate power for export, as this arrangement provides a self-contained water system with no net effluent wastewater and provides a means of destroying odors by using odiferous air for combustion.

After a year and a half of operation, the Hempstead plant was shut down (for a number of reasons), and modifications were made to improve its performance. The ventilation system was revised so that odors would be preferentially captured at the points of generation. A chemical scrubber was added to assure odor destruction at all times, regardless of boiler operation.

The wet system was found to require excessively high maintenance due to the abrasive nature of the glass and grit in the pulped refuse. The process was modified as shown in Figure 4-8.

A primary trommel was installed to remove the major portion of glass and grit from the MSW before feeding it to the Hydropulpers. The trommel unders are now processed in a Hydroclone® without glass-breaking hammers, so that the glass can be removed without entrainment in the organic fraction. These modifications also were incorporated into the Dade County Hydrasposal process.

Wet Separation Equipment

Metals and glass are concentrated in the minus-5-inch portion of MSW, whether shredded or primary trommel thrus. These materials can be further concentrated by removing the major portion of organic materials by trommeling, screening or air classifying, but the concentrate will contain a substantial quantity and bulk of organics, including wood, food and yard wastes. These organics can be readily separated by wet processes and either returned for processing into fuel or landfilled.

Rising Current Separators

Rising current separators (RCS) use a rising current of water to effect an organic/inorganic separation, generally on particles between 20 and 60 mm. The upward water velocity is designed to be greater than the settling velocity of the organics so that they float, whereas the heavier inorganics sink and are picked up by a revolving drum (Figure 4-9).

Typically, about 10% of the circulating water is bled off continually to control grit and the level of suspended solids. This water is then passed through a water cyclone to drop the grit and the water passed to a Hydrocleaner® to remove organics

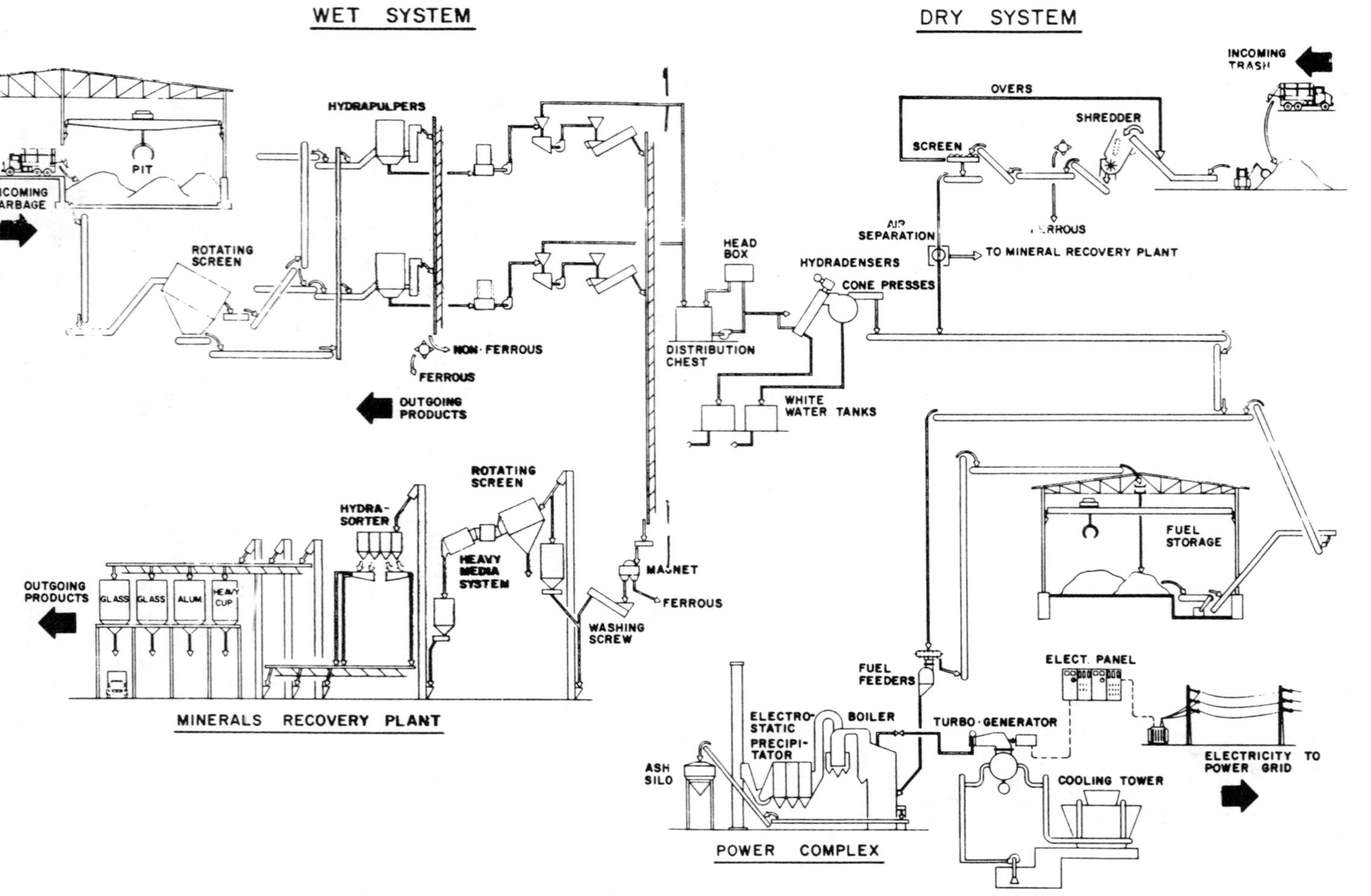

Figure 4-8. Schematic of the revised Hydrasposal process.

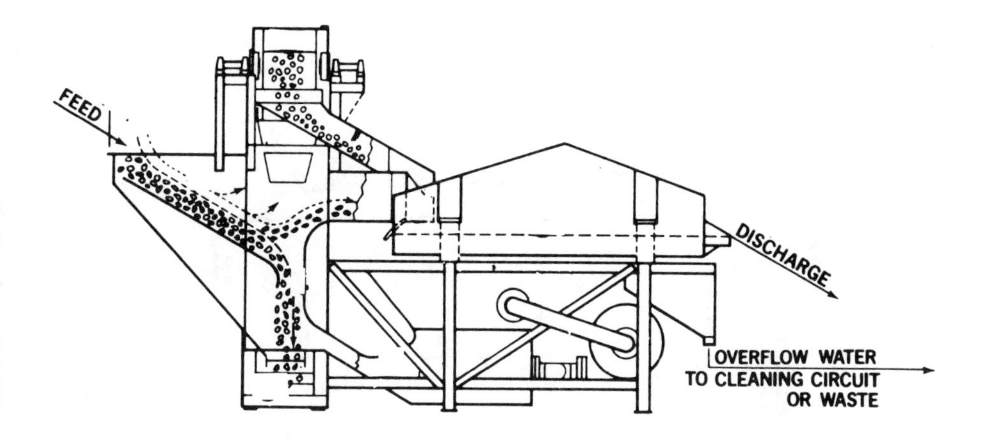

Figure 4-9. Rising current separator.

and oils before returning the water to the process. Due to water removed by the solids, a net makeup of fresh water is required. This can be used to backflush the inorganics so that they exit in a clean state.

Jigs

In jigging, the particles to be separated are usually no larger than 20-25 mm. Jigs have been included in the Americology, Recovery I, Hempstead, Dade County and Monroe County resource recovery systems. The principal parts of a jig are shown in Figure 4-10.

A perforated plate is located 150 mm below the water level of the jig vessel. A pulsating current of water is induced through this plate by an eccentric drive mechanism. Several layers of steel balls held in place by check boards on the perforated plate act as check valves to prevent large particles from going through and getting stuck in the openings.

The feed enters at one end. As it traverses toward the discharge, the pulsations force the lighter organics to the surface from which they are skimmed and taken out the side of the tank. The heavy inorganics pass under a skimming launder and discharge off the end. Extra heavy inorganics, such as copper, brass and stainless are caught by the first one or two check boards and pulled off as separate products.

The organic and inorganic streams come off mixed with water and both have to be dewatered. The organic portion is usually dewatered on a high-frequency, low-amplitude dewatering screen, while the inorganic fraction is dewatered with a spiral classifier. Because the discharge water from the classifier is cleaner, it is returned directly to the jig water in-feed system. The underflow water from the organics dewatering screen has a higher level of suspended solids, which can be cleaned by flotation before reuse.

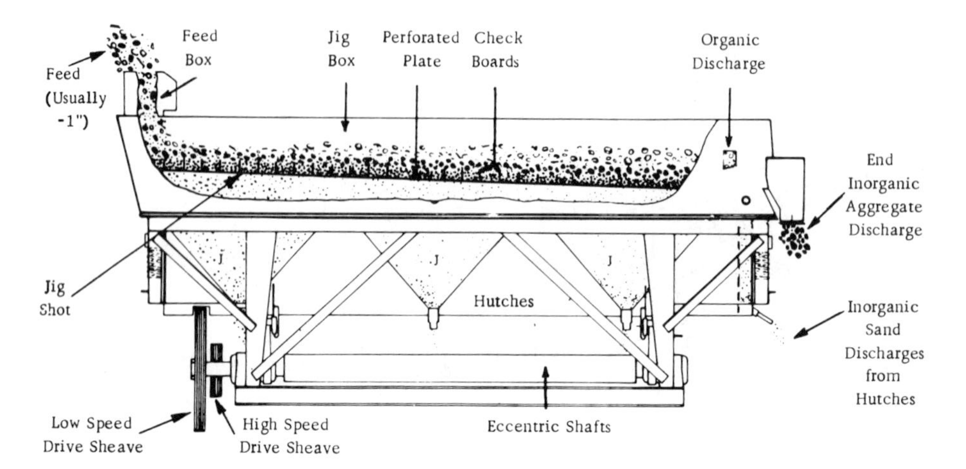

Figure 4-10. Wempco Remer Jig.

Froth Flotation

Froth flotation has been incorporated into the circuits at San Diego, Recovery I and Monroe County as a means of removing ceramic impurities and thus producing high-purity glass. Even very small quantities of ceramics cannot be tolerated in feed to glass plants because they cause faults in products made from glass. Froth flotation also can be used to remove suspended solids and emulsified oils from the process water.

In froth flotation, water containing the solids to be removed or separated is conditioned by the addition of reagents and then fed into a series of tanks. Fine bubbles of air are circulated throughout each tank. When the bubbles come into contact with the solid or liquid to be separated, they become attached. When the bubbles reach the surface they are skimmed off as froth. By the use of suitable chemicals, only the ceramic particles will adhere to the bubbles, permitting them to be selectively removed from the glass particles (Figure 4-11).

Heavy Media Separation

A liquid having a density intermediate between two types of materials can be used to float the lighter particles away from the heavier ones. Ferrosilicon mixed with water can be used to make separation at any point up to 3.6 specific gravity.

Heavy media can be used to separate heavy organics from inorganics and to separate aluminum from heavier nonferrous metals such as zinc, copper and stainless metals. These plants are used to reclaim metals from the millions of automobiles that are junked each year.

The circulating media tends to pick up any fine solids in the feed material. This contamination is removed by passing a portion of the media to a built-in cleaning circuit, where a magnetic drum is used to pull the ferrosilicon out of the liquid for reuse.

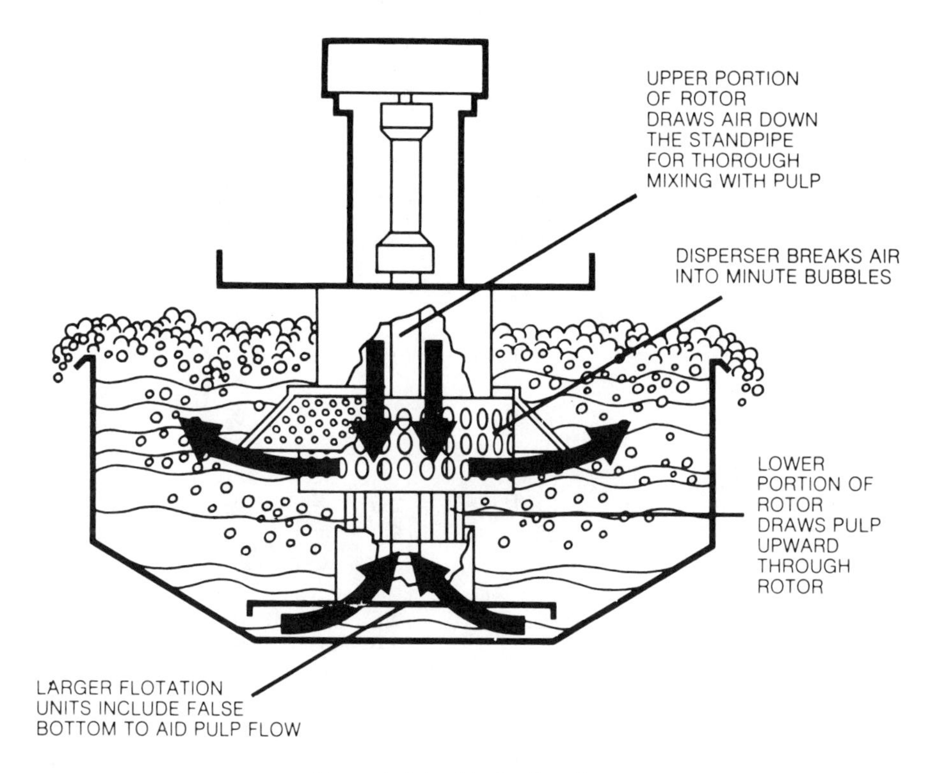

Figure 4-11. Froth flotation.

In applying heavy media, it should be noted that the metals must be properly shredded prior to separation so that no air pockets exist that can change the effective density. Also, it is important to remove most of the paper and textiles from the feed so that they do not absorb the media, downgrading product quality and increasing usage of media.

PERFORMANCE OF METAL AND GLASS RECOVERY EQUIPMENT

Expected typical performance figures for metal and glass recovery equipment used for processing MSW have been given by McChesney and Degner [4], from which Table 4-4 has been derived. This table shows that the rising current separator removed 79/86 = 92% of the organics in the feed, while 0.4/14 = 3% of the inorganics remained as a contaminant of the organics. The inorganics were 13.6/20.6 = 66% of

Table 4-4. Expected Performance of Wet Process Metal and Glass Recovery Equipment

	Feed	–	*Sink*	=	*Float*	
Rising Current Separator						
Organics	86.0	–	7.0	=	79.0	
Inorganics	14.0	–	13.6	=	0.4	
	100.0	–	20.6	=	79.4	
Jig						
Organics	34.5	–	0.0	=	34.5	3/37.5
Inorganics	62.0	–	59.0	=	3.0	= 10%
Heavy Metal	3.0	–	3.0	=	0.0	
Froth						
Glass	85.0	–	2.0	=	83.0	
Other	15.0	–	14.0	=	1.0	
Optical Sorter						
Color sorted	90.0	–	9.0	=	81.0	
Opaques	10.0	–	9.0	=	1.0	
Flint	60.0	–	6.0	=	54.0	(69.4%)
Amber	15.0	–	3.4	=	11.6	(14.9%)
Green	15.0	–	1.5	=	11.5	(14.8%)
Opaques	10.0	–	9.4	=	0.6	(0.8%)
	100.0		20.3	=	80.0	(100.0)

	Feed	→	*Product*	
Heavy Media				
Organics	39	→	39	Low-gravity float
Nonmetal	51	→	11	Low-gravity float
		→	40	Low-gravity float
High-Gravity Float				
Aluminum	9	→	7	
High-Gravity Sink				
Heavy Metal	2	→	2	
	100		100	

the sinks or, in other words, the sinks contained 34% organic contamination. These numbers indicate that this device was efficient in removing the inorganics from the organics, but not vice versa. On the other hand, it effectively removed 79% of the feedstream, increasing the concentration of inorganics from 14% to 66%.

The jig removed all of the organic materials, but the floats were contaminated by 10% inorganics. The sinks were clean inorganics and heavy metals. The froth flotation process floated 98% of the glass, reducing the contamination from 15% to 1.2%.

The optical sorter reduced the "other" contamination from 15% to 1.2% in the opaque sorter. The color sorter reduced the opaques from 10% to 0.6%, while recovering 78% of the colored glass in flint, amber and green.

The heavy media separator was fed 39% organics, which were removed separately from the 51% nonmetals, and recovered the 9% aluminum and 2% heavy metals in separate streams.

Complete plants, such as those described above, have been installed in many scrap-processing plants to recover the metals. The combustibles are generally landfilled. If such a plant is installed as part of a resource recovery plant, the major portion of the combustibles can be recovered and added to the fuel product.

REMOVAL EFFICIENCY INCREASES WITH CONCENTRATION

Other General Considerations in Materials Separation

Separation becomes more efficient as the concentration of the separated materials is increased. MSW contains roughly 50% combustible materials, 25% moisture and 25% metals, glass, dirt, stones and ceramics. The metals and glass, which each constitute only about 10% of the stream, are difficult to remove due to the large weight and bulk of the combustible materials. It is generally helpful to remove as much of the light combustible fraction as possible to maximize removal efficiency separation of the ferrous, glass and aluminum from shredded MSW. In other words, removal is more effective when the concentration of the removed material is high, that is, when the stream is glass or metal rich.

The Contradiction between Capacity and Quality

If the light combustible fraction is destined to be a fuel, it is important to minimize contamination of the combustible fraction by removing as much metal and glass as possible. This results in a large amount of combustibles reporting with the removed contaminants, reducing the recovery of combustibles.

When an effort is made to remove the maximum amount of combustible material from a refuse stream, increasing amounts of noncombustible contaminants are carried with the desired combustibles. Quality and capacity are therefore contradictory.

Multiple Stages Increase Recovery at the Expense of Quality

To minimize loss of combustibles with contaminants, additional processing steps can be used and the combustibles returned to the main combustible stream, which tends to increase the noncombustible content of the fuel.

Likewise, to reduce contamination of the metals and glass streams, additional processing steps can be used to remove the organic contaminants, returning them to the combustibles stream or to landfill.

REFERENCES

1. Rietema, K. "On the Efficiency of Separating Mixtures of Two Components," *Chem. Eng. Sci.* 7:89 (1981).
2. Worrell, W.A., and P.A. Vesilind. "Evaluation of Air Classifier Performance," *Resource Recovery Cons.* 4 (1980).
3. Handler, I., and K. Runyon. "Performance Testing of the Ferrous Recovery System at Recovery I," paper presented at the 1980 ASME SWPC, Washington, DC, 1980.
4. McChesney, D.R., and V.R. Degner. "Metal and Glass Recovery from MSW." in *Energy and Resource Recovery from Industrial and Municipal Waste*, Symp. Ser. 162, Vol. 73 (New York: American Institute of Chemical Engineers, 1977).
5. NCRR. "Aluminum Recovery at Recovery 1," 2 TR-80-15, National Center for Resource Recovery, Inc., Washington, DC (July 1981).

CHAPTER 5

Air Classification

Air classification has long been used in agricultural applications, such as in separating wheat from chaff and stones from grain, to deliver products of greater usefulness or quality. The greater the difference in the properties of the materials to be separated, the more perfect the separation. When the materials have similar properties, the separations are not as decisive, resulting in fractions of the desired and the undesired materials ending up in the wrong compartments. In such cases, additional separations can be made to achieve the degree of purity of the product stream. Usually there are substantial losses of good product in the rejected stream.

The early development of resource recovery from MSW started by applying devices previously used in agriculture and mineral dressing, with varying degrees of success.

HOW AIR CLASSIFIERS WORK

Air classifiers, as applied to processing MSW, use a forced flow of gases, usually air, into which particles are injected in such a way that particles having certain differences in inertial or aerodynamic properties will be displaced, in accordance with these differences, toward separate exits from the system.

When a gas having a velocity higher than that of a particle impinges on the particle, the gas must change its direction to get around the particle, causing it to lose some of its velocity and kinetic energy in its original direction. A portion of this kinetic energy is imparted to the particle, exerting a force called the drag force. The magnitude of this force depends on the relative velocity before impingement, the gas density, the face area of the particle and the shape of the particle.

A blunt-shaped particle, such as a disc, perpendicular to the gas flow, absorbs about 100% of the kinetic energy applied, whereas streamlined objects may absorb as little as 10%. The drag of rough objects may be as high as two times the energy indicated by the projected area. When a drag force is exerted on a particle, it will cause a change, or acceleration in the particle velocity, in the direction of the applied force. In horizontal motion, the particle will be accelerated until it reaches the velocity of the gas, thus eliminating the differential velocity that causes the acceleration.

In vertical motion, when the particle is acted on by the force of gravity, a drag force exerted upward will oppose the force of gravity and tend to reduce a falling velocity or reverse it to a rising velocity. A gas velocity that causes a vertical drag force just sufficient to oppose the force of gravity is called the lifting velocity.

If a particle is falling freely through still air, it will accelerate until its velocity creates a drag force equal to the force of gravity, called the terminal velocity, which is characteristic of the particle as the result of its shape and density, and which is approximately the same as the lifting velocity.

When a mixture of particles of various sizes and shapes is fed into a stream of gas, the various particles will be acted on differently. In a vertical column, the gas can lift all particles having a lifting velocity less than the gas velocity, provided that the particles can be accelerated sufficiently with the energy available in the gas. The need to accelerate the particles is an important factor in the performance of vertical classifiers.

In a horizontal classifier, the horizontal gas velocity will exert a similar force per unit face area on all types of particles; however, as the particles will accelerate in proportion to their mass per unit projected area, light particles will accelerate faster than dense ones. They can thus be "blown out" of the feedstream, provided they do not interfere with each other so that a "light" carries a "heavy" out.

When a large number of particles is fed into the gas stream, some of which will tend to rise and some to fall, they have a chance of interfering with each other, depending on the number of particles present and the way they tend to go. For this reason, the efficiency of separation will decrease as the classifier is loaded.

Also, if a large fraction of the particles have lifting velocities close to the vertical velocity of the classifier, their movement will be indecisive and they will tend to stay within the region of the feedpoint, increasing the particle concentration at this point and reducing the efficiency of separation.

When considering the wide variety of particle types, sizes and densities of materials present in MSW and shredded products, it can be expected that their behavior will be complex and uncertain. Probabilities will govern, not certainties, and good, but not perfect, separation can be expected.

An air classifier will not be able to distinguish between a dense particle and a larger light particle having the same lifting velocity or ratio of drag to inertia. For this reason, sand and paper in certain size ranges cannot be separated. The relative inertia of particles, subjected to the same gas velocity, can be effective in blowing off the lights without changing the trajectory of the heavy particles substantially. The separation achieved in a single step of classification cannot be perfect when the feed is a complex mixture. By arranging more than one stage of separation within the classifier, efficiency can be improved to the point at which particle interference can undo the benefits. Generally, one of the products can be of high quality at the expense of losses of good product to the reject stream. This stream can be processed in another stage or with other equipment to recover the lost product.

NOMENCLATURE

The air classifier has three ports. The feed enters through one and the material exits through two ports. The material exiting with the air stream is called the extract and the remainder the reject. As we see later, air classifiers are not perfect separators because some of the material that has inertial and aerodynamic properties, which should have caused it to exit with the extract, does not and is rejected (for whatever reason—sticking to other particles, caught up in local turbulence, etc.). Thus, there is some fraction of material that should have been removed but was not. Similarly, some of the material that should have been rejected, were the separator doing its job properly, ends up in the extract. We thus make two additional definitions: lights or light fraction is that portion of the feed that should have exited with the extract, while heavier, or heavy fraction, is that portion that should have been rejected.

Finally, air classification literature also confuses the words organic fraction and inorganic fraction with extract and reject, respectively. If we ask an air classifier to separate inorganic materials from organic materials, we must expect the two to have significantly different inertial or aerodynamic properties. It is, in fact, totally unfair to ask an air classifier to separate a grapefruit (organic) from an aluminum can (inorganic). The former will exit with the reject and is, in fact, a heavy, while the latter could very well exit with the extract and act aerodynamically as a light.

Thus we must reemphasize that there has been confusion in the nomenclature and that care must be taken when interpreting such data.

TYPES OF AIR CLASSIFIERS

Air classifiers are made in horizontal, vertical, inclined, zig-zag, air knife and rotary drum configurations. Air classifiers are provided with one or more fans to pull or push the air and a settling chamber or cyclone to separate the entrained light materials from it. The air stream may be vented, or the main portion recirculated and the excess vented. A dust collector is required on the system vent when fine particulate is carried by the air stream. In special cases, the vent gases are used for combustion, to destroy odors. Air classifiers also can perform a drying function by using hot gases.

Air classifiers require a uniform feed of material across the entry port of the unit and a steady rate of feed to achieve efficient performance. Belt conveyors, vibrating pan feeders and rotary drums have been used, with or without rotary air locks, to prevent induction of excessive air with the feed and to assure positive movement of the solids.

Horizontal air classifiers are provided with a high-velocity concentrated stream of air on which the feed material is dropped so that the air propels light materials horizontally, while denser particles fall through the stream to be collected below, in one or more hoppers.

Vertical air classifiers are provided with a rising stream of air into which solid materials are projected with appropriate velocities and angles so that the light materials are carried up and out to the collecting device and the heavies fall to the collecting port.

Inclined air classifiers use a rising current of air that flows, as does the solid material, at various and sometimes alternating angles, so that the solid materials interact with the walls of the column while the lights tend to fly up through the column with little wall contact.

Rotary air classifiers are provided with appropriately shaped flights to lift the material as the drum rotates and drop the material into an axial air stream so that aerodynamic separation takes place. The drum may be horizontal if the flights progress the material or inclined either way to discharge the rejects at either end. Screens can be provided to remove fines simultaneously.

Air knife classifiers use a strong jet of air to separate paper and other light materials from dense materials such as wood and metals.

Air-assisted screens are used to remove light materials from a bed of material moving along the screen.

The features of these various types of classifiers are often combined within a single unit. In general, classifiers have a feed zone, a dispersing zone to separate agglomerations, an acceleration zone, one or more discriminating or separating zones and paths by which the heavies can be removed without reentrainment with lights, so lights can be relieved of heavies.

Air classifiers are sensitive to the weight of solids feed and to the weight and velocity of the air. Their separation function changes considerably as loading changes, as does the quality and yield of product. Efficient separation is accomplished by proper selection or adjustment of the air velocity and the shape of the device to optimize the complex operations taking place within.

AIR CLASSIFIERS APPLIED TO RESOURCE RECOVERY

The main types of air classifiers that have been used in resource recovery plants are illustrated in Figure 5-1 [1-3].

Vertical (Rader Pneumatics) (Figure 5-1a)

The feed is dropped (usually from a vibrating pan feeder or drag chain to level the material and discharge steadily) through a rotary air lock, which discharges it at an angle into the air stream entering from the solids discharge. The side plates are adjustable to create the optimum shape in the accelerating and separating regions. High velocities are created at the lifting throat, after which the outlet area increases to reduce lifting velocities, thus allowing entrained heavies the opportunity to fall back.

Zig-Zag Vertical (Figure 5-1b)

The feed, with or without a rotary air lock, as desired, drops down a chute inclined to suit the zig-zag shape and into the rising current of air. As the solids flow over the edge of the "knees," the rising air current forces the lights to separate, at least partially, and fly up the column, while the heavy materials fall to the opposite wall and slide down the next lower ramp to repeat this process. As a result, many stages of classification can be carried out, both above and below the entry point. The heavies are discharged out the bottom, where the column air enters.

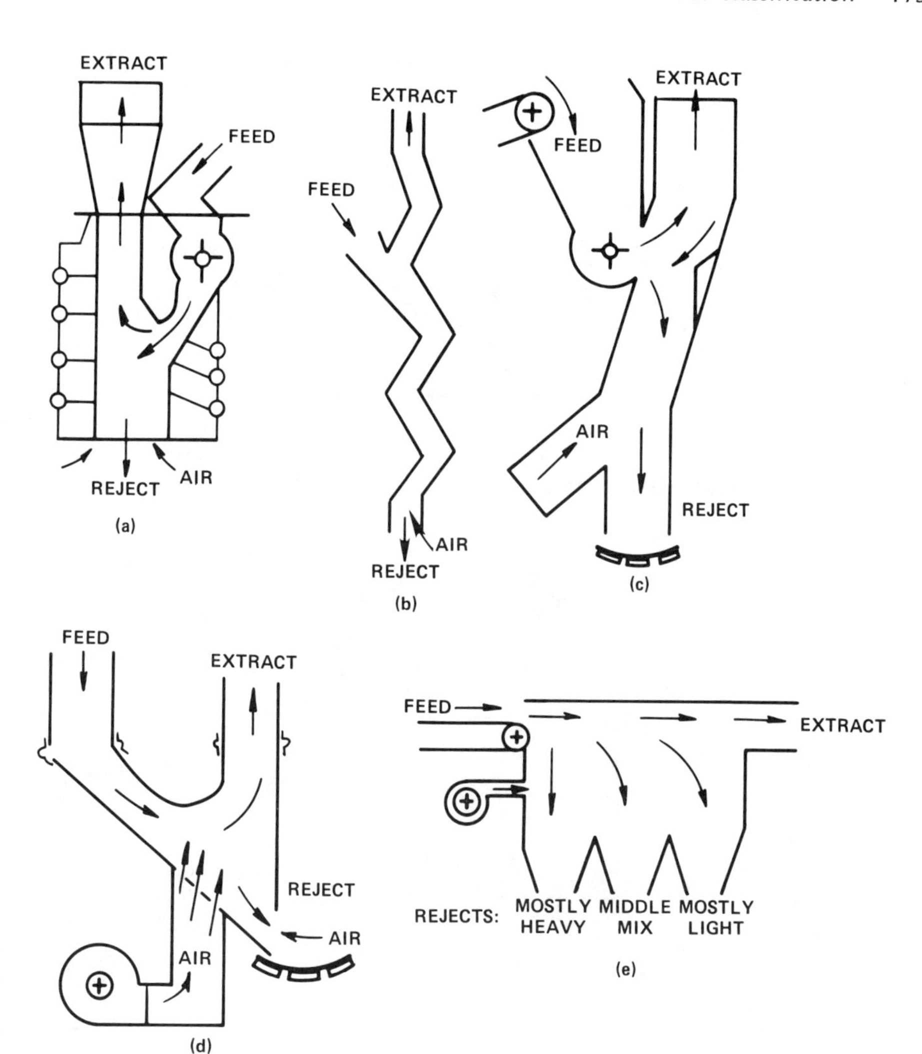

Figure 5-1. Air classifiers used in resource recovery [1-3]:
(a) vertical (Rader Pneumatics);
(b) zig-zag;
(c) vertical inclined (CEA – Carter-Day);
(d) "Vibrolutriator" (Triple/S Dynamics);
(e) horizontal (Bureau of Mines);
(f) horizontal (Boeing);
(g) ferrous concentrator (Triple/S Dynamics)
(h) nonferrous separator (MAC)
(i) rotary drum (Raytheon Service Co.).

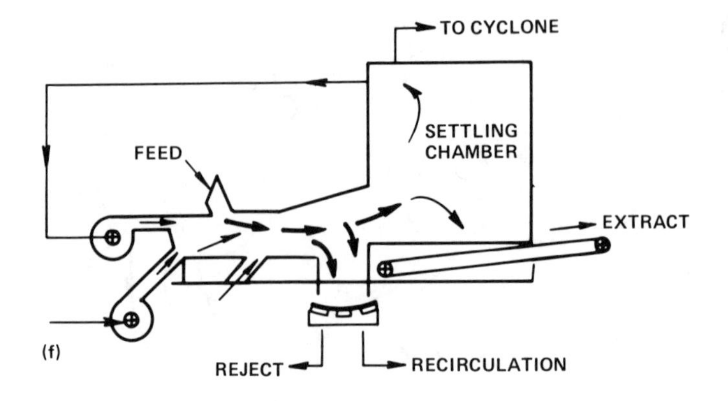

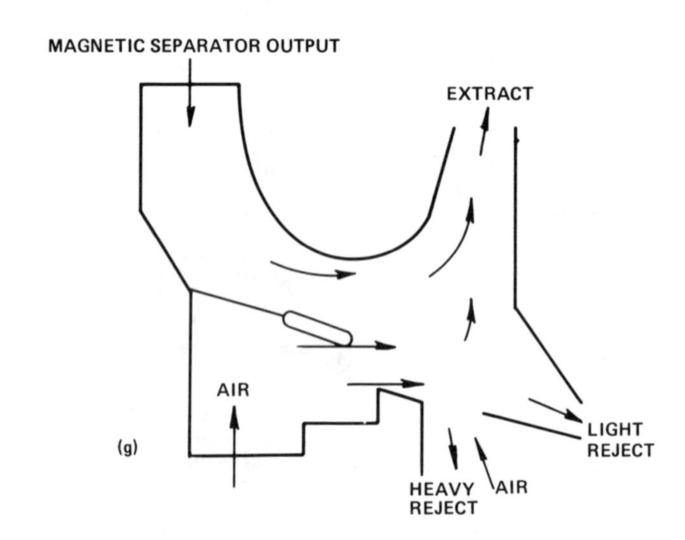

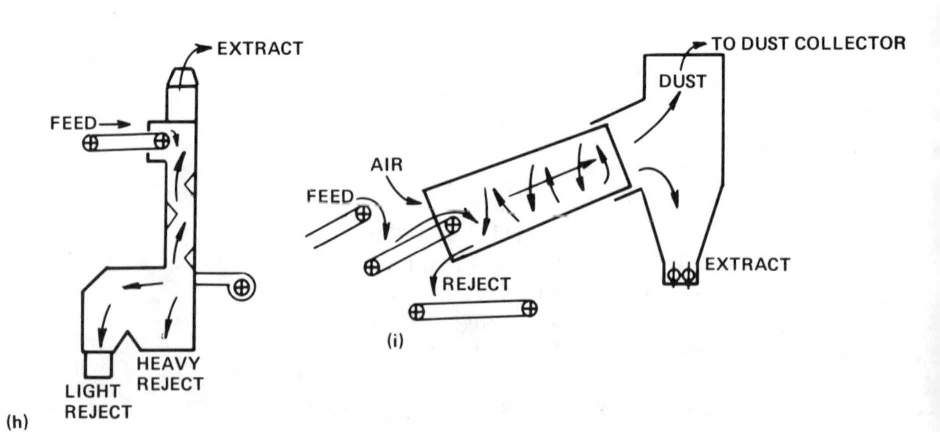

Figure 5-1. (Continued)

Inclined Vertical (CEA Carter-Day) (Figure 5-1c)

This design uses the rotary air lock to propel the solids in a direction angled upward to assist acceleration, and one or more knees to provide additional stages of separation of lights from the heavies, which fall to the conveyor below. The air enters through a vane-controlled port below the rotary valve. The inclined back wall assists the heavies in finding a separate path to the bottom without reentrainment.

Vibrolutriator® (Triple/S Dynamics) (Figure 5-1d)

This device uses a vibrating pan within the unit to progress the feed material over forced air. This forced air is delivered by a booster fan under the pan, which blows the light materials out of the bed into a substantially larger current of air that is drawn by an induced air fan beyond the classifier lights outlet. This air is drawn in with the feed, aerating it, and from the end of the pan, carrying the lights up the vertical column.

Horizontal Classifier (Bureau of Mines) (Figure 5-1e)

The feed material is discharged from a belt conveyor onto a high-velocity air stream, which accelerates the light materials along a horizontal path to the outlet, while the heavier materials take different trajectories to the hoppers below. The first hopper catches the heaviest materials, the next the intermediate materials, and so forth.

Horizontal Classifier (The Boeing Company) (Figure 5-1f)

This horizontal classifier has three air jets to permit closer control of the particle trajectories, and three points of solids discharge. The first heavies stream is collected by a "batter board" onto a discharge conveyor. The second stream is similarly collected and recycled back to the feed for a second pass. The lights are carried to the settling chamber from which they fall to a discharge conveyor. The air stream is drawn from the top of the chamber and most of it returned to the forced draft fans. A vent is drawn to a dust collector to maintain negative pressures within the unit.

Ferrous Concentrator (Triple/S Dynamics) (Figure 5-1g)

This unit is similar to the Vibrolutriator except that no vibrating pan is needed for metals. It provides the air sweep needed to remove light combustible contaminants from the magnetic pickup product, and an air knife to separate the light metals from the heavies and separate discharges for them.

Nonferrous Separator (MAC Equipment) (Figure 5-1h)

A zig-zag column is used to remove combustible lights from the nonferrous-rich stream, at the bottom of which is an air knife to separate light (aluminum) from other nonferrous metals. An induced-draft fan carries the lights to a cyclone collector.

Rotary Drum (Raytheon) (Figure 5-1i)

Solids are fed from belt conveyors into the center, lower end of a rotating drum angled at about 25°. The upper end of the drum discharges into a settling chamber with a twin discharge screw at the bottom. The upper part of the settling chamber is connected via a bag filter to the induced-draft fan, which draws air into the feed opening of the drum and axially through the drum. The flights in the drum lift the material, drop it into the air stream and carry the heavies down to the feed end discharge conveyor. The lights progress up the drum each time they are dropped until they fall into the settling chamber. Intermediates will tend to accumulate if they are not decisively light or heavy. Fan volume can be used to control classification to some extent.

HISTORY OF AIR CLASSIFICATION IN THE RESOURCE RECOVERY INDUSTRY

U.S. Bureau of Mines, Salt Lake City

The Bureau of Mines, recognizing the potential value of recovering materials from MSW, embarked on extensive research on separation of metals and nonmetals from urban refuse in 1969 at the Salt Lake City Metallurgy Research Center [4].

The shredded refuse obtained from Madison, Wisconsin was screen sized to determine the potential for mechanical size separation. The tests showed that the metals were in the plus-4-mesh fraction and most of the dirt and glass in the minus-4-mesh fraction. The paper was mostly between 4-mesh and 1-inch fractions, and the cardboard in both the 4-mesh X 1-inch and plus-1-inch fractions.

They concluded that screening offered possibilities for good separation of materials, but that air classification should be done first because the paper fractions interfered with screening and hindered the direct application of wet beneficiation processes such as jigging, sink-float, tabling and water elutriation. On the other hand, the low densities of paper, plastics and cardboard were found to be advantageous to air winnowing.

Both vertical and horizontal air classifiers were tested and proven to be effective. The horizontal classifier was found to have a much higher tonnage throughput than the vertical separator when treating low-density urban refuse and, hence, was selected for initial tests on separation of the shredded MSW.

The horizontal classifier (Figure 5-1e) could process about 1.25 TPH with a 1300-scfm blower, separating the metals, plastics, hard rubber and other high-density materials into the nearest compartments, while collecting the paper, textiles, fibers and light materials in the downstream compartments. A bag filter was provided to collect the dust and other fines. The reject was magnetically separated to recover the iron, and the sand and glass collected under the 4-mesh screen. The extract was also screened on 4-mesh to remove the sand and glass that was carried with it, and the two plus-4-mesh fractions fed to the vertical elutriator at a rate of 0.4 TPH with 500 scfm.

These simple steps recovered 92% of the metal in a clean product and 97% of the combustibles (retaining only 3% noncombustible impurities), and the minus-4-mesh residue contained only 3% of the combustibles.

The plus-4-mesh elutriated products were further treated by passing through the vertical classifier, obtaining a clean fraction of 85% light paper, 13% heavy paper and only 2% other combustibles, film plastic, fabric and wood, representing 41% of the total combustibles in the original refuse. The reject of this classification process was 94% combustible, with only modest ash, and was suitable as fuel. The minus-4-mesh fraction contained 84% glass and recovered 83% of the glass in the original refuse.

The horizontal classifier was operated at an air/solids ratio (ASR) of 0.43 pounds of air per pound of solids—a solids/air ratio (SAR) of 2.3—and the vertical classifier was operated at 2.8 pounds of air per pound of solids (0.36 lb solids/lb of air). The air/solids ratios used in the Salt Lake City tests have persisted in later designs, with or without basis.

These separations were based on light loadings of the air classifiers, which gave highly efficient separation and, hence, gave optimistic estimates of the potential of resource recovery. While they revealed the nature of the urban waste, they did not provide a basis for scaling up these operations to full-sized equipment or any indication of the relationship between efficiency and loading.

These findings were widely reported and led to construction of the College Park pilot plant, which continued in the same directions and expanded the process of refining to recover the glass, paper and nonferrous fractions.

With special emphasis on recovery of paper and elimination of metals and glass, a series of experiments were carried out with a Scientific Separators pilot air classifier at the Bureau of Mines laboratory in Salt Lake City. This classifier was a ten-stage zig-zag unit with a 6-inch × 12-inch throat, similar to that in Figure 5-1b.

Samples of MSW were obtained from five sources: the PHS-TVA composting project at Johnson City, Tennessee; the SRI, from the Los Angeles County Sanitation Scholl Canyon Landfill; the Bureau of Solid Waste Management, Cincinnati, Ohio; and material supplied by Eidal International of Albuquerque, New Mexico.

The purpose of these tests was to determine the following data:

- velocity required to effect potentially useful separations;
- weight percentage yields of the separated fractions for process design material balances;

- maximum allowable feedrate for column throat-size calculations; and
- combined influence of feedrate and air velocity on the degree of separation obtainable.

Runs were made at low feedrates to establish the separating velocities for splitting the paper fractions. Higher rates of feed then were used until column operation became unstable and the column became choked because of overloading, thus establishing the maximum allowable feedrate. Tests were then run within these limits. As higher feedrates were used, the sharpness of separation decreased, as expected. The number of experimental runs was insufficient to establish limiting column loadings with any degree of reliability, although a total of 45 runs were made on the five samples.

Runs were made at column velocities ranging from 300 to 1800 ft/min, separating the raw refuse into lights and heavies, then further classifying the lights into additional fractions of paper. Column loadings ranged from 3.7 to 14.3 pounds of solids per pound of air. Table 5-1 summarizes the observations made regarding the types of materials lifted within successively higher ranges of vertical column velocities.

The experimenters noted that it was helpful to have the heavy fractions in the feed because they prevented column clogging by a sort of reflux action. They also suggested that the corrugated and other strength-grade papers that left with the bottom product of the air classifier could just as well or better be removed by other separating methods, such as screening, ballistic techniques or even by a simple air cascade.

They visualized a simpler process in which a coarse shredder would be used to produce a paper component particle size in the 6- to 12-inch range, followed by a screen to remove the minus-6-inch material that would largely comprise cans, broken bottles and dirt, with a ballistic device to remove large plastic bottles as an appurtenance

Table 5-1. Materials Separated at Various Lifting Velocities

Velocity (ft/min)	*Materials Separated*
300– 400	Light fines, dust balls, film plastics
700– 800	Newspaper and magazine stock (groundwood papers), toweling, tissue paper, writing paper, bleached board and bag paper (chemically pulped), film plastic and Styrofoam; more groundwood than chemically pulped
900–1000	More chemically pulped than groundwood, some corrugated
1100–1200	Less newspaper and more corrugated cardboard
1200–1300	Same as above
1300–1400	Kraft and corrugated paper, aluminum, tire fragments, inner-tube fragments, glass, tin cans

in the air classifier feeding mechanism. They recommended that the characteristics of shredders be investigated to determine, for a given mill, the effects on product characteristics of rotational speed, grate or screen size and feedrate to the mill, to be able to specify the grinding principles to be employed or the characteristics desired in the shredded product. Because of the empirical nature of the problem, they believed that only full-scale tests would yield significant information.

The Bureau of Mines pilot plant in College Park, constructed to fulfill the needs for demonstration of resource recovery, incorporated a coarse shredder, horizontal classifier and a vertical classifier. The latter was not the ten-stage zig-zag, but a simpler agricultural "aspirator" with three stages.

National Center for Resource Recovery

The National Center for Resource Recovery established a pilot plant in Washington, DC, that served as a testing ground for a variety of unit operations, including a zig-zag classifier similar to the original one tested at Salt Lake City, and a Triple S classifier [2]. These tests were oriented toward removal of aluminum cans with the heavy fraction of the air classifier, and showed that too much heavy material is dropped when the aluminum recovery is effective.

University of California, Berkeley

Tests at the University of California showed the performance of air classifiers under pilot-plant conditions, where the operating conditions could be controlled and the feed and products analyzed [5]. Trezek determined the particle size of the major components in the feed, lights and heavies, and concentrated attention on the ways and means of recovering the clean combustible paper and plastics.

The work of Savage, Diaz and Trezek [6] gives the weights of lights and heavies, the constituent splits, and data on loadings through a wide range of conditions in seven full-scale installations, where the air flow could be varied above the design level and feedrates controlled to permit testing of a wide range of conditions in several full-scale installations. Unfortunately, in most installations the air flow cannot be varied beyond the design level, nor can the feedrate be sustained at levels that permit testing of overload conditions.

Duke University, Durham, North Carolina

The fundamental aspects of air classification were studied at Duke University. Pierce and Vesilind [7] reported that it was feasible to estimate the performance of air classifiers using a "drop test," in which the individual particles are dropped from a height sufficient to have them attain terminal settling velocity. This velocity is calculated and the light and heavy particles identified. It was further found that air classification seems to be inherently unstable, with slugs of material (lights and heavies) moving downward, acting as single particles, while other materials carry upward, acting as air foils.

Other Installations and Research Programs

The St. Louis demonstration sponsored by the EPA to prove the feasibility of burning RDF in Union Electric's utility boilers was equipped with a Rader air classifier after several types had been tested.

Meanwhile, the American Can Company tested its proprietary zig-zag air classifier, which was ultimately installed at Americology in Milwaukee. A zig-zag was also seen in the Occidental Research facility in San Diego. Further, Triple S (inclined), Rader, CEA Carter-Day and Allis-Chalmers (vertical), General Electric and Raytheon rotary and Boeing horizontal classifiers have been in operation in full-scale plants, offering various configurations for application in resource recovery.

The Rader classifier was installed at the Ames and St. Louis demonstration plants. The Triple S was installed at Recovery I in New Orleans. CEA, at its East Bridgewater facility, installed one similar to Aspirators, made by Carter Day for agricultural applications, and additional ones at Bridgeport, Connecticut. Rotary classifiers were installed by Raytheon in Monroe County, New York, in a facility developed in cooperation with the Bureau of Mines.

Conclusions

Air classifiers have been disappointing to those who expected the high separation efficiencies achieved in pilot-scale units or splits between materials with aerodynamic properties that are too much alike. They have not been effective in removing the aluminum fraction as unders and have the unfortunate characteristic of flying the sand and fine glass. The reasons for this will become apparent when we examine the fundamental principles governing their operation.

BASIC PRINCIPLES OF AIR CLASSIFIER OPERATION

When particles are subjected to a stream of gas, the gas applies forces to the particles, which result in an acceleration or change in their velocity. This change depends on the difference in velocity between the gas and the particle, the shape and the weight of the particle. The behavior of individual particles can be predicted and described by mathematical relationships based on fundamental principles, which are summarized below.

Basic Equations

Conservation of Energy

This principle states that the total energy of a system will remain constant, including work, heat, kinetic energy, potential energy, flow work, internal energy and friction. Bernoulli's simplification omits the internal forms of energy, work and heat, and includes potential and kinetic energy and friction. It states that

$$\text{(Potential + kinetic) energy + work = friction + heat}$$

$$PE_1 + KE_1 + \text{work} = PE_2 + KE_2 + \text{heat} + \text{friction} \tag{5-1}$$

where PE = static pressure or elevation

$$KE = \frac{(V)^2}{2g} = \text{Velocity Head} = h \tag{5-2}$$

Conservation of Momentum

The sum of the momentum, M (mass(m) × velocity(V)) of entering streams, a, equals the sum of the momentum of the leaving stream, b, or the momentum sum remains a constant:

$$M_aV_a + M_bV_b = M_aV_{a'} + M_bV_{b'} \tag{5-3a}$$

or

$$M_aV_a + M_bV_b = (M_a + M_b)\,Vm \tag{5-3b}$$

Newton's Laws

A body in motion tends to stay in motion (in the absence of friction). A force, F, exerted on an object causes an acceleration, a, in the direction of the force, which is inversely proportional to the mass of the particle:

$$F = M \times a \tag{5-4}$$

The velocity, V, resulting from constant acceleration will increase with time, t:

$$V = a \times t \tag{5-5}$$

The distance, S, traveled under constant acceleration will increase with the square of time:

$$S = \frac{a \times t^2}{2} \tag{5-6}$$

Drag Force

A fluid exerts a drag force, F_d, on a particle when the fluid has a velocity different from the particle with which it is in contact. This force is the sum of forces due to the viscosity of the fluid and dynamic forces resulting from the velocity of the fluid. The kinetic energy of the fluid is converted to force as it impacts on the object. The drag coefficient, C_d, represents the fraction of the kinetic energy, $\rho(V_2/2g)$, per unit of "face" or projected area, A_p, which has been converted to force:

$$F_d = C_d A_p \rho \left(\frac{V^2}{2g}\right) \tag{5-7}$$

where ρ = fluid density

Choking Velocity

As the amount of solid material being transported by a fluid stream is increased, a point is reached at which the fluid stream is no longer able to maintain the velocity of the solid and its flow starts to decrease. At this point the density of the material dispersed in the fluid becomes the bulk density of the solid. This fluid velocity is called the choking velocity.

$$V_{ch} = \frac{W}{A \times \rho_b} \tag{5-8}$$

where W = mass flow
A = cross-sectional area perpendicular to flow
ρ_b = bulk density of the solid material

Dynamics of Particles under Gravity, without Drag

The trajectories of particles acting under the force of gravity and with an initial velocity, V_o, are given by the following expressions, referring to Figure 5-2.

Free Fall:

$$V_x = V_o \cos\theta \qquad V_y = V_o \sin\theta - gt \tag{5-9}$$

$$X = (V_o \cos)t \qquad Y = (V_o \sin\theta)t - \frac{gt^2}{2} + h \tag{5-10}$$

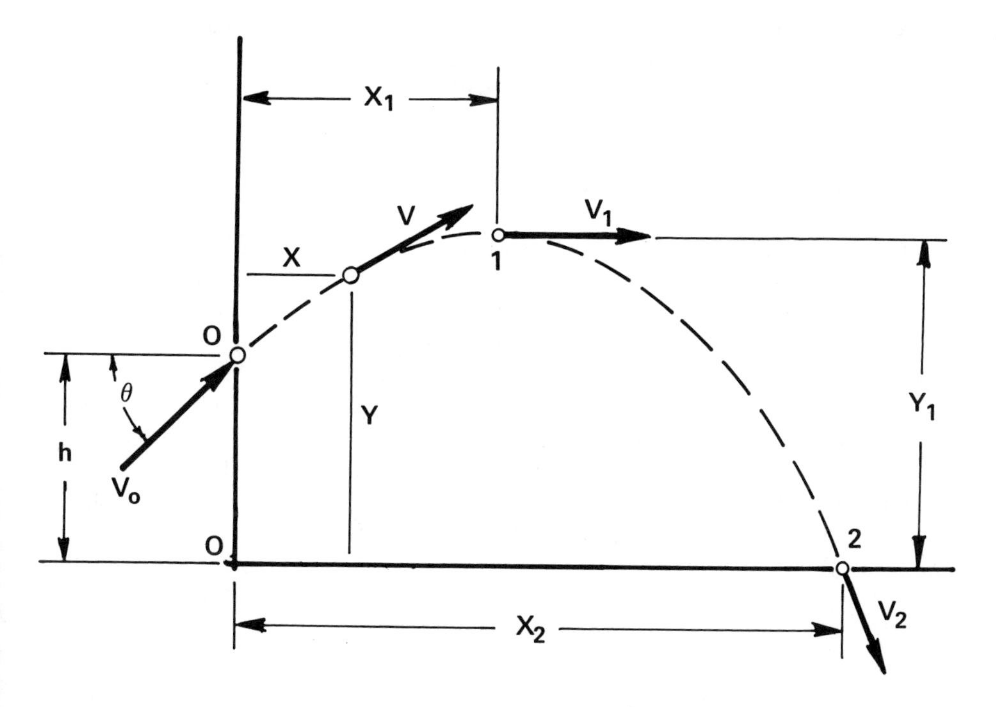

Figure 5-2. Trajectories of particles without air drag.

Vertex:

$$X_1 = \frac{V_o^2 \sin 2\theta}{2g} \quad Y_1 = h + \frac{V_o^2 \sin^2\theta}{2g} \tag{5-11}$$

Terminus:

$$X_2 = \frac{V_o^2 \sin 2\theta}{2g} + V_o \cos\theta \sqrt{\frac{2h}{g} + \left(\frac{V_o \sin\theta}{g}\right)^2} \tag{5-12}$$

where θ = angle to horizontal
V_x = horizontal component of velocity
V_y = vertical component of velocity
t = time
X_1 = initial vertical component
Y_1 = initial horizontal component
h = horizontal distance

These expressions do not take into consideration drag forces exerted during the flight of the particle. These forces can be taken into account by complex mathematics or differential equations, which are easily used with programmable calculators.

Particle Acceleration Caused by Drag Forces

When a particle falls through a fluid or gas under the force of gravity in a fluid, it will accelerate in speed until the drag force created by its velocity relative to the fluid is equal to the force of gravity. Under these conditions it is said to be falling at its terminal velocity. In another way of expressing this situation, if the particle is maintained at a constant relative velocity by a fluid exerting a drag in a direction opposite to that exerted by gravity, this relative velocity is called the lifting velocity.

When this equilibrium has been reached, the kinetic energy of the gas impinging on the face area of the particle, A_p, creates a force that is balanced by the buoyant force resulting from the volume of the particle, ¥, multiplied by the displaced volume of gas. This balance can be expressed as follows:

$$\frac{(V_t^2)\,A_p}{2g} = \frac{¥(\rho_p - \rho_g)}{\rho_g} \tag{5-13}$$

Since the volume, V, = Mass/Density = M/ρ_p,

$$\frac{V_t^2\,A_p}{2g} = \frac{M_p(\rho_p - \rho_g)}{\rho_p\,\rho_g} \tag{5-14}$$

where V = volume of particle
ρ_p = particle density
ρ_g = gas density
V_t = terminal settling velocity

The general expression for terminal velocity, V_t, can therefore be expressed as follows:

For any shape:

$$V_t^2 = \frac{2g\,M_o(\rho_p - \rho_g)}{C_d\,A_p\,\rho_p\,\rho_g} \tag{5-15}$$

where C_d is a correction factor for other particle shapes.

For spheres:

$$V_t^2 = 3\left(\frac{g\,D_p(\rho_p - \rho_g)}{\rho_g}\right) \tag{5-16}$$

where D_p = particle diameter

Another way of determining V_t is by measuring the free-fall velocity or finding the vertical gas velocity that will just suspend the given particle. This velocity is a useful property of the particle and can be determined for particles of all shapes. It is also the property that actually determines its aerodynamic properties in air classifiers and other devices.

When a stream of gas impinges on a particle with a velocity, V_g, the particle will accelerate in accordance with the differential velocity, V_r, which is the gas velocity minus the particle velocity, or $V_g - V_p$. The force created by this differential velocity will create an acceleration in the same line as the force. When several planes are involved, the resultant forces can be used to determine the resultant acceleration.

Acceleration of a Particle in a Horizontal Air Stream

Under the conditions in which only dynamic forces are applied and the drag is proportional to the square of the relative velocity (under Newton's law of drag), the following expression may be derived:

$$\text{Drag force} = F_d = M \times a \text{ (Newton's law)}$$

under gravity,

$$F_g = M \times g$$

Dividing,

$$\frac{F_d}{F_g} = \frac{a}{g}$$

and

$$a = g \frac{F_d}{F_g}$$

Because

$$F_d \sim V_r^2$$

and

$$F_g \sim V_t^2 \quad \text{(in free fall } V = V_t\text{)}$$

Substituting back in Equation 5-4 (Newton's law), then

$$a = \left(\frac{V_r}{V_t}\right)^2 = \left(\frac{V_a - V_p}{V_t}\right)^2 \qquad (5\text{-}17)$$

The following equation can be derived to describe the distance traveled by a particle being accelerated in a horizontal direction by a horizontal air stream (without gravity), as a function of the particle velocity, V_p, for a given air velocity, V_a:

$$S = \frac{V_t^2}{g}\left[\frac{V_p}{V_a - V_p} - \ln\left(\frac{V_a}{V_a - V_p}\right)\right] \qquad (5\text{-}18)$$

Acceleration of a Particle in a Vertical Air Stream

When the particle is moving vertically in a vertical air stream, the effect of gravity must be added, as follows:

$$a = \frac{d^2y}{dt^2} = g\left[1 - \left(\frac{(V_a - V_p)}{V_t}\right)^2\right] \qquad (5\text{-}19)$$

where a = acceleration, ft/sec^2
t = time (sec)
g = gravitational constant
y = distance traveled (ft if g is ft/sec^2)

APPLICATION OF BASIC PRINCIPLES TO AIR CLASSIFIERS

Newton's laws state that particles, once in motion, tend to stay in motion, and that dense objects accelerate at a rate of 32.2 ft/sec in our gravitational field. Knowing this, we can calculate the position of any dense object in free fall, given its initial velocity and direction, and the time elapsed after release.

Horizontal Air Classifier

Figure 5-3 shows the trajectories of material flowing off the head pulley of a conveyor belt with horizontal velocities of 200 and 400 ft/sec and zero vertical velocity. These trajectories would apply to any dense material that is not significantly affected by wind resistance or drag.

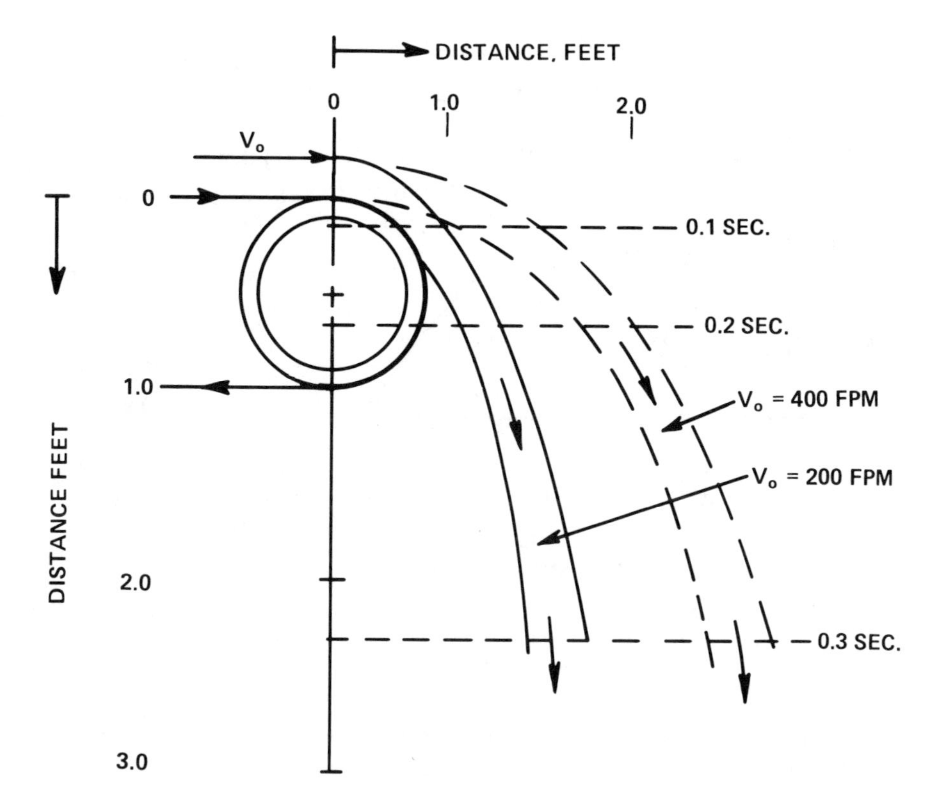

Figure 5-3. Particle trajectories at various conveyor belt velocities.

As the conveyor speed is increased, the particle is projected farther, horizontally, from its starting point, but would reach the same vertical level at any given time as it would have if the conveyor speed were reduced. The elevations reached at increasing times give a good visual indication of the rate of acceleration.

Trajectories of this type can be used for horizontal air classification by the application of air streams, which apply a greater vertical drag force to some of the particles than to others. Whatever angle is used, horizontal and vertical components can be evaluated to determine the trajectory of the particles.

Figure 5-4 shows the calculated trajectories of various particles accelerated by a horizontal air stream while falling 12 feet. The distance traveled depends on the air velocity and the type of material. This principle has been used in the Bureau of Mines and Boeing horizontal classifiers, as well as the General Electric, Raytheon and AENCO rotary air classifiers. In the rotary classifiers, the material has a new opportunity to advance each time it is dropped. By sloping the drum upward, the heavies can be removed at the opposite end from the lights. When the drum is sloped away from the

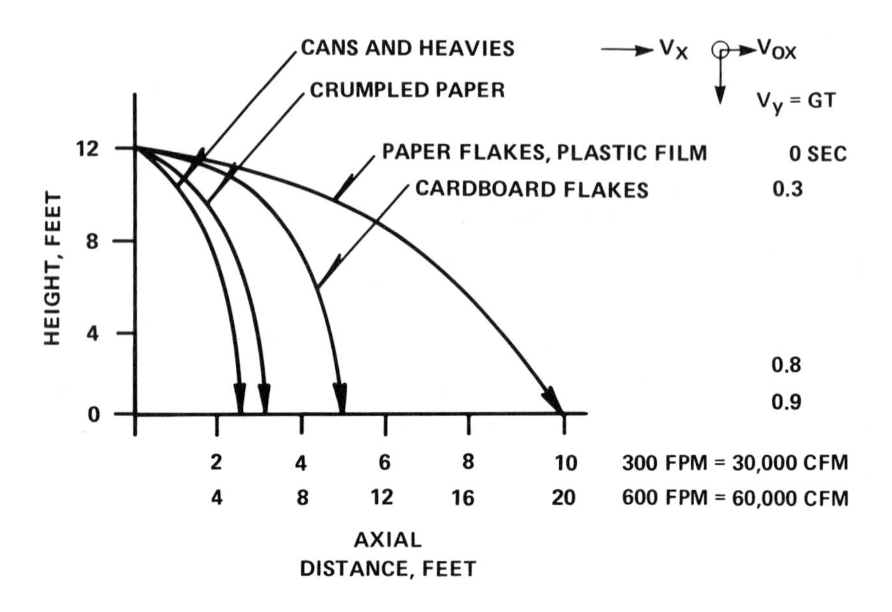

Figure 5-4. Particle trajectories in a horizontal air classifier.

feed end, the heavies can be removed in a falling trajectory at the exit end while the lights remain airborne.

Vertical Air Classifier

When a particle is injected into an air stream having a relative velocity, V, it will be affected by the drag force exerted by the air and, hence, will be accelerated toward the air velocity. If the air stream is moving horizontally, without friction, the particle will actually reach the air velocity after overcoming its inertia. If the air stream is moving vertically upward, the particle will be influenced by the air drag and the opposing force of gravity. The vertical velocity that balances these forces is called the lifting velocity (V_l). If the particle falls in still air, it accelerates until the drag force equals the force of gravity, at which point its velocity is called its terminal velocity (Vt). Lifting velocity and terminal velocity are the same unless the particle is unstable when lifted from below.

Figure 5-5 shows the velocity of a particle having a lifting velocity of 25 ft/sec, starting from zero velocity relative to a rising current of 33 ft/sec. At the outset, the differential velocity is 33 ft/sec, but as the particle velocity increases the relative velocity is reduced until it reaches the lifting velocity. This acceleration takes a significant amount of time. If it had continued at its initial rate of acceleration, it would have reached the terminal velocity in 0.4 seconds. Actually, it only approaches it.

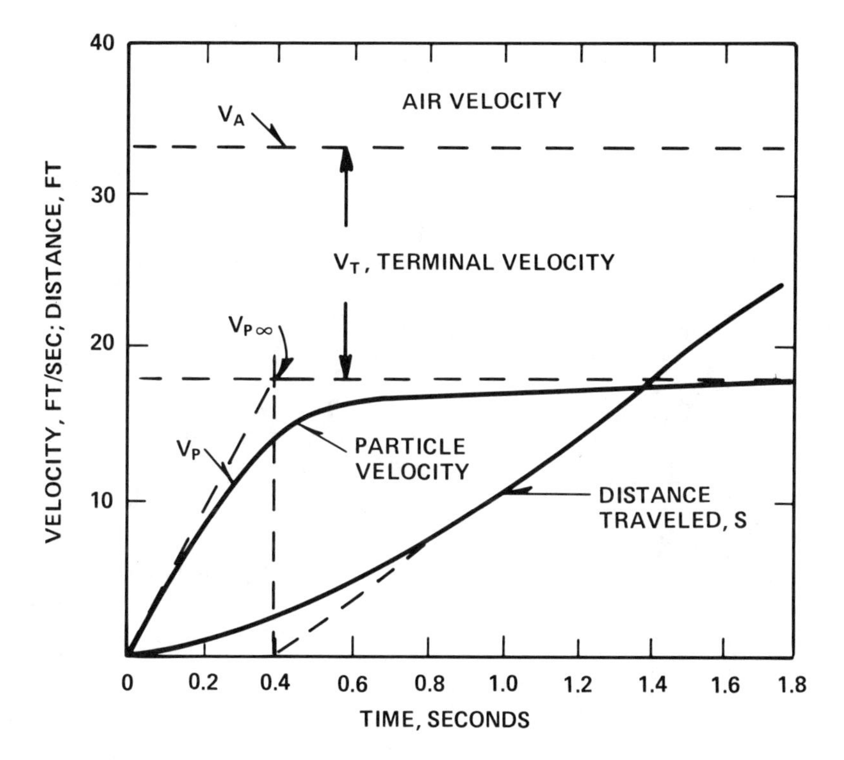

Figure 5-5. Acceleration of particles in a rising air stream.

The distance traveled, S, is also plotted: this increases slowly as the particle velocity increases. Note that the extrapolation of the ultimate straight line is also about 0.4 seconds. The acceleration lag is an important aspect of air classification and other situations in which particles fall in air. The density of a flowing material depends on the area perpendicular to its velocity and its velocity. At low velocity it is more dense, potentially causing jam-ups. For this reason, acceleration zones must be provided to allow for the acceleration lag.

The behavior of particles moving horizontally while falling must be considered in classifier design. Figure 5-6 shows a series of trajectories of particles resulting from horizontal entry velocities of 100 to 600 ft/min. Horizontal lines show the vertical position at 0.1 to 0.3 seconds. The "throw" depends on the horizontal velocity.

Also shown are the trajectories that would result if the material had a lifting velocity of 650 ft/min and air was applied vertically at 1000 ft/min. Again, the 0.1- to 0.3-second lines are shown. At this differential velocity, the particles are carried in an upward trajectory.

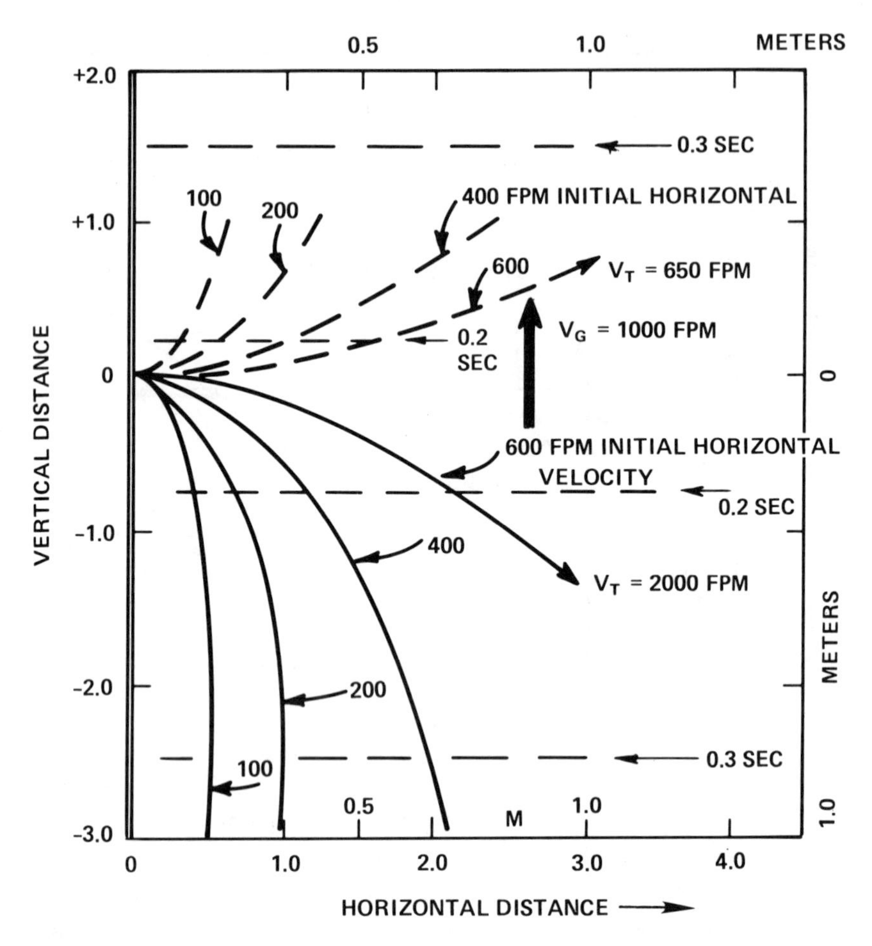

Figure 5-6. Particle trajectories at various initial horizontal velocities if the particles had a terminal velocity (V_t) of (a) 2000 ft/min (b) 650 ft/min.

Determining Lifting Velocities

Lifting velocity can be determined under atmospheric conditions by experimental means or calculated on the basis of theoretical principles with a knowledge of the fluid properties and the shape, size and density of the particle. The lifting velocity of a given particle can be found by placing the particle on a screen and drawing air upwards under the screen at increasing velocities until the object lifts off the screen and floats in the air stream. This air velocity is the lifting velocity and is essentially the same as the terminal velocity (the velocity reached when a particle is in free fall and the drag forces are equal to the force of gravity), provided that the particle floats in a stable manner.

Figure 5-7 shows lifting velocities of "irregularly shaped" particles (below 10,000 μ, or 0.4 inch) of various densities from 2.65 (for stone) to 0.64 (for wood). The line representing the lifting velocity of Eco-Fuel powder shows that it behaves as though its density were 1.5, although it has a bulk density of about 25 lb/ft^3 (density of 0.4). This illustrates the fact that a fibrous material has entirely different lifting qualities than a spherical or irregular particle.

Note that at 200 mesh and 10 mesh the curves change their shape. The drag forces applied by air and other gases have three major regimes, depending on the size of the particle. Small particles (under 200 mesh) have lifting velocities that are proportional to the particle diameter squared. In the size range up to 10 mesh, the lifting velocities are roughly proportional to the diameter. Larger particles have lifting velocities proportional to the square root of the particle diameter. In this region,

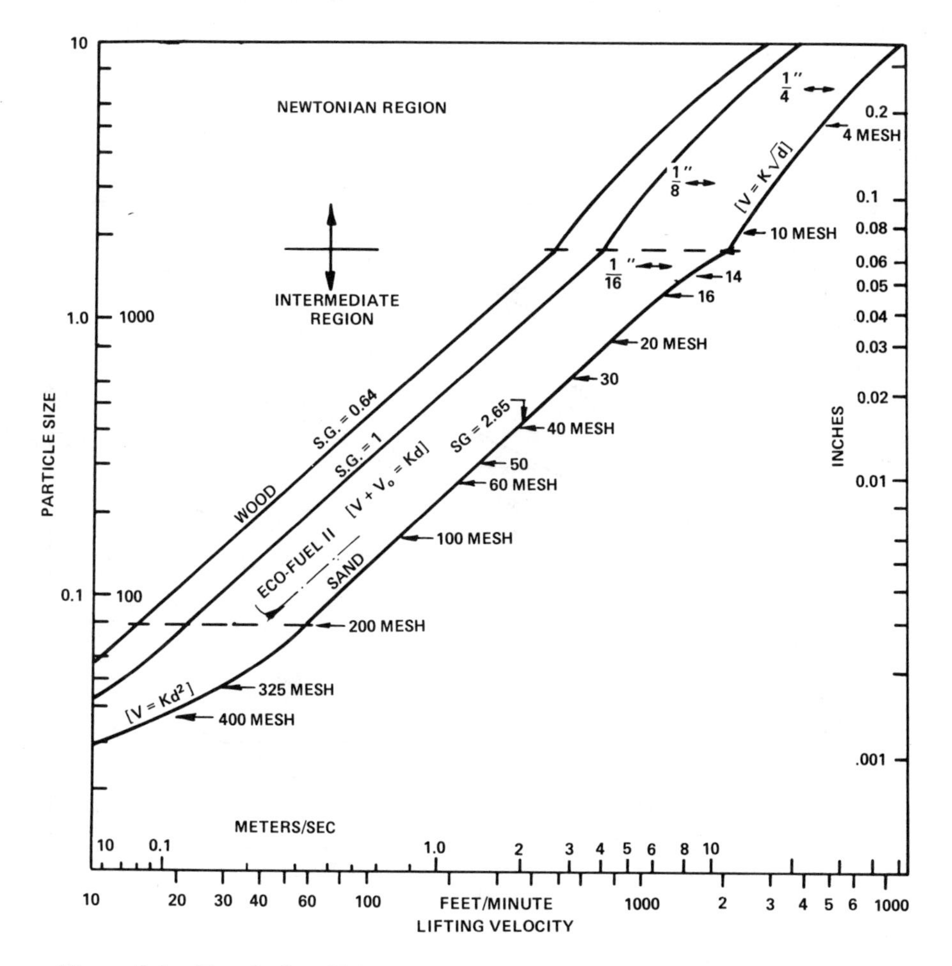

Figure 5-7. Terminal or lifting velocities for sand and other irregular particles.

where most resource recovery operations take place, drag resistance is proportional to the differential velocity squared.

Effect of Density on Lifting Velocity

For a given lifting velocity, the size particle that will be lifted depends nearly directly on the density, so that a particle of half the density and twice the diameter would be lifted with the same velocity. This causes difficulties in separating particles of different densities and sizes.

A velocity of 150 ft/min, for instance, will lift a 1-inch flake of paper, a 30-mesh sphere of wood and a 100-mesh sphere of stone. A velocity of 2000 ft/min will lift a 1/8-inch piece of glass, a 1/4-inch piece of wood and 6-inch flakes of heavy paper.

Figure 5-8 shows the upper range of particle sizes, where Newton's law applies. Here we note that the lifting velocities of flakes diverge radically from those of spherical shapes. These features may be noted:

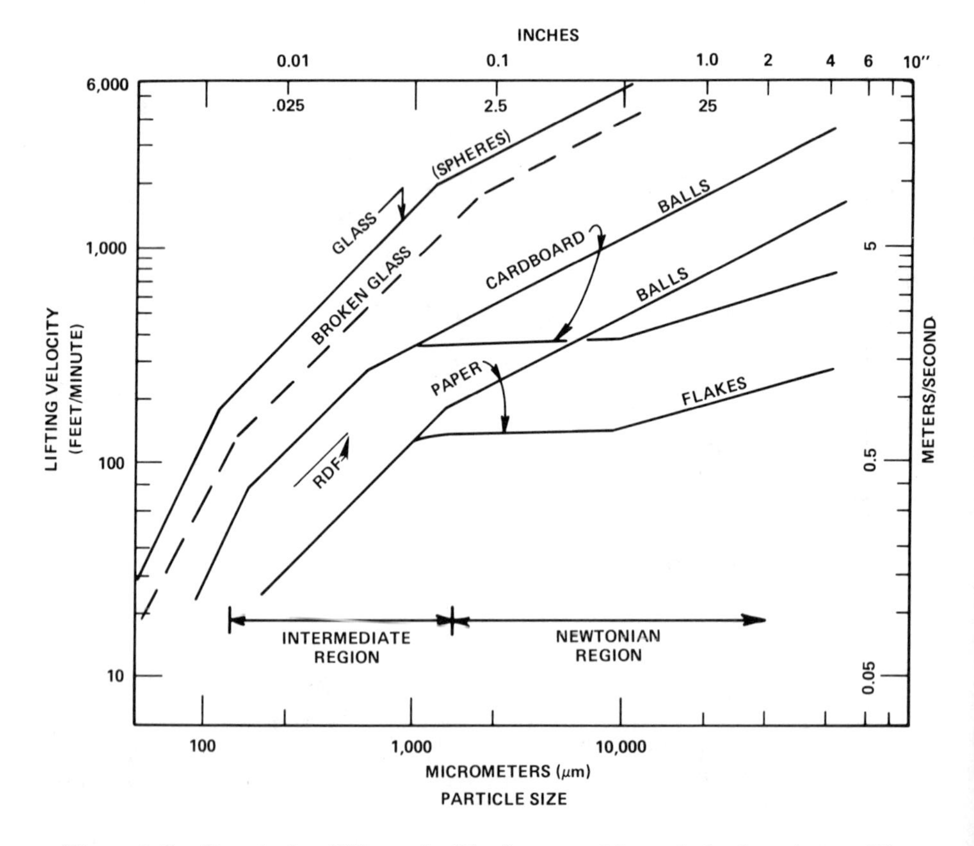

Figure 5-8. Terminal or lifting velocities for several irregularly shaped materials.

- Cardboard and paper flakes follow parallel lines, reflecting the difference in thickness of the material. Flakes or discs of regular shape do not suffer much change in lifting velocity within a wide range of particle sizes because their drag depends on their frontal area and their weight on thickness.
- Cardboard and paper spheres follow parallel lines of an entirely different slope, diverging from flakes and discs as they enter the Newtonian region. These curves have been calculated on the basis that the paper was crumpled into a spherical shape, containing much more mass than a disc of the same frontal area. Test data seem to confirm these calculated curves, although the irregularity of such particles defies accurate testing.

As most materials found in MSW are not spherical or merely angular or irregular, their aerodynamics are more complicated, requiring actual test. An elutriator can be used, but for unstable particles, such as flakes, the measurements are difficult and inexact. A simple method for determining the actual lifting velocity of a particle is to drop it a given distance, about 8 feet or more, and measure the time of fall. This method is especially useful for determining the terminal velocity of the odd materials found in MSW, including paper, cardboard, aluminum cans (whole, crumpled or dented), rags, plastic bottles and clumps of material in their natural form. It is not obvious how the time required to fall a given distance can be related to the terminal or lifting velocity.

The following expression has been derived to obtain the distance traveled by a particle in free fall in air, versus time, given the value of the terminal velocity, V_t:

$$y = \frac{1}{\alpha} \ln \left[\frac{\exp(\sqrt{\alpha g}\, t) - \exp(-\sqrt{\alpha g}\, t)}{2} \right] \tag{5-20}$$

or

$$y = \frac{1}{\alpha} \ln \left[\frac{e^{\sqrt{\alpha g}\, t} - e^{-\sqrt{\alpha g}\, t}}{2} \right] \tag{5-21}$$

where $\frac{1}{\alpha} = \frac{V_t^2}{g}$

t = time (sec) to travel distance y

g = gravitational constant

V_t = terminal velocity (ft/sec)

y = distance traveled (feet if g is ft^2/sec)

Figure 5-9 shows a graph of alpha versus V_t with measured values of alpha and V_t for a number of typical materials found in MSW. Table 5-2 lists additional materials and their terminal or lifting velocities determined by this method. Equation 5-21 is plotted in Figure 5-10 and gives typical values for cardboard, irregular (crumpled) paper

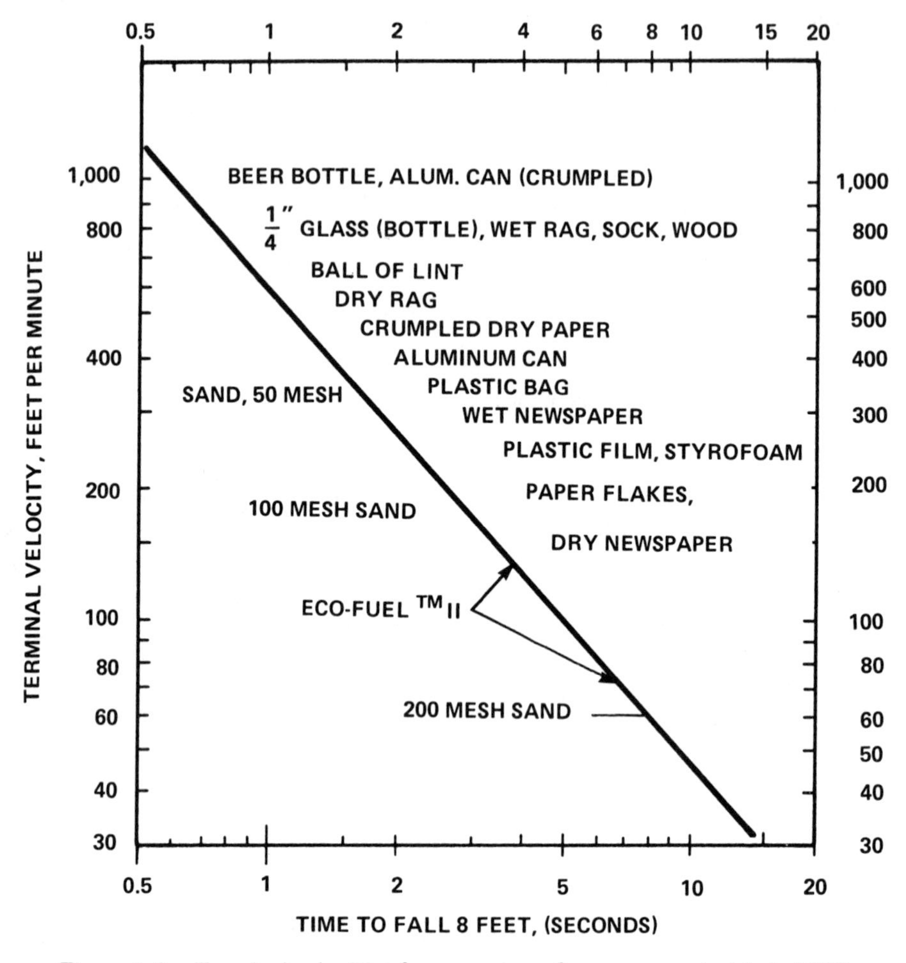

Figure 5-9. Terminal velocities for a number of common materials in MSW.

and flakes. Dense materials, such as chunks of wood and denser substances, follow the straight line for acceleration without (significant) drag.

Some of the particles found in MSW have fairly consistent lifting velocities (LV) for given sizes, such as shredded dry paper and glass. Others, such as clumps of rags or paper/cardboard, have highly erratic lifting velocities, falling through a much wider range. Aluminum cans, when whole, have different lifting velocities depending on whether they are oriented sideways or lengthwise. When crumpled or flattened, another set of values for lifting velocities is obtained. As it happens, the cans will lift if the cardboard clump lifts and both would fall at lower velocities.

Particles of shredded materials are distributed over a wide range of sizes as well as shapes. Hence, such a mixture will also have a wide range, or distribution, of lifting velocities. Figure 5-11 shows the ranges of lifting velocities that have been

Table 5-2. Lifting Velocities of Materials in MSW vs. Size

Velocity		*Material and Size, μ*			
(m/sec)	*(ft/min)*	*Sand*	*Eco-Fuel*	*Wood*	*Paper*
25	4,800	5,000		15,000 (0.6 in.)	
12	2,400	2,000		8,000	
6	1,200	1,100		4,400	
3	600	600	1,100	2,200	
1.50	300	300	550	1,100	300,000 (12-in.)
0.75	150	150	300	600	75,000 (3-in.)
0.38	75	90	160	350	20,000 (3/4-in.)
0.18	38	50	80	160	
0.10	18	35	50	90	
0.05	10	30	35	55	

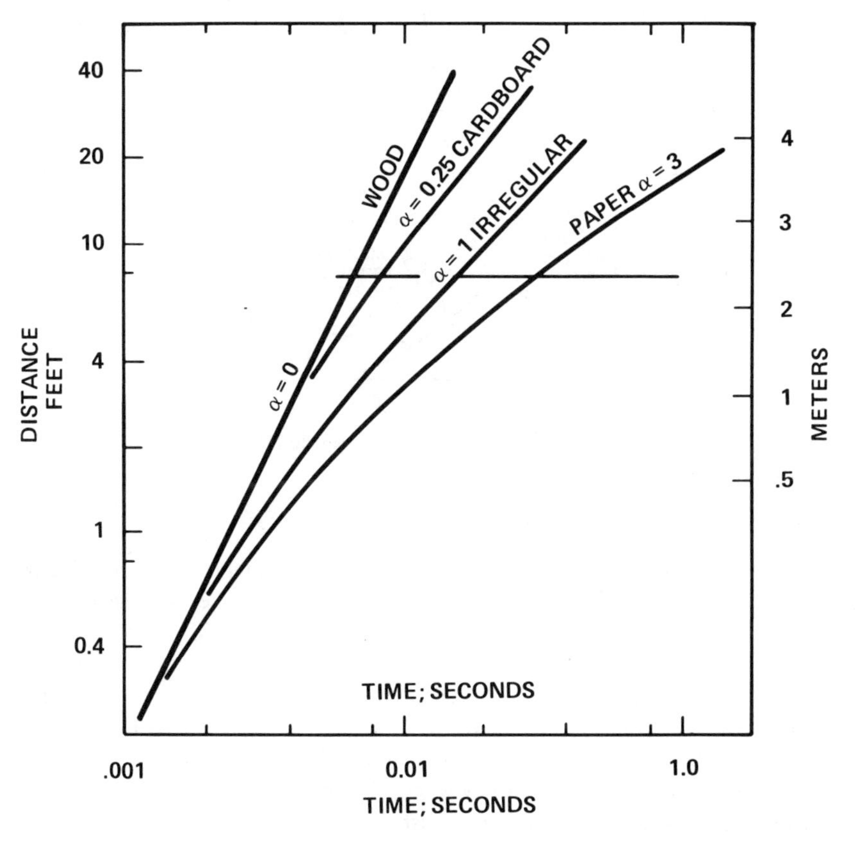

Figure 5-10. Time of fall for various particles.

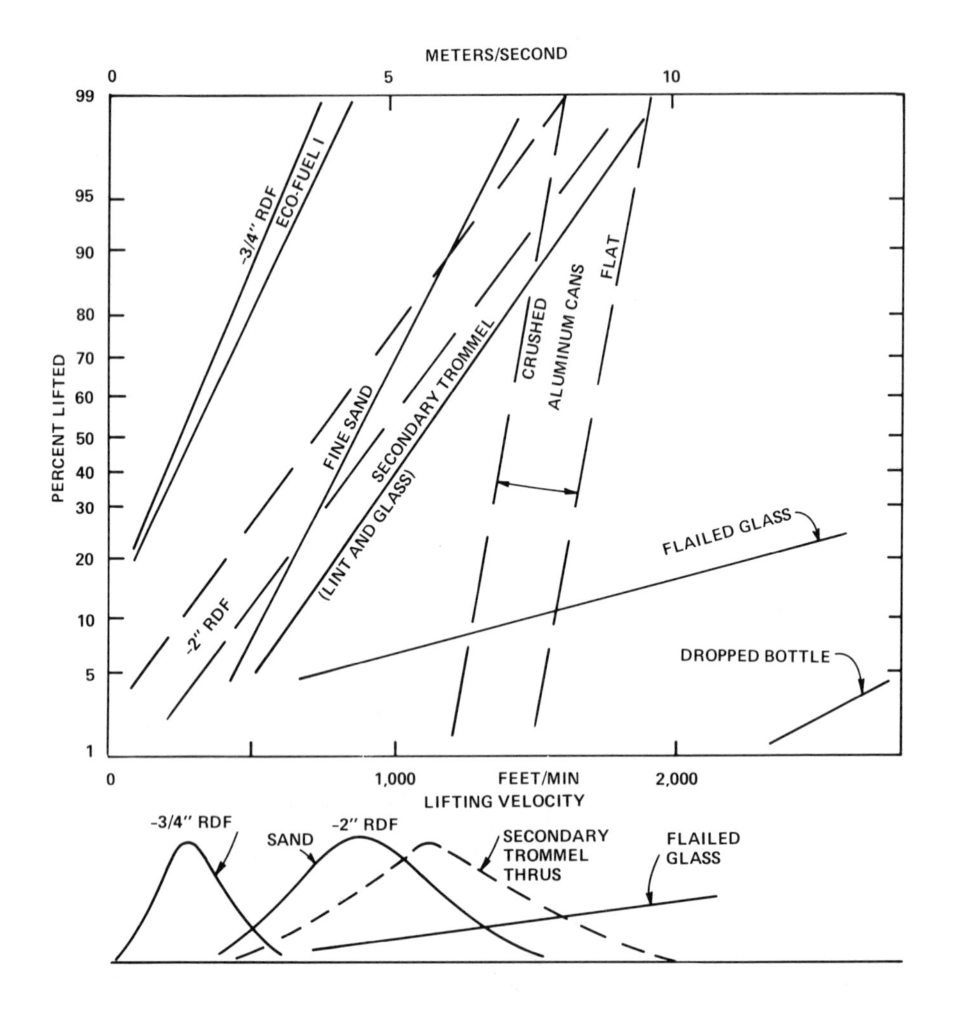

Figure 5–11. Lifting velocity distributions of shredded MSW components (courtesy Bureau of Mines, Salt Lake City).

measured for a number of discrete particle types, with histograms to help visualize the distribution of velocities for some of them. The aluminum types have very steep lines on probability coordinates and sharp histograms. A narrow cut of ground wood is likewise steep, but as a wider range of sizes is present the overall distribution widens.

The velocity distribution of broken glass depends on the size distribution. The distribution for a dropped bottle is shown for comparison with that of glass found in flail-milled RDF and with sand. It is clear that the shredded glass is in the same region marked for lights and that of raw glass is in the heavies region.

The mean velocity of a velocity distribution of shredded MSW depends on the degree of shredding: the double-shredded RDF lifts at extremely low velocities, under 300 ft/min, below that of sand. The velocity distribution of air classified lights of single-shredded MSW is broadly dispersed, depending on the character of the MSW. The same is true of the heavies.

THEORETICAL VERSUS ACTUAL PERFORMANCE

The ideal behavior of single particles injected into an air stream with uniform velocity can be described mathematically by the methods presented above. Particles will fall, float or rise according to whether their lifting velocities are greater than, equal to, or less than the vertical component of the air stream. The rate at which they rise or fall will depend on the difference between their lifting velocity and the gas velocity, and the weight per unit face area of the particle.

The conditions in a real air classifier, wherein a large number of particles of various sizes, shapes and densities enter from the side of the rising air columns, are more complicated. The particles enter from more than one point, must enter with a velocity and a direction, then will either hit the wall opposite the entrance or be accelerated to fall or rise. At the entry zone, the particles that may be densely concentrated must be separated so that the air can act on them more or less separately. Ideally, the lightest and heaviest particles are removed from the feed stream rapidly, leaving the particles that will not accelerate rapidly due to the low differential between their lifting velocity and the gas velocity, or to their high inertia.

It is apparent that the feed and separating zone is a critical area. The happenings within it will greatly influence the effectiveness of the classification process.

As the particles enter the classifier, there is a significant chance that the particles will misreport, so that lights fall and heavies are carried upward due to variations in the point and angle of particle entry and in the gas velocity across the classifier vertical cross section. By providing more than one stage of classification, most of the misreporting particles can be made to change their direction and report properly.

The discussion and analysis that follows will consider what happens in a single stage of classification, then in a multistage classifier, which returns misreporting lights to the heavies, and vice versa.

Single-Stage Vertical Classifier

The simple vertical classifier illustrates the behavior of single particles entering the classification zone. Particles that are not rapidly removed during the horizontal traverse will splash against the wall, going sideways or vertically upward or downward. Lights may happen to be splashed downwards and may or may not be dragged back upward, and vice versa for heavies. Heavy particles may be carried by lights, and vice versa. The majority of the particles will be accelerated by gravity and the air column and move toward the upper or lower exits through which they will leave the system.

The behavior of the particle is thus seen to be a probability function. Indeed, when a number of identical particles are fed to such a classifier, with the vertical velocity set at the lifting velocity of these particles, about half the particles will rise and half will fall. Obviously, they cannot all stay at the center of the classifier, although they will tend to congregate there.

To test a classifier, a set of identical particles can be fed to the classifier at a series of lifting velocities and the fraction or percent determined; also, the residence time of each particle can be determined at each lifting velocity. The results of such a test, performed by Senden and Tels [8] with styrene spheres of 4.5 mm in diameter, fed to a 1-meter vertical classifier, are shown in Figure 5-12. The data, plotted on probability paper, show a "separation curve," which follows a straight line, indicating almost complete probability control.

The steeper the slope of the separation curve, the "sharper" the "cut" of the classifier. In this case, the mean lifting velocity was 1.645 m/sec, and the velocities ranged from 1.4 to 1.8 m/sec for 96% of the spheres, a range of 4/1.645, or about 24%.

The mean residence time of the particles is a measure of the degree of oscillation of the particles within the classifier before they exit. During the tests of the polystyrene spheres, the residence time was measured for each particle and averaged at each column velocity as a function of the mean particle velocity. Those particles that happened to acquire higher velocities either up or down tended to stay in the classifier for shorter periods, as would be expected. Particles with very low exit velocities had average residence times of more than 9 seconds, compared with 3 seconds for others (Figure 5-13).

The ratio of actual particle residence time to the gas residence time gives an indication of the concentration of particles within the classifier, which results from "indecisive" particles, or particle interference. In this case, the gas moving at 1.645 m/sec would traverse half the classifier in 0.5/1.645 = 0.3 seconds, whereas the mean particle residence time was about 4.5 seconds. The ratio of 4.5/0.3 = 15 is called the nondimensional residence time.

Senden and Tels [8] have developed a theoretical concept for the description of the separation process in a vertical gravitational air classifier, which was confirmed by the tests described above. The model assumes that particle inertia is negligible and that the transport parameters are constant along the height of the classification zone. The removal of the particles from the classification zone is described by rate equations. At the heavy fraction exit, the removal rate is assumed to be linearly proportional to the fall velocity of the particle and the relative particle concentration at the heavy fraction outlet. At the light fraction exit, the removal rate is assumed to be proportional to the superficial air velocity and the particle concentration at the light fraction exit.

They derived expressions for the separation curve and for the mean residence time of the particles, which is an indirect measure of the throughput capacity of the classifier. The residence time was calculated and found to influence the sharpness of the separation curve. They concluded on the basis of theory that suppression of particle mixing and accelerated removal of the particles from the classification zone yields the highest separation efficiency for a given particle residence time, and

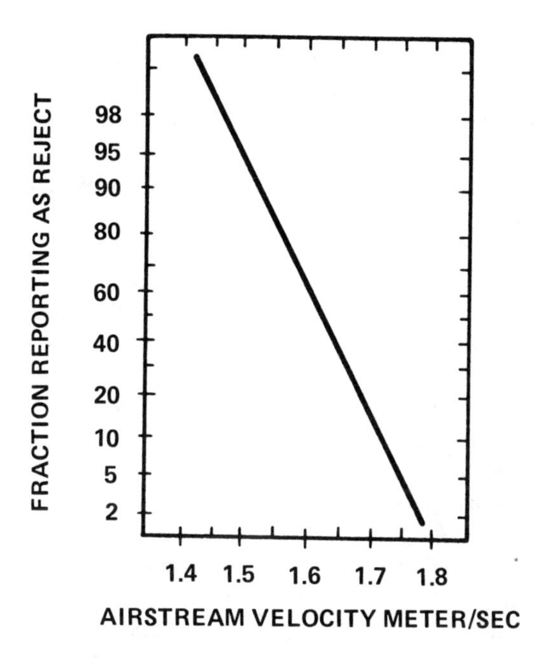

Figure 5-12. Separation curve for porous polystyrene balls [8].

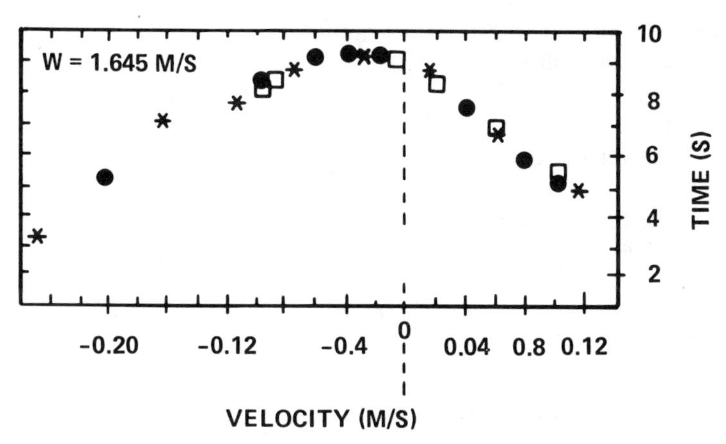

Figure 5-13. Relationship between column velocity and theoretical residence time (Source: Senden and Tels).

that increased residence time increases the sharpness of the classification. Their experimental concepts were verified experimentally for particles with small inertia at low particle concentrations in the zone of classification, in tests such as those described above. The theory makes it possible to predict performance of vertical classifiers under different process conditions and with different feed material properties.

This study indicates that high separation efficiencies may best be achieved by suppressing particle mixing in the classifier and by using high rates of particle removal from the classifier exits. This results in the shortest residence time for the floating particles that have fall velocities equal to the superficial air velocity, and creates conditions favorable to achieving high throughput capacities.

Some conclusions can be drawn as to the conditions desired in an air classifier (including horizontal designs):

- The air velocity profile over the classifier cross section should be as flat as possible.
- The classification zone should be shallow to avoid developing a changed air velocity profile along the length of the classifier.
- Removal of the particles at the product exits should be accelerated.
- The solid particles should be fed homogeneously over the classifier cross-sectional area in a constant stream.
- Particle-particle and particle-wall interactions should be minimized to avoid flow disturbances.

Obviously, high feedrates interfere with these idealized requirements and reduce the effectiveness of classification. On the other hand, as there may be a clear distinction between the light and the heavy materials in MSW (especially in regard to their terminal or free-fall velocities) and a high ratio of light to heavy materials, actual conditions may approach the ideal.

The effect of residence time on capacity and separation is clearly shown by the theoretical analysis and is the nemesis of full-scale, practical air classifiers. Materials that have falling velocities close to the superficial gas velocity will accumulate in the classifier and can cause plugging.

Multistage Vertical Classifier

The zig-zag classifier provides multistage classification. Commonly built in ten stages, it achieves very sharp cuts when suitable materials are processed. This type of classifier was tested with MSW by the Bureau of Mines in Salt Lake City, and subsequently by Americology, the NCRR and Occidental Research (ORC).

The theory of the zig-zag classifier has been developed by Kaiser [9] and Senden and Tels [8] and will be outlined here as an illustration of the principles involved, although many of the assumptions of the theory are violated by MSW, especially at high feedrates.

Zig-zag classifiers use a current of air rising in a single column to carry up light particles while allowing heavier particles to drop out the bottom. The column is made with a number of zigs and zags having a 120° angle over the intersections of the sides, as illustrated in Figure 5.14. Other angles, such as 90°, may be more suitable for other materials.

These classifiers have been developed to separate granular particles of many types from minerals to wood and some types of flaky materials. For instance, they

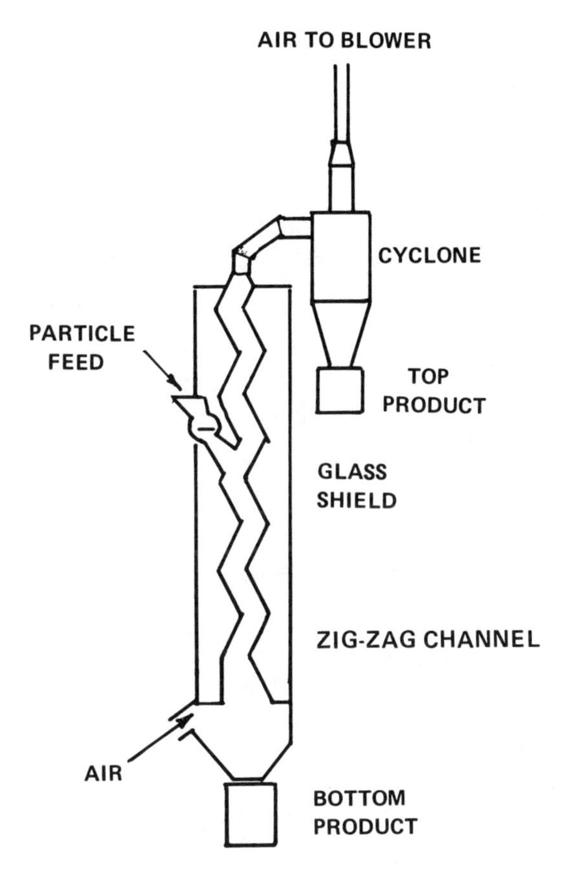

Figure 5-14. Typical zig-zag air classifier.

are very effective in separating wood flakes of different sizes and shapes in the manufacture of particle board and flakeboard. They are used to achieve a fine cut of materials from 0.1 to 10 mm, while using velocities from 1.5-10 m/sec (270-1800 ft/min).

Figure 5-15 shows a portion of a zig-zag classifier. A falling stream of heavies and a rising stream of lights is carried by a rising air flow. The rising particles are almost entirely airbound, moving through a central zone without hitting the walls, whereas the falling particles travel down the sloping walls, making trajectories across the rising gas column, striking the opposite wall, then accelerating again down the steep slope to make another trajectory across the gas stream. Each time the falling materials pass across the rising column, another opportunity is offered for light particles that may be entrained in the heavies to break loose and rise. Repeating this action results in a clean separation between the two types of materials.

Modeling of zig-zag classifiers has examined the behavior of particles of uniform size and under conditions in which there is little particle interference. Neither of these conditions is valid in practice, especially in processing wastes. However, the

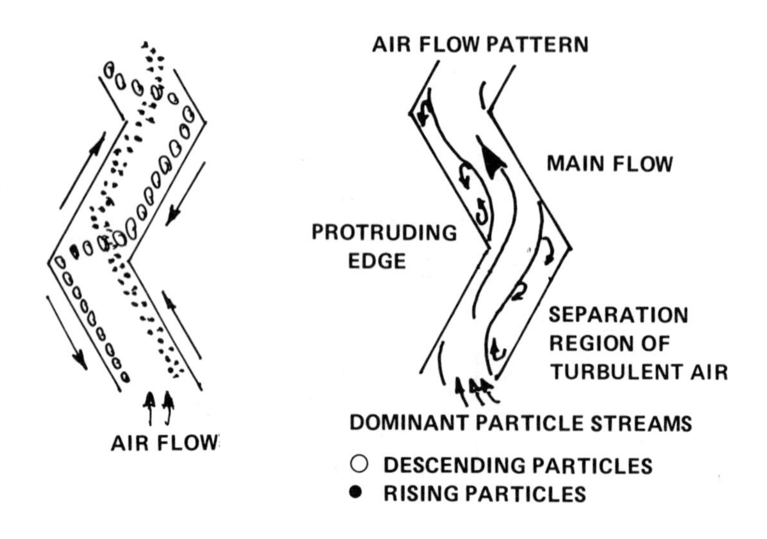

Figure 5–15. Particle and air movement in a zig-zag classifier.

physical happenings described in the mathematical models can serve as a basis of comparison of the real against the ideal and help us see to what extent there is room for improvement. The application of theory to multistage air classifier operation has been studied by several research teams.

University of Eindhoven, The Netherlands

One of the first attempts to understand and model air classification was made by Senden and Tels [8] at the University of Eindhoven in the Netherlands. Some of their results, using single, ideally shaped particles, have been discussed previously (Figure 5–16). They argued that perfect separation, where all particles with the same falling velocity would end up in the same stream if the air stream velocity were slightly greater or smaller than that of the particles, is only a concept, and was never achieved under real conditions. A curve showing recovery versus air velocity thus has to have a slope that is characteristic of the classifier.

Figure 5–15 shows the separation for a typical classifier versus the vertical velocity of the air, or column velocity. For a feed containing essentially identical particles, any slope in the curve reflects imperfection of the classifier itself. The sharpness of separation is the slope or steepness of the curve relating the quantity lifted to the velocity of the air stream. It is commonly evaluated at the point where percentage reporting is 50%, that is, half the material by weight lies above and below this point. This also defines the probability, P_c, that a particle has a 50–50 chance of going either way at the velocity, V_t, the cut velocity.

Senden and Tels [8] defined the amplification factor, I_r, for a classifier having r stages, the ratio between the separation sharpness of the multistage classifier, S_r, and a single-stage classifier, S_o.

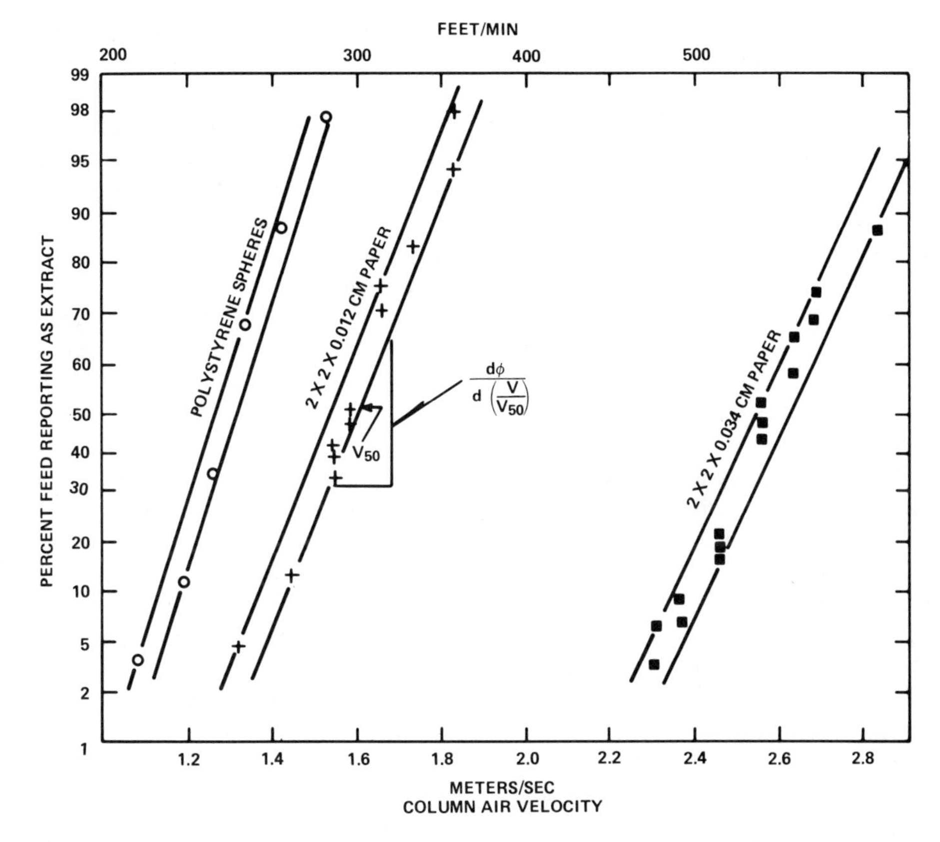

Figure 5-16. Percentage reporting as extract for three feeds; 120° zig-zag [8].

$$\text{Amplification factor} = I_r = \frac{S_r}{S_o}$$

The amplification factor, based on theoretical calculations, indicates that it could be five for a ten-stage zig-zag, with the feedpoint five stages below the top, with material having a 50% chance or probability of going up or down. However, other work suggests that the amplification factor for materials that do not need to be broken apart may be close to 1.0 [9].

The data in Figure 5.16 were obtained in a series of tests using three types of uniform particles: (1) 2 cm × 2 cm, 0.0121-cm-thick paper; (2) 2 cm × 2 cm paper, 0.034 cm thick; and (3) 0.45-cm polystyrene spheres. The particle behavior in a laboratory zig-zag was photographed and the fractions rising and falling determined versus the column velocity. Figure 5-16 illustrates some important principles. First, it shows that even uniform particles have considerably less than infinite sharpness. The spherical particles in a ten-stage classifier were lifted to some extent at 1 m/sec, and all at 1.5 m/sec. The slope of the separation function in the center can be expressed as a differential of 12% of the average column velocity.

The heavy paper particles, on the other hand, were lifted to some extent at 2.2 m/sec, and all at 3.1 m/sec, the slope again being about 13% of the average column velocity. By comparison, the heavy paper showed a velocity range of 1.6 to 3.8 m/sec in a single stage, a spread 80% of the average. The mean lifting velocity of the heavy particles versus that of the light particles was found to be proportional to the square root of the thickness of the particle. The lifting velocity of the 1 cm X 1 cm particle was not significantly different from that of the 2 cm X 2 cm particle.

The effect of zig-zag geometry has been studied by Senden [10]. The findings show that the 120° classifier is more suitable for shredded paper and that the optimum ratio of depth of throat to length of side is about 1:2. Greater sharpness can be achieved with a narrower throat, but only at the expense of reduced classifier throughput.

Salt Lake City, Utah

A ten-stage zig-zag column built by Scientific Separators was tested by the Bureau of Mines in Salt Lake City. The throat was 15 cm X 30 cm, considered large enough to test shredded MSW with a maximum particle size of 10 cm. The suction blower provided could produce column velocities of 100-770 m/min. The column was operated with six stages above and four below the feedpoint. MSW was tested that had been shredded by five different methods, including a Gruendler hammermill, Dorr-Oliver rasp, Williams rigid-arm paper shredder, Williams swinging-arm refuse shredder and an Eidal Model 400 coarse-grind shredder. The MSW was collected from five locations associated with the five shredders, namely those in Johnson City, Tennessee, Los Angeles, California, Cincinnati, Ohio and Albuquerque, New Mexico. The average particle size from these sources varied from 1.25 to 3.5 cm.

The general procedure was to make two runs on a sample to recover the paper-rich middle fraction: (1) a dedusting run to remove light fines and film plastic; and (2) a run to remove heavy constituents such as metal, rocks, glass, rubber, heavy plastic and wood. The middle paper fraction was then split by air classification into the chemically pulped fraction (containers that would include cardboard boxes, grocery bags, milk cartons and similar items) and a mechanically pulped fraction, essentially newsprint.

Runs to establish separating velocities for splitting the paper fractions were made initially at low feedrates. As the feedrate was increased, the sharpness of separation decreased until the column operation became unstable and choking occurred due to overloading. Subsequently, overloading conditions were avoided. Typically, the runs started with a dedusting velocity of 90-120 m/min, removing a small quantity of very light fines, dust balls and a major portion of the film plastic contained in the raw sample. Next, the dedusted sample was run at about 215-250 m/min, again at 275-300 m/min, 340-370 and 370-400 m/min. At 400 m/min, the heavy fraction contained almost entirely corrugated stock, contaminated with more than 50% shredded metal, wood chips, heavy plastic, glass and other high-density materials. These contaminants were effectively removed by air classification at 550 m/min. They could have been removed by a 5-cm screen.

Air-dried material was used for most of the tests, but wetted material also was tested to discern the effect of moisture. Higher separating velocities were needed to get the same species separation, but the percentages removed were essentially the same, as would be expected because only the density of the paper fractions was changed.

Data from these tests are plotted in Figure 5-17 with percentage lifted plotted against the corresponding column velocity. The fraction lifted during a test with a given sample consistently follows a linear relation with velocity. The light and heavy fractions show similar straight lines intersecting at a common apex.

The data taken during runs at light loadings reflect the nature of the material rather than the characteristics of the classifier because the sharpness of separation is high for a ten-stage unit. At heavier loadings, the sharpness would be expected to fall. The heaviest column loadings tested were at ASR of 3.1-4.7 pounds of air per pound of solid. The highest feedrates were at a column loading of 0.9 TPH/ft^2 of throat area.

National Center for Resource Recovery

The NCRR has reported tests on a zig-zag classifier and on a Triple S classifier [2]. The NCRR pilot plant was set up to demonstrate the recycling of metals, especially aluminum, by passing the air-classifier heavies through an aluminum magnet. The gas velocities were adjusted accordingly. As the lifting velocities of aluminum cans are close to those of corrugated cardboard, both materials tend to fall together, increasing the burden from which the magnet has to pull the aluminum and reducing the recovery effectiveness of the process.

Figure 5-18 shows NCRR data relating the percentage of aluminum dropped to the percentage of the feed reporting to the heavy fraction. The aerodynamic shape of the cans greatly influences the collection efficiency: flat cans reached 50% collection only when 30% of the feed was dropped; crushed cans were totally collected with only 20% heavies. This illustrates why the material to be processed must be

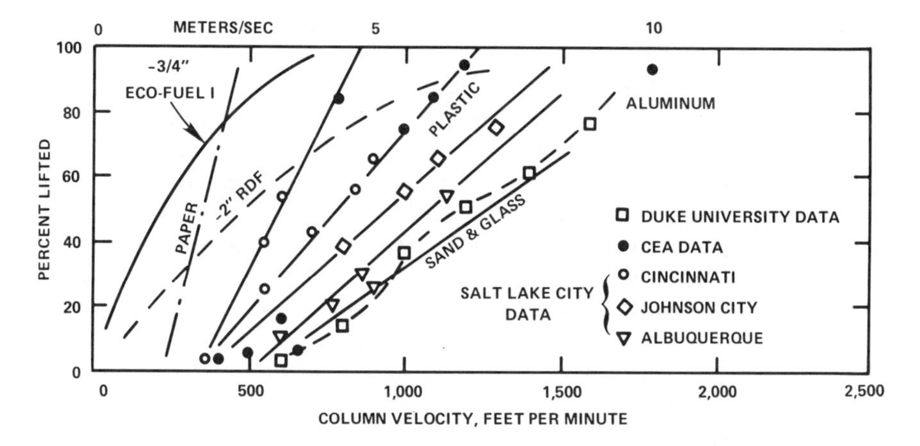

Figure 5-17. Bureau of Mines tests, Salt Lake City (and other data).

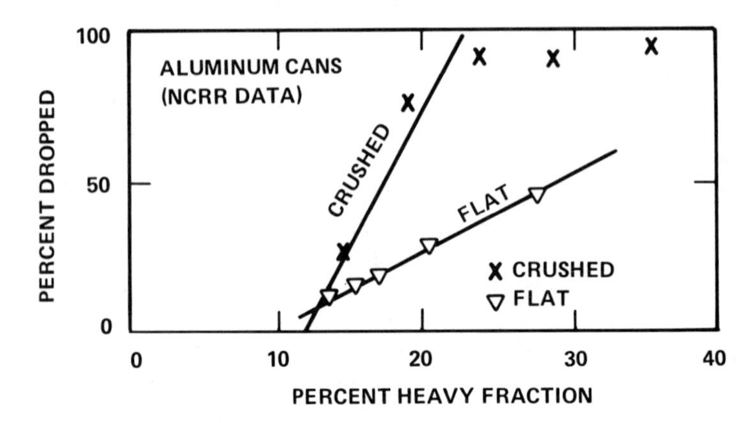

Figure 5-18. NCRR tests on aluminum cans [2].

tested to find out whether the classifier can make a desired separation and achieve acceptable recoveries at practical rates of flow without loss of otherwise valuable products.

Duke University

A comprehensive set of tests on air classification were performed at Duke by McNabb [11]. A method for determining the terminal velocities of components of shredded MSW was developed, tested and its validity confirmed by tests performed with the same samples in a laboratory air classifier.

To evaluate air classifier effectiveness in separating noncombustibles from the combustible light fraction, a sample was prepared that included only paper, plastics, aluminum and steel material, and shredded to minus 3/4-inch size. Portions of this sample material then were subjected to drop tests to determine terminal velocity and then tested in the air classifier to see to what extent the classifier lifting velocity agreed with the terminal velocities.

The terminal velocities of individual particles were determined by timing drops from 6-, 9- and 13-foot heights, plotting the times and extrapolating to obtain the terminal velocity. The lifting velocities were determined by running the air classifier at a series of velocities and weighing the portion of the sample that dropped and flew, thus obtaining the percentage lifted as a function of lifting velocity. Lifting velocity distributions were thus obtained for the paper, plastics and metals (as shown in Figure 5-19), for comparison with the drop-test data.

The lifting and terminal velocities of the paper were in complete agreement, confirming the validity of the drop test as a method for determining terminal velocities. Those for the plastics also were in close agreement, but the air classifier data showed a wider spread of velocities.

The aluminum particles did not behave the same way in dropping as in lifting, as clearly shown in the percentage of lifted curves and dramatically pictured by the histograms. The air classifier divided the aluminum particles into a light and a heavy

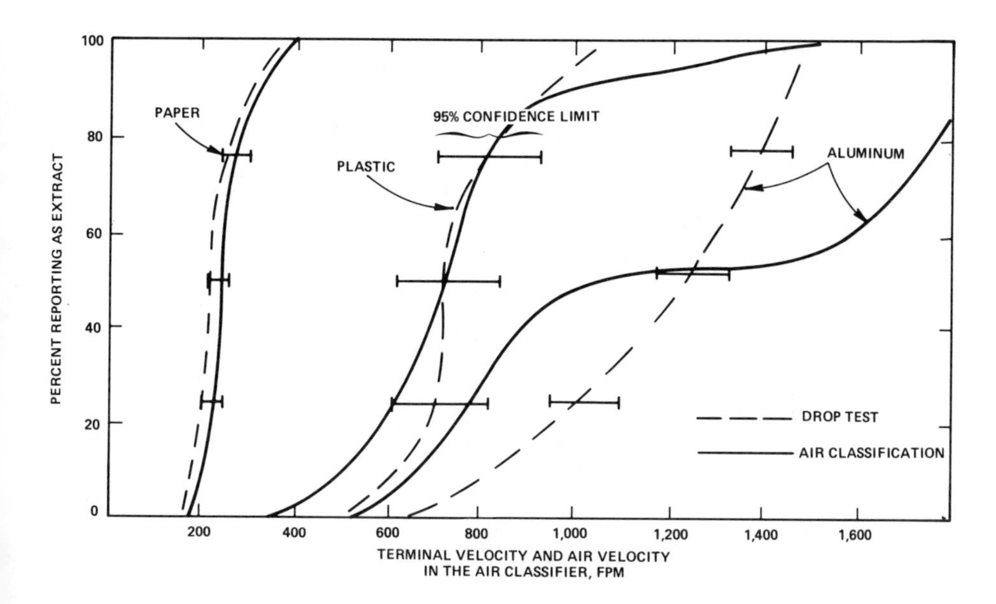

Figure 5-19. Recovery (percentage lifted, or fraction as extract) compared to drop test velocities [11].

concentration. The light concentration follows the peak of the plastics, indicating interaction between the plastics and the aluminum, whereby the plastics tend to carry up more aluminum than without interaction. Another explanation is that the zig-zag causes the aluminum to accelerate sideways back and forth between the walls, emphasizing the higher inertia of the aluminum, requiring that it be reaccelerated frequently, causing it to lose ground and fall as a heavy. This is the more plausible explanation and demonstrates a potentially valuable feature of the zig-zag. The depletion of material caused the heavies to be concentrated at 1700 ft/min, whereas the terminal velocity peak was at about 1400 ft/min.

The steel particles could only be lifted at velocities far above those needed to lift all other materials, around 2000 ft/min. Therefore, they can properly be classed as heavies. The percentage lifted versus velocity curves when plotted on probability paper shows that the data for each species plot as a straight line, proving the probabilistic nature of both lifting and terminal velocities.

The Worrel definition of efficiency (p. 148), defined as the product of the percentages of lights and heavies lifted, was used to compare the drop test with the results of air classification.

Figure 5-20 shows the comparison between the efficiencies of the dropping and lifting classifiers. The air classifier reaches 70-75% and maintains this level up to 1600 ft/min. The drop classifier reaches 85%, which must be considered the best possible, or ideal.

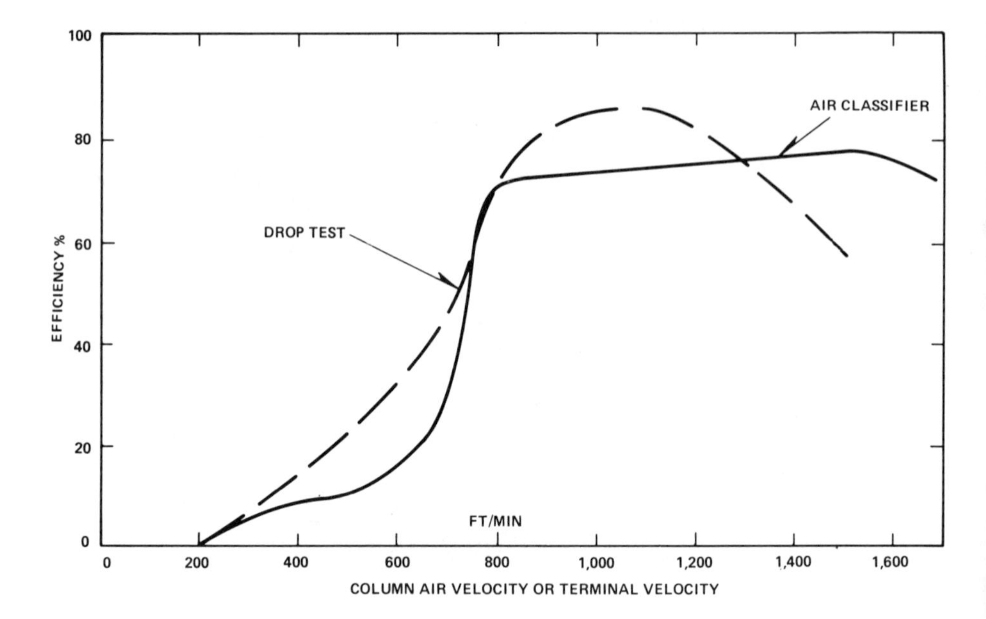

Figure 5-20. Comparison of the drop test with actual air classifier performance [11].

ACTUAL PERFORMANCE OF AIR CLASSIFIERS

Very little information is available on full-scale air classifiers, especially comprehensive data covering a wide range of operating conditions with various types and amounts of feed. There are many reasons for this lack of data:

- The fan capacity is normally limited closely to the design capacity and cannot be increased significantly because the horsepower required goes up as the cube of the gas flow.
- The feed equipment or, more often, the discharge equipment, is limited in capacity.
- Methods of weighing the flow on an instantaneous basis are usually lacking.
- Equipment is close coupled so that there is no way to interpose a weighing device.
- The labor required to run tests and to analyze the products is extremely costly.
- Due to the highly variable nature of the feed materials and the variability of flowrate, the number of tests and samples required to achieve statistical confidence in the results is excessive.

Some test results are available, however, including a series of tests performed at Chemung County by Americology and a series of field tests sponsored by the EPA.

Americology Tests at Chemung County

A large assemblage of data was obtained in a series of tests performed by Americology. A statistical analysis of these data, reported by Lawler [12], has been made available for publication here. Due to the nature of the test facility, it was possible to obtain at least a small amount of data under high air flows and heavily loaded conditions, beyond design conditions. In addition, the many bits of data could be analyzed by regression analysis, yielding a high level of precision by virtue of many data points, despite the high variability of the MSW feed material.

The analysis of these data illustrates the methodology that can be used to analyze and interpret performance data, with application to many devices other than air classifiers. A total of 55 sets of data were collected, of which 40 included moisture, ash and heating value in addition to feedrate, air flow and percentage of feed lifted. Lawler's findings included these:

- Air flowrate has a major effect on the amount of extract produced—three times as much influence as feedrate—raising air flow from 21,000 to 29,000 ft^3/min and increasing the lights from 35 to 70%.
- Feedrate has a significant effect on the percentage of feed exiting as extract, raising feed from 20 to 60 TPH and reducing lights from 60 to 50%.
- The moisture in the lights had a direct effect on the ash in the lights as moisture went from 10 to 30% and ash increased from 12 to 18%.
- Higher air velocities are needed to achieve a given percentage of lights as the feedrate is increased: TPH—20, 40, 60; ft^3/min—50% lift: 22,800, 24,200, 25,500; 60% lift: 21,200, 22,400, 24,400.
- At low feedrates, 1900 ft/min nominal lifting velocity was sufficient to lift essentially all available lights, and at high feedrates 2100 was needed to be equally effective. Lesser velocities might not be effective and reliable, and higher velocities might be wasteful or not effective in separation of heavies.

Most of these findings are readily explained after the data have been analyzed or the reasons investigated. The method of analysis was as follows:

The data were analyzed by computer regression analysis to find which variables had the strongest influence on trends. This produced the information that air flowrate was the most important variable, with feedrate second in influence.

Moisture was found to have no influence other than the obvious effect on the weight of the material and the amount of air required to lift the additional weight. Figure 5-21 shows how the moisture affected the light fraction expressed on an as-received basis and on a dry basis. The scatter in the points was fairly extreme, but regression analysis of the data for runs at fixed air velocities showed strong trends in straight lines, as shown in Figure 5-22, with percentage of feed lifted plotted against the feedrate. Clearly there is a limit to the amount of the feed that could be lifted, and even an apparent drop-off when the optimum velocity was exceeded, due perhaps to turbulence in the classifier.

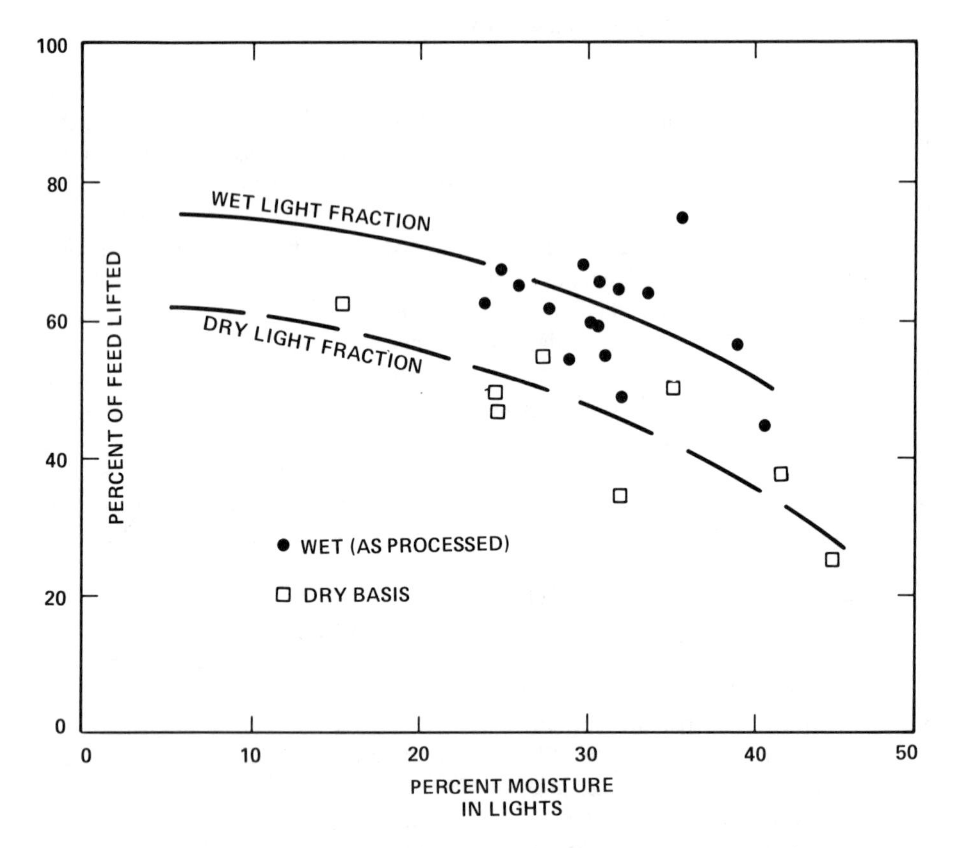

Figure 5-21. Effect of moisture on recovery of organics [12].

The straight lines did not make much sense until the lines of constant ASR were added, resulting in a third dimension and a delineation of the optimum condition. On average, all these data showed the percentage of available lights to be close to 66%.

A plot of the nominal column velocity versus the percentage of lights at various rates of feed is illuminating. Figure 5-23 shows clearly that more velocity is needed as the load is increased to achieve a given level of available lights lifted. The data show that above a given velocity the percentage lifted actually falls off due to deterioration of column performance.

Table 5-3 shows the results of a momentum analysis of the data. The theoretical velocity that would have to be applied to the solid stream to achieve the measured exit gas velocity is determined by the momentum balance. As the classifier column has a constant cross section, resulting in constant average gas velocity, the accelerating energy has to be obtained at the expense of static pressure. This can be calculated by determining the static pressure needed to create the theoretical velocity. Additional static pressure is needed to accelerate the solids and air to the conveying velocity and to overcome duct and cyclone resistance.

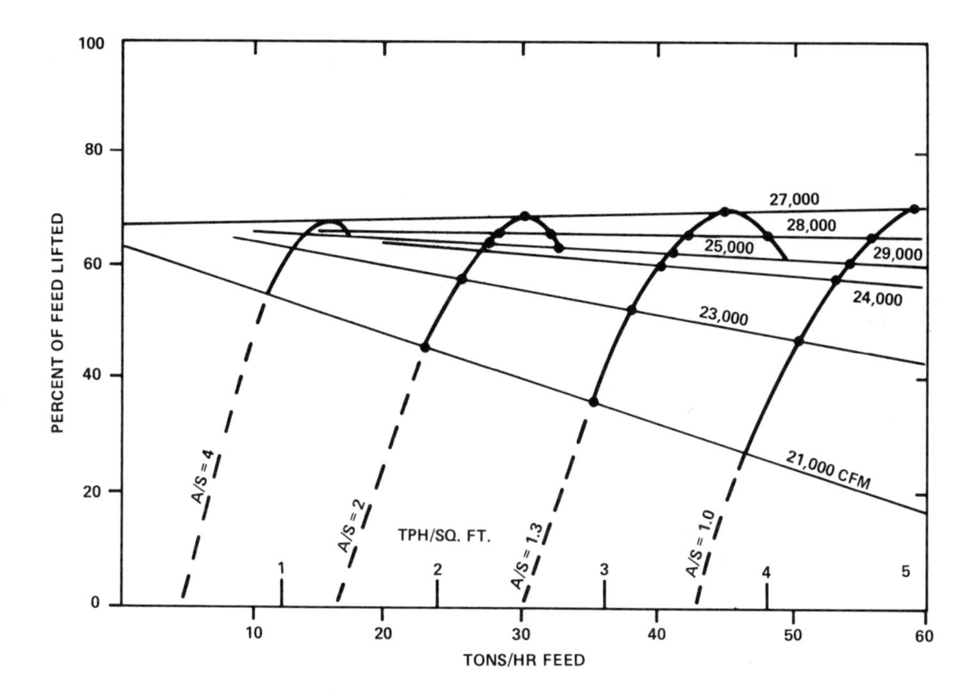

Figure 5-22. Effect of feedrate on performance [12].

Figure 5-24 shows the generalized curves that result when the percentage of available material lifted is plotted against ASR and various air velocities. These trends explain the test data.

Field Tests of Various Types of Classifiers by Midwest Research Institute and Cal Recovery

Seven air classification systems were field tested and evaluated under a two-year program sponsored by the EPA to characterize and compare the operation and performance on an equivalent basis. They included horizontal, vertical and vibratory inclined types, as listed in Table 5-4.

These comments can be made, after inspection of Table 5-4:

- The critical throughput was less than, but close to, design in cases 3, 4 and 6, indicating that the units were expected to be operated under a loading that would drop some lights as the price for product quality. In cases 2 and 6, the critical was much less than design, possibly the result of the method for determining the critical.

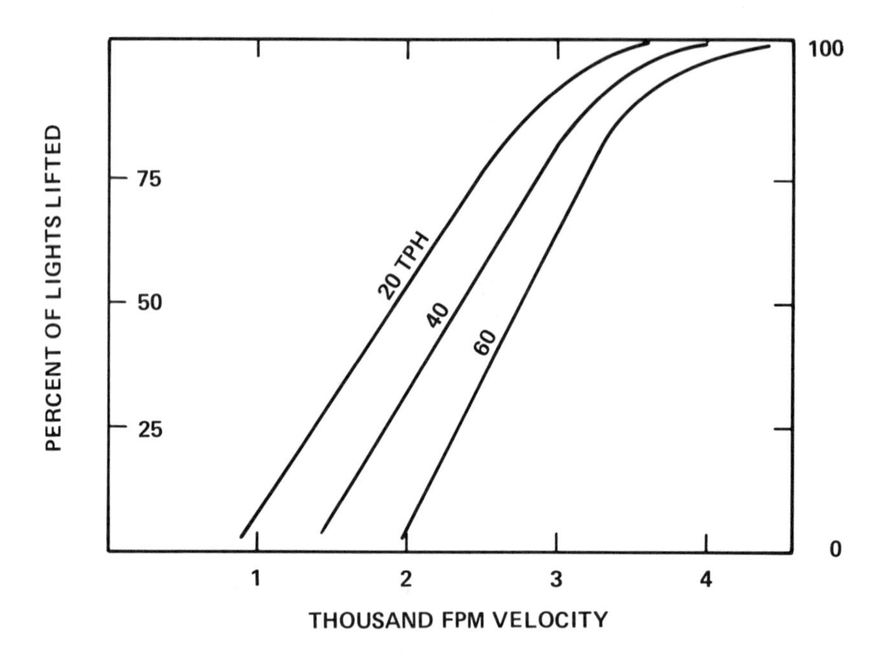

Figure 5-23. Effect of air velocity in column on recovery of organics [12].

Table 5-3. Calculation of Resultant Material Velocity

MSW Feed, TPH	20	40	60
Lights Lifted (65% of feed)	13	26	40
Column Air Flow, TPH	51	54	55
Outlet Air Velocity, ft/min	1,900	2,000	2,050
Theoretical Inlet Velocity, ft/min	2,384	2,963	3,541
Air/Lights Ratio	3.9	2.1	1.4
Air/Feed Ratio	2.5	1.4	0.9
Static Pressure to Produce Inlet Air Velocity, in. H_2O	0.4	0.6	0.8

- The ASR varied from 1.6 to 6.6, while the column velocity varied inversely. The product of V × A/S, however, was nearly the same for units 1 to 4 for the Bridgeport classifier tested by the author, showing that the momentum is the controlling parameter, not the mass or velocity alone.
- The units with high column loadings, 1, 4, 6 and 7, had relatively high column air velocities, apparently by necessity.

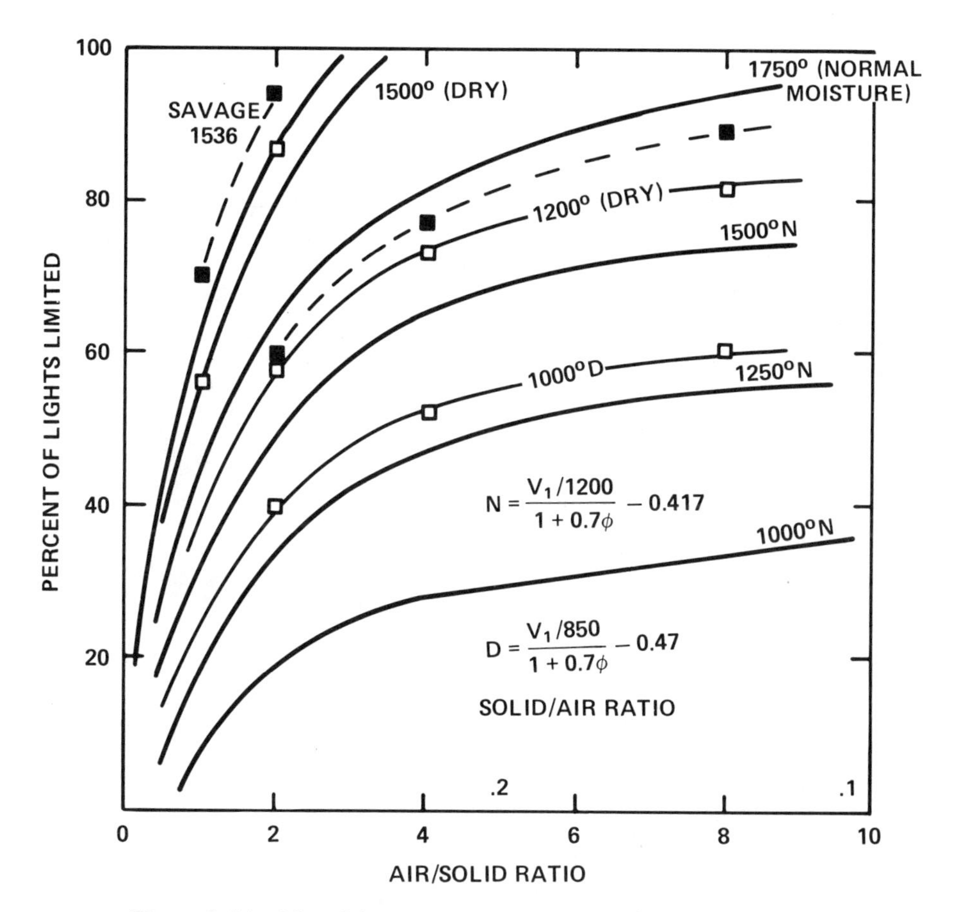

Figure 5-24. Material recovery as extract vs air/solids ratio [12].

The properties of the MSW fed to the classifiers varied widely. Ash content varied from 12.5 and 13.7% for Pompano Beach and Ames facilities (which had grit-removing screens installed prior to air classification) to 31.5 to 39.7% for the other systems without prescreening. Air-dry moisture content ranged from 11.2 to 15% for the screened materials and 14.4 to 27.1% for the others. The paper and plastics content of the feed ranged from 30 to 81%, ferrous metals from 2 to 9%, and characteristic particle size from 1.7 to 3.6 cm.

In an effort to discount the effect of variable feed properties, the performance was evaluated in terms of ratios of output to feed properties and quantities. As a result, many of the performance curves show clear and understandable trends. Tests were run at high, medium and low air flows in all cases. The medium was generally the normal operating point. Samples taken from stopped belts were used to measure flowrate, particle size, physical properties and heating value. Figures 5-24 to 5-29, constructed from reported test data depicting the results, can be discussed as follows:

Table 5-4. Air Classifiers Tested in EPA Field Test Program

				Critical Parameters				
No.	*Location of Air Classifier*	*Type*	*Design Throughput (mg/hr)*	*Column (mg/hr)*	*Throughput (mg/hr)*	*A/S*	*Velocity (ft/sec)*	F^a
1	Tacoma, WA	Horizontal	73	>46.0	>120	1.6	3,200	51
2	Baltimore City, MD	Vertical	91	9.8	36	3.5	1,250	44
3	Richmond, CA	Vertical	4	8.8	3.3	5.1	1,250	64
4	Ames, IA	Vertical	45	>18.6	> 21.0	<2.6	2,248	58
5	Los Angeles, CA	Vertical	4	8.8	3.2	6.6	1,722	112
6	Akron, OH	Inclusive	64	16.7	19.9	6.4	4,400	281
7	Pompano Beach, FL	Inclusive	6	11.8	4.4	5.6	2,180	122
8	Bridgeport, CT[b]	Vertical	50	20	45	2.3	2,200	55

[a]F = (critical velocity) × ASR.
[b]Tested by the author, not part of EPA test program.

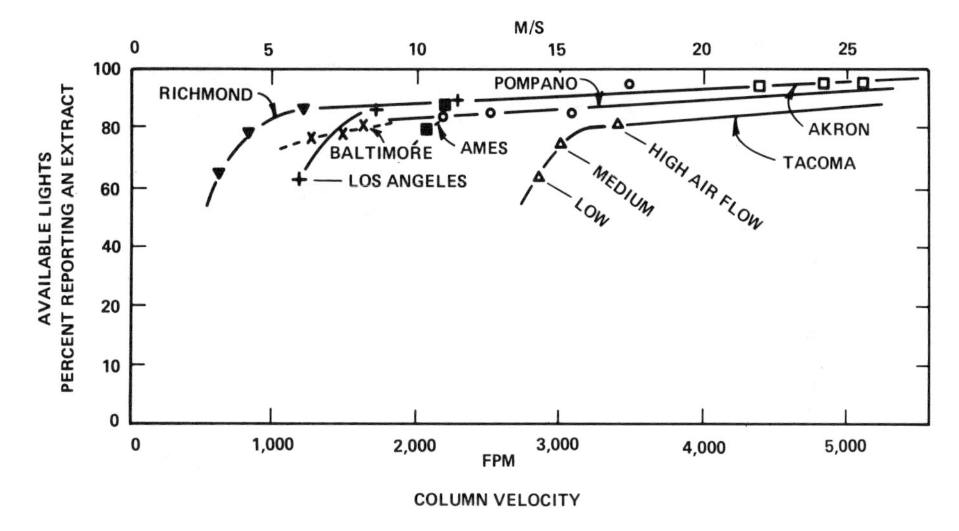

Figure 5-25. Fraction of available lights reporting as extract [6].

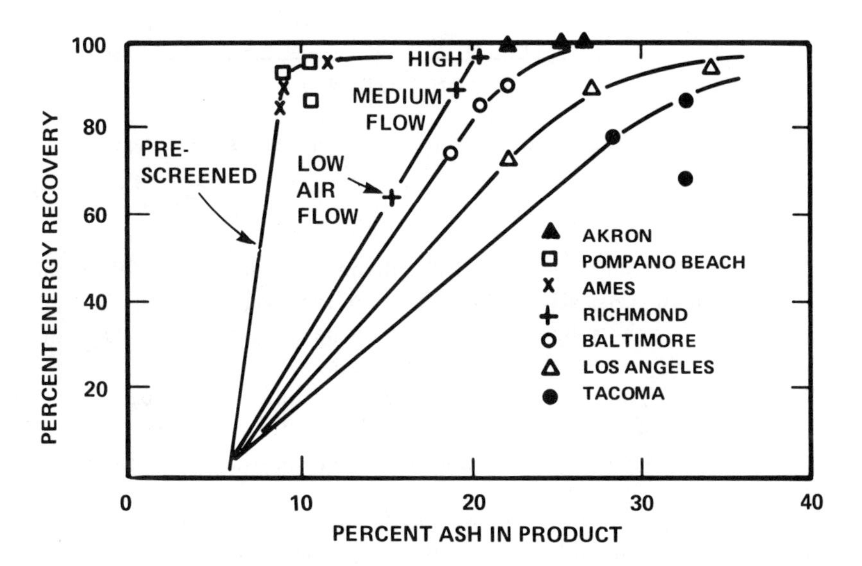

Figure 5-26. Energy recovery vs ash in the extract [6].

Figure 5-25 shows how the column velocity affected the split to the extract, as a percentage of the constant light fraction or of the available lights. The three test points in four cases show the "knee," which represents the region in which higher air velocities no longer influence the split. A velocity 1% below the constant velocity was called the critical velocity. If velocity x area had been used as the plotting parameter, the curves would have converged closer to a single line, making it harder to see the individual curves.

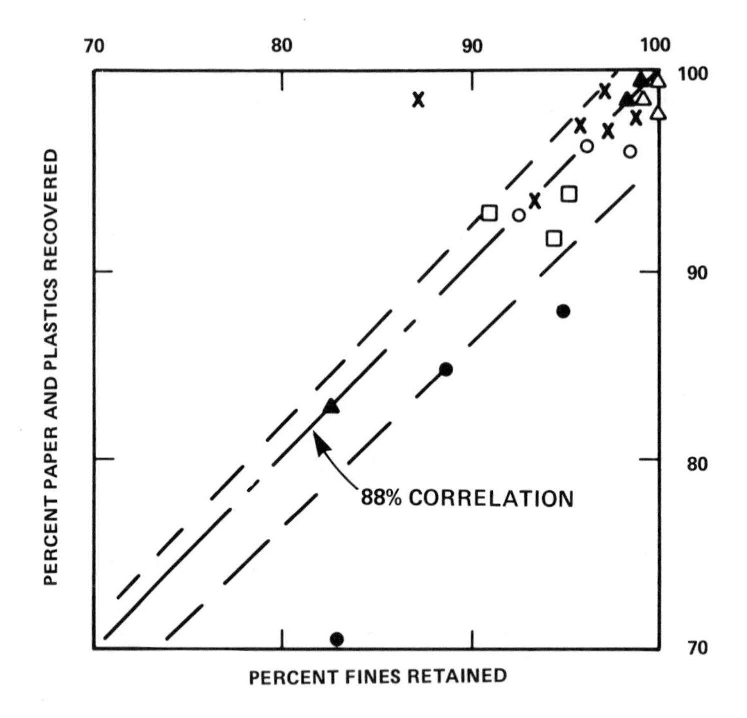

Figure 5-27. Recovery of paper and plastics vs percentage fines in extract [6].

There is a wide range in the critical column velocities, from 1000 ft/min (50 m/sec) for Richmond (vertical) to 3000 ft/min for Tacoma (horizontal). These velocities were measured at the column cross section or the gas discharge and have different significance, depending on the classifier configuration.

The momentum, or product of column velocity times air flow per unit of solids (factor F in Table 5-4) is nearly constant for four of the classifiers, indicating that momentum controls this critical condition. The others may have been assigned critical points far above the actual "knee."

The performance of the classifiers must be evaluated in terms of the fraction of lights lifted and the resultant quality of the light product. Quality could be defined by the paper fraction, but when a RDF is produced the ash content is the more critical parameter, measured versus either paper and plastics or energy recovery.

Figure 5-26 shows the energy recovery versus the ash in the product. Here, Ames and Pompano Beach plants show low ash due to precleaning of the feed. The other classifiers show similar trends but different ash levels, which can be attributed to both the nature of the feed and the classifier itself.

The higher air flow produced both higher energy recovery and more ash in the product. The curves "point" at 6% ash, which is the inherent ash of the light fraction. Additional ash is the "extraneous" ash, sand, glass and fines that could be separated

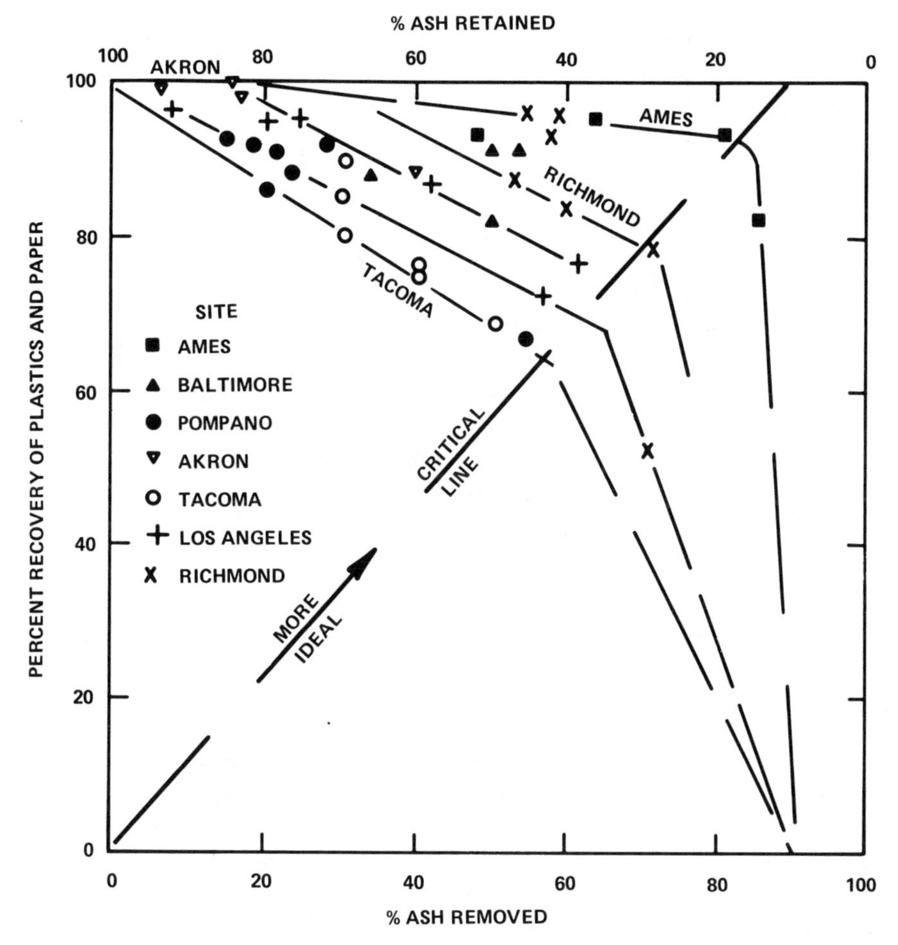

Figure 5-28. Recovery of paper and plastics vs ash retained [6].

but that flew with the light fraction. These had been removed in the Ames and Pompano Beach plants by prescreening.

In an effort to cancel out the effect of various feed compositions, recovery of paper and plastics can be plotted against the percentage of feed ash retained in the light product. Figure 5-27 shows trends similar to the energy/ash curves, but has a much clearer definition, resulting from the large number of data points. The sharpest classification is the Ames, which shows low ash retention at recovery rates as high as 90%. The inclined Akron unit was set for high recovery but poor ash separation, approaching that of the horizontal unit.

Plotting recovery of paper and plastics versus the percentage of ash removed shows not only the interdependence of recovery and ash, but also the classification effectiveness (Figure 5-28). The horizontal classifier shows a 60% recovery/removal efficiency, whereas Richmond showed 80% and Ames 90%.

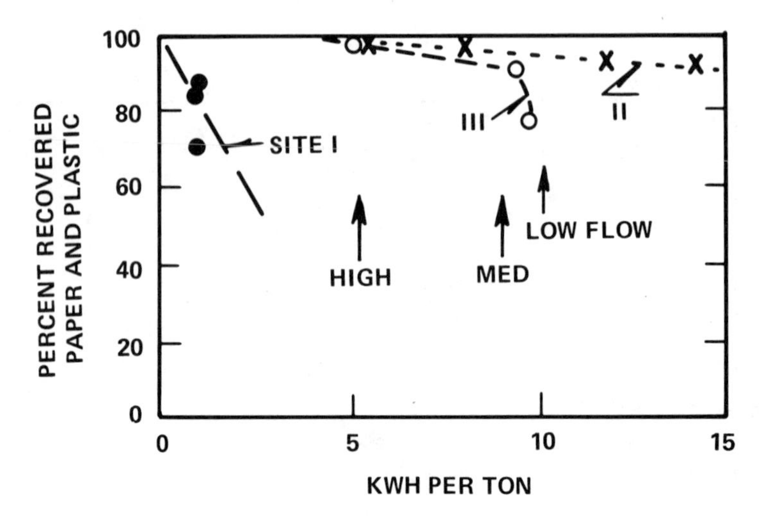

Figure 5-29. Recovery of paper and plastics vs energy per ton.

The various classifiers used substantially different amounts of energy while achieving fairly similar quality and recovery, as shown in Figure 5-29. For a facility with a given fan system, the lowest specific energy (kWh/ton) is achieved at the maximum throughput or feedrate. The quality appears to be a matter of adjusting the classifier to the best operating point, not the power usage.

The close relationship between the fines retained and the paper and plastics recovered is shown in Figure 5-27, where the points for all seven tests are correlated with an 88% coefficient of correlation. The ratio of (energy recovery)/(ash retained) versus percentage ash (Figure 5-30) does not correlate as well as recovery versus fines.

Specific energy, the power required by the classifier per unit of feed capacity, varied from about 1 kW for the horizontal classifier to about 6 kW/ton for the vertical units when they were loaded to their critical points. When loaded less heavily, they drew almost the same power; hence, the specific power increased. Figure 5-31 shows that neither the quality of the product at critical load nor the recovery varied much with the power input for the seven units tested.

In summary, the results of this comparison illustrate the factors and the wide range of consequences, but lead one to believe that the moisture, ash and inert fines in the feed are the major factors in the quality of the product, and that screening out dirt and fines results in a major improvement in RDF quality. Also notable is the large amount of energy required for vertical classifiers compared with the apparently less efficient horizontal unit.

There is no absolute means of comparing air classifier performance per se because other factors such as feed properties and the importance of product quality dominate. The broad picture of classifier performance is given in Table 5-5, showing high, low and typical ranges of performance.

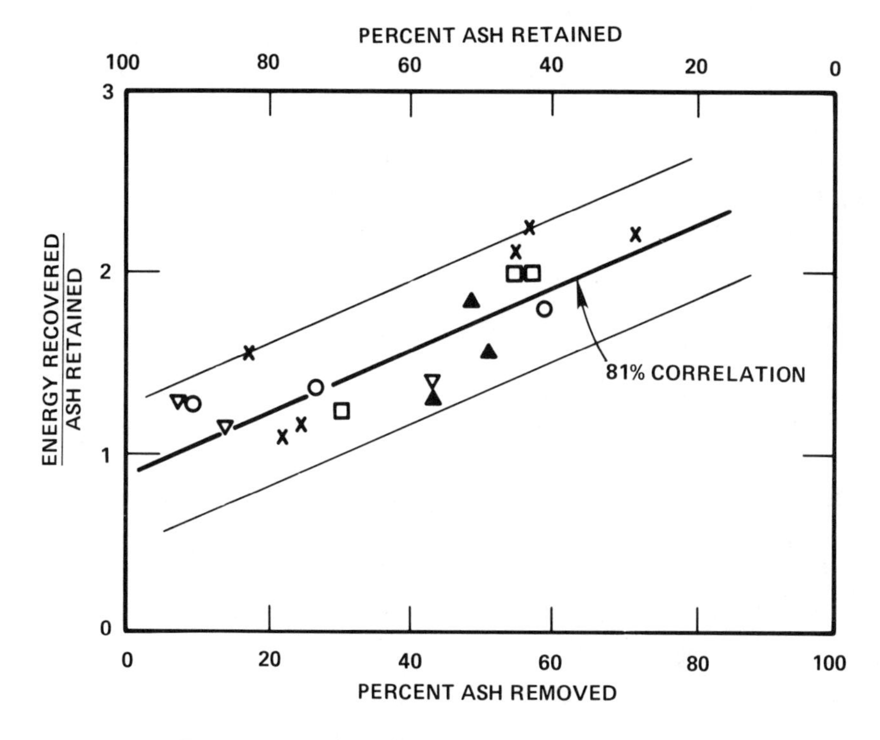

Figure 5-30. Energy recovery and ash content [6].

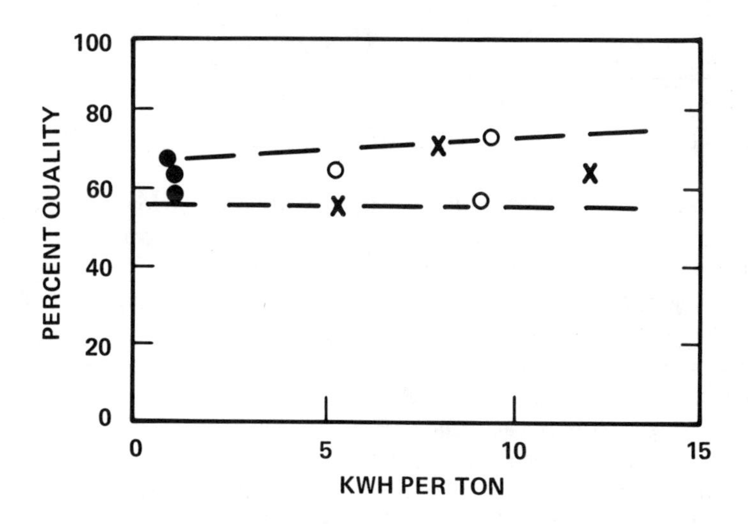

Figure 5-31. Specific energy of air classification [6].

Table 5-5. Summary of Operating and Performance Parameters of Seven Air Classifiers

	High	*Low*	*Typical Range*
Critical Air/Solids Ratio	8.5	< 1.5	2–7
Paper and Plastics in Heavies	42.8	0.8	5–30
Light Fraction Composition, %			
Ferrous metals	3.3	0.0	0.1–1.0
Nonferrous metals	1.5	0.0	0.2–1.0
Fines	37.5	4.0	15–30
Paper and plastic	87.8	22.9	55–80
Ash	34.3	5.8	10–35
Percentage of Component Retained in Light Fraction			
Ferrous metals	32.1	0.0	2–20
Nonferrous metals	96.7	0.0	45–65
Fines	100.0	68.4	80–99
Paper and plastic	99.9	69.5	85–99
Ash	85.2	29.3	45–85
Recovered Energy	99.9	65.2	73–99
Specific Energy, kWh/mg	15.0	0.9	1–11
Column Loading, mg/hr/m^2	>46.0	2.0	5–40
Recovered Paper and Plastics/ Retained Fines	0.9	1.1	1.0
Recovered Energy/Retained Ash	2.3	1.1	1.2–2.0
1980 $/(Mg/hr)	< 14,000	4,300	5,800–8,200

The following are some conclusions from this study:

- Best classification and product quality are achieved at low densities or high ASR. As this results in more capital investment, some compromise is to be expected to achieve good operation.
- Increasing the energy recovery rate generally results in higher product ash.
- Inert fines are not separated well by air classifiers because they have low lifting velocities similar to those of heavy or high moisture paper. They must be removed prior to air classification if a low-ash product is desired.
- High lifting velocities can compensate for low ASR but tend to carry more inerts and high-moisture materials.
- The ash content of the lights increases with moisture content because the inerts stick to damp paper, and higher velocities, needed for heavy paper, lift more inert fines.

- No correlation was found between specific energy consumption and capital cost with product quality. Simpler or better designed classifier systems may be more effective.
- The product of column velocity and ASR appears to be a constant for many classifiers, despite variations of these parameters.
- The single-stage horizontal classifier tested operated with very low energy at the expense of somewhat lower energy recovery than inclined or vertical classifiers. The vertical classifiers appear to give sharper separation but are more sensitive to operate and adjust to variations in feed.
- The quantity and velocity of air flow to an air classifier are the fundamental adjustments that can be made under normal operation, by changing fan flow and/or the flow cross section to minimize the amount of lights in the heavies or by varying the amount of combustibles in the heavies to maintain a desired product ash quality.
- The ability to vary the cross section of a classifier as well as air quantity is important to maintain classification effectiveness with varying feed moisture content and composition.
- It is more difficult to operate close to the critical point at high classifier efficiencies.
- To maintain classifier efficiency, it is necessary to keep the feed constant or to vary air flow and velocity.
- The test data reveal what the variety of air classifiers can do and under what conditions they operate best. With this information it is possible to determine, for a given unit, how well it is being operated, that is, how close it is to its point of best operation.
- The data show what can and cannot be expected from various types of classifiers, given the type of feed material, its moisture, ash and extraneous inerts content.
- The selection and use of a classifier must be evaluated to determine what is "good enough," how important quality is, what investment is justified to assure and maintain quality and what capacity can be achieved without sacrificing quality. When loss of combustibles to the heavy fraction is the price of a quality product, what is the cost of processing and/or disposal of the heavy fraction?

REFERENCES

1. Boettcher, R.A. "Air Classification for Reclamation of Solid Wastes," *Compost Sci.* 46 (November–December 1970).
2. Chrismon, R.L. "Air Classification in Resource Recovery," NCRR, Washington, DC (October 1979).
3. Finnell, R.N., and W.J. Larson. "Garbage In = Garbage Out? – Not Always," ASCE Preprint 3685, Atlanta, GA (1979).
4. Dean, K.C., C.J. Chindgren and L. Peterson. "Preliminary Separation of Metals and Non-metals from Urban Refuse," Tech. Progress Report 34, US BuMines, Salt Lake City, UT (June 1971).

5. Trezek, G.J., et al. "Significance of Size Reduction in Solid Waste Management," EPA-600/2-77-131 (July 1977).
6. Savage, G.M., G.F. Diaz and G.J. Trezek. "Comparative Study of Seven Air Classifiers Utilized in Resource Recovery Processing," in *Proc. Seventh Annual Research Symp.*, EPA/MERL, Philadelphia, PA (March 1981), pp. 67-84.
7. Pierce, J.J., and P.A. Vesilind. "Fundamental Aspects of Air Classification: Final Report," DOE, NTIS (1981).
8. Senden, M.M.G., and Tels.
9. Kaiser, F. "Der Zick Zack Sichter—ein Windsichter nach neuen Prinzip," *Chem-Ing-Tech.* 35 (4):273-282 (1963).
10. Senden, M.M.G. "Performance of Zigzag Air Classifiers at Low Particle Concentrations," paper presented at the 1980 NWPC, ASME, New York, 1980.
11. McNabb, M.B. "Fundamental Aspects of Air Classifier Performance," MS Thesis, Duke University, Durham, NC (1980).
12. Lawler, S.P. "A Statistical Analysis of the Chemung County Americology Air Classifier, on Tests Made October to November, 1975," American Can Company, Greenwich, CT, unpublished results.

CHAPTER 6

Size Separation

SCREENING

Screening devices are used to separate mixed materials containing particles of different sizes and types into groups or grades as the result of passing or not passing apertures of selected sizes.

Screens sort materials in accordance with their size: as the material passes along a surface containing holes, the particles having a controlled dimension less than that hole size may pass through, thus dividing the feed into overs and unders, or accepts or rejects.

The feed may be a single type of material, which is to be divided into two or more size "cuts," or it may consist of several materials that can be separated from each other by virtue of their size differences. For instance, referring to Figure 1-5, a screen with 5-inch holes theoretically would separate out almost all ferrous, nonferrous, glass and inerts, as well as most of the plastics, rubber, wood, garbage and yard waste, while reclaiming all the paper and cardboard.

NOMENCLATURE

The screen through which the material has passed may be called a limiting screen. The material retained on the screen is called the oversize, or plus (>), material of that screen, and that passing is the undersize, or minus (<). The terms overs and unders are convenient abbreviations for oversize and undersize. The materials that have passed through the screen holes are called thrus. Screen performance can be evaluated on the basis of the parameters described in the following five sections.

Yield

Yield (Y) can be defined as the amount or, alternatively, the percentage of the feed that passes through the screen, hence is the same as the split. It may refer to the total flow, or to a particular species, such as the glass, ash, combustible material or moisture.

Capacity of a Screen

Capacity must be specified: it may be the feed, the thrus or the overs, and may be either weight or volume flowrate. Test curves, showing the weight passing through the screen, plotted against time, are "thrus" capacity curves.

Screen Efficiency

The efficiency or effectiveness (E) of a screen can be defined in many ways, all of which compare in some way that fraction of the possible that has been achieved, something that varies with the objective.

A number of methods are used to express screening efficiency, each suited to a specific purpose. For instance, some efficiencies are concerned with the quality of the product, whereas others concentrate on the characteristics of the screen itself, independent of the material being processed.

When the quality of the product is our concern, we may be interested in the amount of undersize in the oversize product, for removal of undesirable fines, such as glass and sand removal from larger material; or the removal of oversize material from an under product, when removing metals, combustibles and stones from fine glass. Taggert [1] lists the following efficiencies used in the mineral dressing industry, each having different application and significance:

For undersize recovery:

$$\text{Recovery efficiency, } E_1 = \frac{\text{weight of undersize obtained}}{\text{weight of undersize in feed}}$$

If samples of the feed and oversize are available, this recovery formula can be used:

$$\text{Recovery efficiency, } E_2 = \frac{100 < 100\,(e - v)}{e(100 - v)}$$

where e = undersize in the feed (%)
v = undersize in the oversize (%)

On this basis, efficiencies on well-operated vibrating screens are reported to be 80-90% with 4 mesh and larger, and 65-70% at 10 mesh with dry materials; however, with moisture, they may fall to below 60%, according to Taggert.

When the efficiency of the screen itself is of interest, the Warner company suggested that the following expression should be used, taking into account the number of difficult grains present:

$$\text{Recovery efficiency, } E_3 = \frac{U(F - O)}{F(U - O)}$$

where F = difficult grains in the feed (%)
U = difficult grains in the undersize (%)
O = difficult grains in the oversize (%)

Difficult grains are defined as those held on the next finer Tyler standard testing sieve whose aperture is less than 83% of the aperture of the screen under investigation.

According to Taggert [1], Siden offered another expression, defining difficult materials as 0.75 to 1.5 times screen aperture:

$$\text{Recovery efficiency, } E_4 = 100\,\frac{(U-(ZC))}{U}$$

where C = difficult grains in the feed (%)
U = true undersize in the feed (%)
Z = fraction of true undersize based on DG remaining in the oversize

$$\text{Percentage removal} = \frac{\%\text{ true oversize in the feed}}{\%\text{ feed going into actual oversize product}}$$

Removal efficiency, for removal of undesirable fines, is expressed as

$$E_5 = \frac{100 \times (O)}{(O + 2C)}$$

where O = percentage of true oversize in the feed

The screen efficiency may be expressed this way to consider difficult grains only:

Removal efficiency:

$$E_6 = \frac{100(C - ZC)}{(C)} = 100\,(1 - Z)$$

where Z = fraction of true undersize, based on the difficult grains only remaining in the oversize product

Tkaing into account undersize in the feed, overs and unders, the definition commonly used for flat-deck screens is

$$E_7 = (100 - b) - \frac{(a - b)}{(c - b)}\,(100 - b - c)$$

where E = efficiency (%)
a = undersize in feed (%)
b = undersize in oversize (%)
c = undersize in undersize (%)

In its simplest terms, screen efficiency is the fraction or percentage of the material in the feed that passed, divided by the fraction that COULD have passed, through the screen. The important concept is "the fraction of the possible." To evaluate this, it is necessary to measure the possible in some way. In the case of screening, the amount of material that can possibly be passed can be measured with the actual machine by running it "forever," or for a reasonably long time, which we may call the ultimate time.

Another way is to take samples of the feed and product of the machine material and screen them on the same mesh screen in a laboratory for a sufficiently long time to find how much fines remain in the product, as compared with the fines in the feed.

To be certain that the screening has been done for a sufficient time, the weight passed versus time may be plotted as shown in Figure 6-1. The use of semi-log paper or doubled time coordinates makes it easier to see that the screening has continued well into the depletion range (see Figure 6-5).

In this text, unless otherwise noted, screen efficiency will be defined as

$$\text{Screen efficiency} = E = \frac{\text{actual flow}}{\text{possible flow}} = \frac{\text{undersize in thrus}}{\text{undersize in feed}}$$

Specific Capacity

The amount passed through the screen per unit time divided by the screen area (usually the open area, but sometimes the total or nominal screen area) can be called the specific capacity, as TPH/ft^2 or TPH/m^2 of open area. For a given screen, the higher the flowrate, the lower the screen efficiency. Hence, the higher the specific capacity, the lower the screen efficiency.

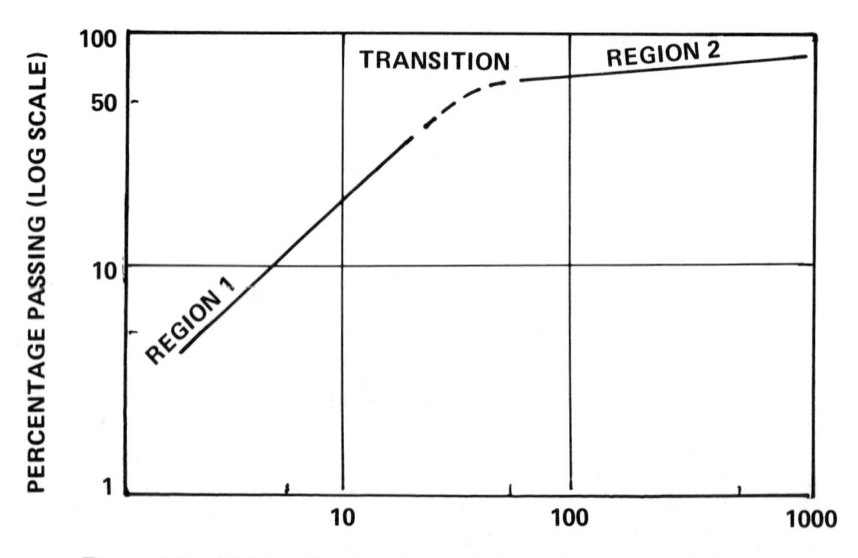

Figure 6-1. Weight of material passed through a screen with time.

Specific Area

Another term that is convenient to use is the area required to pass a ton of material per unit time. This is the reciprocal of the specific capacity and can be called the specific area, expressed as ft^2/TPH or m^2/TPH. The higher the specific area, the higher the screen efficiency.

TYPES OF SCREENS

The major types of screening devices, shown in Figure 6-2, have these basic characteristics:

- **Flat screens**: Vibrating, shaking, inclined, air assist. Good for dry granular, not moist, linty or flaky materials. Tend to blind as material fills the holes.
- **Disc Screens**: Good for scalping off oversize materials and dropping out fines and removing grit from flaky and moist materials.
- **Rotary (trommel) screens**: Good for flaky and moist materials, due to tumbling action, and for materials that tend to blind holes, due to reversal of gravity forces.

SCREENS USED IN RESOURCE RECOVERY

These are some of the main applications of screening devices in processing refuse:

- Separate (scalp) oversize material from the main stream, possibly for further size reduction.
- Separate undersize material from the main stream to avoid shredding material that is already small in size.

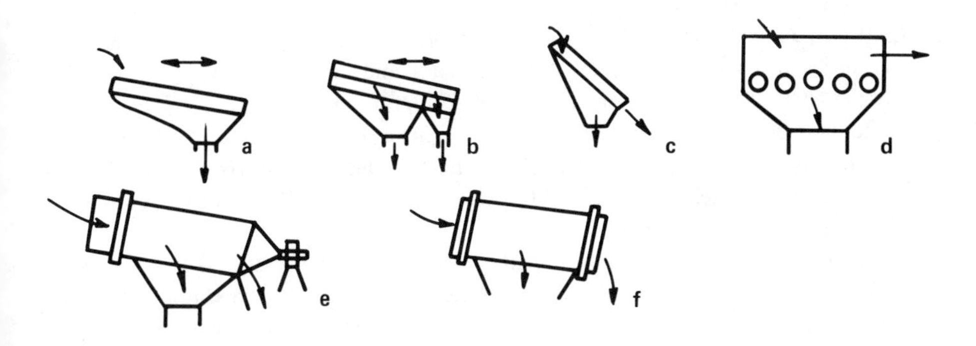

Figure 6-2. Screen types: (a) vibrating flat screen; (b) vibrating double flat screen; (c) inclined flat screen; (d) disc screen; (e and f) rotary or trommel screens.

- Divide a stream into light combustibles and heavy, moist and incombustible materials.
- Recover paper, plastic and textile fraction as overs, while concentrating glass and metals in the thrus.
- Separate glass, grit and sand from combustible materials.
- Remove can stock from incinerator residue.

Figure 6-3 shows several systems used for processing waste-to-energy and for recovering by-products, showing the locations where screening operations have been effective, such as the following:

- A primary trommel can be used to remove glass from MSW before it is shredded and to remove small-sized particles that do not need further shredding.
- A scalping screen can be used to remove oversize materials from the shredder discharge, such as rags and other objects that may be troublesome downstream.
- A screen can be used to process the primary trommel unders to remove glass and inerts from the combustibles and to concentrate metals in the middling stream.
- Before feeding shredded material to the air classifier, the glass fines should be removed because they will fly in the classifier, increasing the fuel ash content.
- A screen may be used to remove the glassy fines in the air classifier lights.
- The screens may have more than one hole size to remove several types of contaminants or products while passing through a single unit.

HISTORY OF SCREENING IN THE RESOURCE RECOVERY INDUSTRY

As early in 1924, efforts were made in Britain to sort MSW by trommel screening. In a revival of interest, new studies were initiated at the U.S. Bureau of Mines by Sullivan et al. [2] in 1973; at Warren Springs, U.K. by Douglas and Birch [3, 4]; by Woodruff and Boles [5]; by Savage and Trezek [6] at the University of California, Richmond Laboratory; and recently by Glaub et al. [7] at Cal Recovery. Extensive research on a 10-foot-diameter laboratory trommel screen has been done by Menon et al. [8] at CEA R&D in Minneapolis, and production trommels have been tested by Hasselriis at East Bridgewater, Massachusetts and Bridgeport, Connecticut [9] and by the NCRR in New Orleans, LA, and Washington, DC [10, 11].

Rotary trommel screens have been put into operation in resource recovery plants such as those in Ames, Iowa (1973), East Bridgewater, Massachusetts (1975), Recovery I, New Orleans (1977), Cockeysville, Maryland, Bridgeport, Connecticut and Monroe County, New York (1979) [12, 13].

In the Bureau of Mines pilot plant, a trommel screen was used to remove the glass fraction from air-classified heavies using 3/4-inch holes. This application was demonstrated full scale at East Bridgewater and Cockeysville. At Ames, Iowa, a trommel was installed to process the air classifier heavies into several size fractions, separating glass and nonferrous metals, as was done at San Diego in the Occidental Research Company pyrolysis process [14].

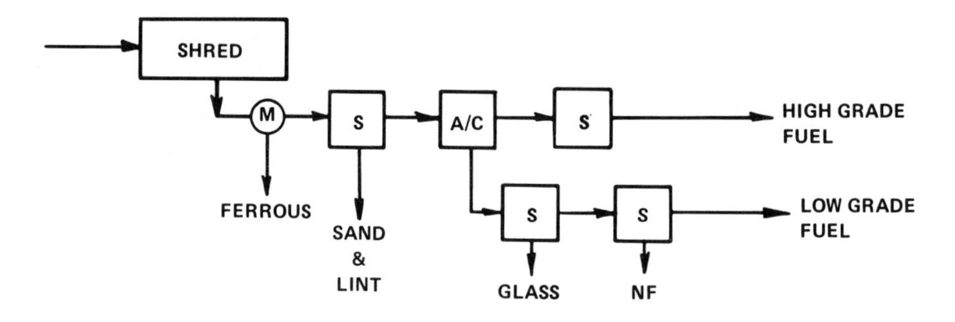

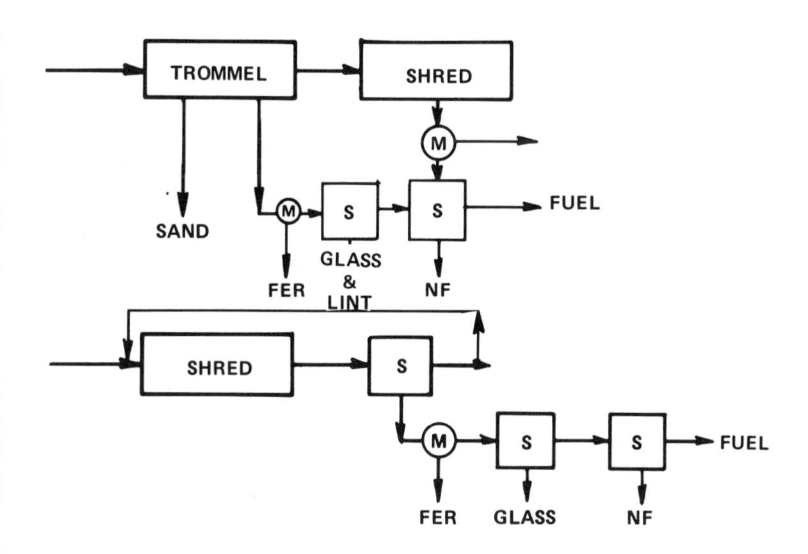

Figure 6-3. Application of screens in resource recovery.

Recovery I and Bridgeport were the first commercial applications using the trommel to process raw MSW, achieving effective glass separation prior to shredding operations, reducing glass in the fuel fraction, power consumption and wear on the shredder.

- **Bridgeport, Connecticut**: Primary trommels are fed with unprocessed MSW. Oversize material is fed to a hammermill; after shredding, both streams go to an air classifier and then to a secondary trommel to remove grit. The other primary trommel feeds oversize to a flail mill, which discharges back to the feed: only undersize material is processed.
- **Cockeysville, Maryland (Maryland Environmental Services)**: Processes shred the air-classified light fraction to remove minus-32 mm (minus-1-1/4 in.) grit and organics.

Table 6-1. Dimensions and Capacities of Various Trommels [12]

No.	*Diameter × Length (ft)*	*Hole Size(s) (in.)*	*Feed Capacity (TPH)*	*Power (hp)*	*Speed (rpm)*
1	11 × 40	4-3/4, 2-1/2, 3/4, 1/4	75	40	11
2	12 × 28	1-1/4	45	20	9
3	12 × 38	2-1/2	100	120	10.5
4	10 × 51	2-1/4	65	60	10
5	10 × 45	4-3/4	62.5	80	11
6	10 × 26	1-9/16 8-1/4 × 10	10	6	9-30
7	10 × 21	Various	5	–	0-20
8	9 × 30	4-3/4	25	17	14
8	8 × 20	5/8	25	25	14
9	9 × 30	3/4	40	20	11
10	8 × 44	3/8 × 2, 1 × 4, 8 × 10, 6	33	8	8
11	8 × 35	2-3/4, 1	50	50	–
12	7 × 20	Various, 1/2	45	2	
13	7 × 20	4, 2	50	10	13
14	7 × 15	7/8	10	12	28
15	6 × 24	3/4	20	15	11
16	5 × 20	4, 1-1/4, 5/8	5	15	25
17	4 × 18	3/4 or 3	15	15	20
18	4 × 8	1/2	4.5	7.5	8-16
19	4 × 12	5.2	7.5	7.5	22
20	4 × 11	3/4	2.5	7.5	10-30
21	4 × 8	3, 3/4	15	15	1-20
22	3 × 16	3/8, 1/2	10	2	12-36
22	3 × 10	4, 2	4	1.5	4-43

- **Hempstead, New York (Hempstead Resource Recovery Corp. and Dade County, Florida)**: Primary trommel diverts glass and other small particles to a nonattrition pulper to reduce abrasion in main pulpers and improve RDF quality.
- **Houston, Texas (Reynolds Metals/Browning Ferris Industries)**: Trommel separates raw MSW into three streams: overs, middlings (aluminum can-rich) and unders.
- **New Orleans, Louisiana (Waste Management, Inc.)**: Primary trommel separates MSW into oversize (landfilled) and undersize, from which metals and glass can be recovered.
- **Doncaster and Byker, United Kingdom**: Flailed MSW is trommeled into three fractions: (1) oversize goes to paper recovery; (2) middlings are air classified, shredded, dried and made into fuel pellets; and (3) undersize goes to glass recovery.
- **Stevenage, United Kingdom (Warren Springs Laboratory)**: Experimental trommel used to separate products from MSW.

- **East Bridgewater, Massachusetts (CEA)**: MSW flail-milled before trommel; oversize returns to mill; undersize air classified and lights screened to remove grit in secondary trommel.
- **Rochester, New York (Monroe County)**: Heavy fraction from rotary air classifier is trommel screened to separate out concentrated glass and noncombustibles.
- **Byker, United Kingdom**: Described above.
- **Madison, Wisconsin (City of Madison)**: Trommel processes flailed MSW after ferrous metal removal; undersize and middling material rejected (potential fuel); oversize shredded to RDF.
- **Los Gatos, California (Vista Chemical)**: Shredded MSW degritted.
- **Milwaukee, Wisconsin (Americology)**: Shredded, air-classified heavies are separated into glass and aluminum-rich concentrates for materials recovery.
- **Eastborne, United Kingdom**: Shredded MSW trommeled to remove grit; oversize air classified, light processed to d-RDF for cofiring with coal.
- **Toronto, Ontario, Canada (Ontario Ministry of the Environment)**: Removal of inorganic fines to beneficiate compost product.
- **Ames, Iowa (City of Ames)**: Four-way separation of air-classified heavy fraction to remove glass, nonferrous metal and grit.
- **Pompano Beach, Florida (Waste Management)**: Three size separations; middling is air classified; lights are mixed with sewage sludge in digester to form methane gas fuel.
- **Jacksonville, Florida (U.S. Navy)**: Flailed MSW is trommeled to remove inorganics, glass and grit; oversize burned in modular incinerator or used to make d-RDF.
- **Salem, Virginia (City of Salem)**: Trommel unders are used for aluminum can recovery and beneficiation of MSW prior to incineration.
- **Avondale, Maryland (U.S. Bureau of Mines)**: Air classifier heavies are screened to separate glass and metal-rich material from combustible fraction.
- **Richmond, California (Cal Recovery)**: Experimental trommel used to remove grit from shredded refuse.
- **Upper Marlboro, Maryland (NCRR)**: Research trommel.

Disc screens have been installed at the City of Ames facility [15], Americology [16], Dade County and Lakeland, Florida, and Akron, Ohio.

PRINCIPLES OF SIZE SEPARATION BY SCREENING

The basic function of screening is to pass undersize particles through the holes or apertures of a screen and reject oversize particles. The maximum rate of flow of undersize materials through screens is limited by access to the screening surface, approach velocity, rate of acceleration and friction.

In processing applications, at economic capacity the large number of particles present interfere with each other. Those that are too large to pass through have a motion parallel with the screen, tending to prevent undersize particles from passing through it. This is particularly true of flaky particles.

When a sample of material is placed on a screen and some degree of agitation or shaking is applied, undersize material will flow through the screen. The nature of this flow is complex, depending on the size distribution (PSD) of the material, its shape and density, the height of the material bed on the screen, and the nature and degree of the agitation. As the size of the particles approaches the screen dimensions, interference increases. The rate of flow will change as the amount of undersize material above the screen decreases and oversize material becomes more concentrated. There are three phases discernible in graphs of flowrate: constant flow, depletion phase and probability phase.

Constant flow occurs when there is an ample supply of fine material above the screen and the screen itself is limiting the flow. The product under the screen increases linearly with time under these circumstances.

In the depletion phase, the flowrate declines as more and more fine material is screened out and its concentration becomes significantly reduced. The rate of flow becomes dependent not on the screen, but on the concentration of fines above it, causing a logarithmic decrease in flow and in the accumulation of product under the screen.

In the probability phase, the material that is near to screen size begins to accumulate. As it has a lesser chance of passing through the screen and interferes with the flow of finer material, the flow decreases in accordance with probability.

Plotting screening test data reveals the nature of this flow. Figure 6-4 shows test data obtained from screening a sample of minus-4-mesh shredded wood particles through an 8-mesh rotary screen and weighing the discharged material after each two turns of the drum. These data reveal some of the screening characteristics of relatively homogeneous materials having a distribution of particle sizes. The weight of the material accumulated under the screen is plotted versus the number of turns in this case, which is equivalent to time and also to the area of screen surface that has passed. Curve A shows the performance of the original mixture, and curve B shows the behavior of the undersize material from run A. It is evident that the undersize material flowed through much more readily.

Also plotted are the rates of flow versus number of turns of the screen. The feed material, A, has a fairly constant flowrate for 10 turns, after which it falls off due to depletion. The flow of undersize material, B, was much more rapid than that of the feed material, increasing during the early stage, then falling off logarithmically as the undersize material in the overscreen fraction was depleted.

The probability phase of screening is more easily discerned by plotting the data on semi-log paper, as shown in Figure 6-5. The depletion phase plots as a straight line, characteristic of flow, with a declining potential force. The data fall away from this line after more than 90% of the undersize material has passed through the screen, following a new line which, as we shall see later, is a straight line on probability paper. Curve "C" shows a lengthy delay before starting the declining period, indicating that this sample had a large quantity of undersize material in the feed, the flow of which was delayed for various reasons, such as by the time required to break bags and other containers, or to break bottles.

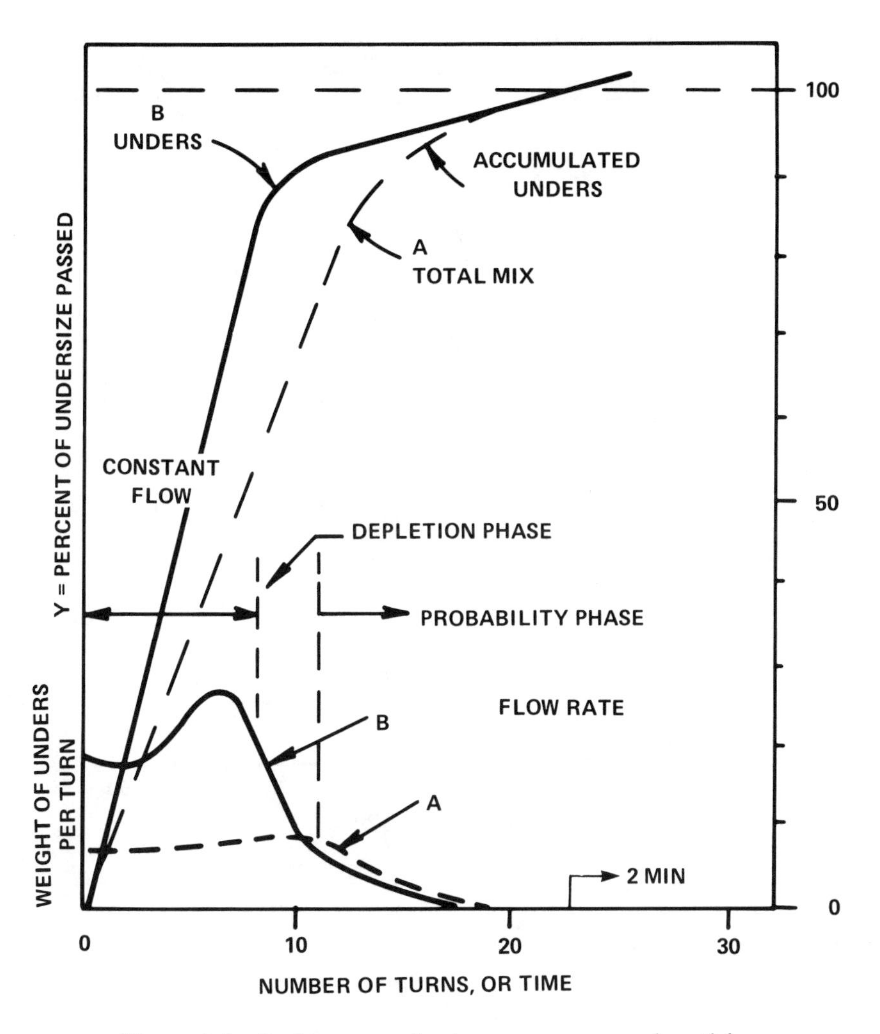

Figure 6-4. Performance of rotary screen on wood particles.

Figure 6-6 shows the percentage of undersize material that has passed through the screen (the screen efficiency) and the corresponding flowrates of glass and combustibles versus screening time, when screening raw MSW in a laboratory trommel. The characteristics of the flow are similar to those of the wood particle data plotted in Figure 6-4.

The actual data points for these tests are plotted in Figure 6-7 on semi-log paper to show how closely straight lines describe these data, and the remarkable similarity of slope in the depletion phase prior to the probability phase (not shown).

In the constant flow range, the relationship between accumulated underflow can be approximated by a simple arithmetic (linear) relationship with time:

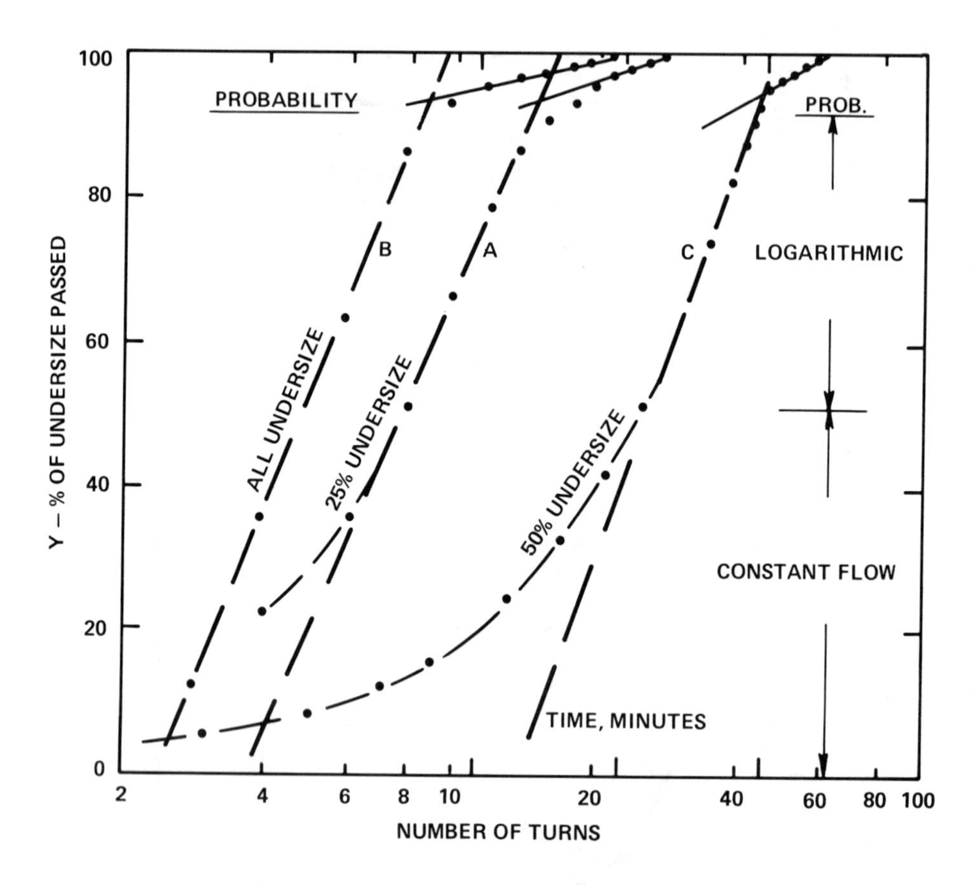

Figure 6-5. Curves from Figure 6-4 as a semi-log plot.

Percentage of undersize passed = screen efficiency = E

$$E = B \times t \tag{6-1}$$

where t = time
B = the flow rate constant

In the decreasing flow range, the following logarithmic expression fits the data well:

$$E = A + B \times \ln t \tag{6-2}$$

where E = passible feed passed (screen efficiency) (%)
t = time to pass E
$A = E - B \times \ln t$
$B = \dfrac{(E_2 - E_1)}{\ln (t_2/t_1)}$

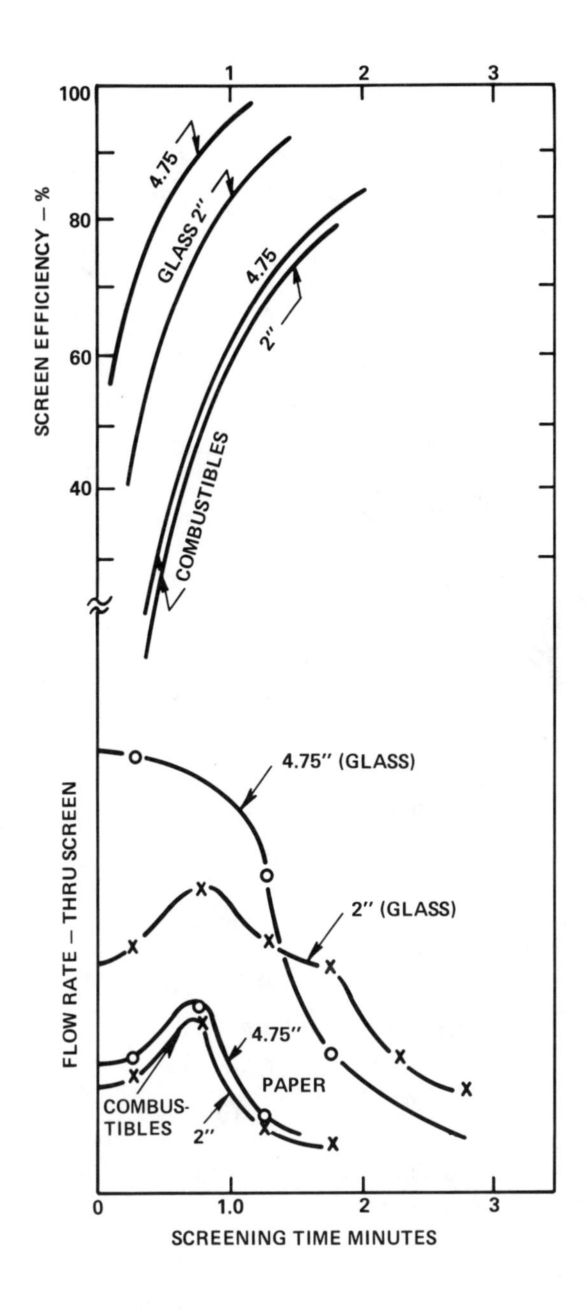

Figure 6-6. Screening time, flowrate and efficiency of a trommel screen.

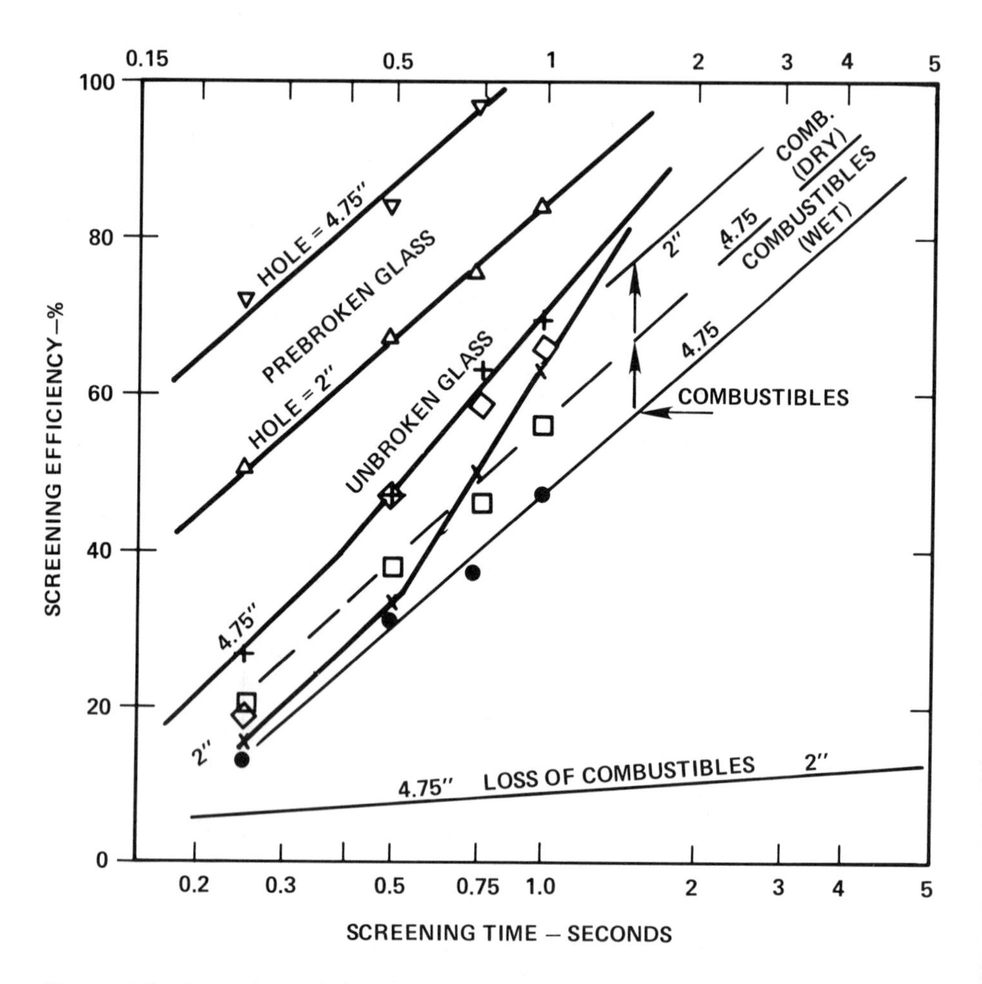

Figure 6-7. Screening efficiency and loss of combustibles for 4.75-inch and 2-inch hole trommels, against logarithmic time scale.

The virtual intercept of this line with E = 100% is

$$T_{(100)} = \exp\left(\frac{(100 - A)}{(B)}\right)$$

The constants A and B can be determined from two data points in the 50 to 80% range of the curve. A larger number of data points can be used to determine the goodness of fit by regression analysis, provided no data from the constant flow and probability phases are included.

Analysis of large amounts of trommel and screen data has shown that, for many materials found in MSW (and even properties of materials), this equation is the best fit for E from 50–80% passed.

Using the constants A and B and the value of $T_{(100)}$ thus determined, we can determine the relative amount of time, the number of turns or shakes of the screen, or the screen area required to achieve desired values of E, up to the point where the slope changes.

Effect of Particle Size Distribution on Screening

The PSD of the material to be screened is a determining factor in the selection of screen hole size and the area of screen needed to achieve a given screening efficiency and capacity.

To anticipate how a given material will be separated in a screening operation, it is first necessary to characterize the material by determining the PSD, which is done by screening the material.

Particle size analysis may be done by placing the sample on top of a succession of screens and providing manual or mechanical agitation to provide the material with ample opportunities to pass through the screen apertures. When screening has been completed, each screen in the series will retain all the particles within the sample that lie between the various screen sizes, plus the pan at the bottom.

The first time this is done with a new material, it is advisable to pass the material through each screen manually so that nearly 100% screening efficiency is achieved in each stage. The amount of material that can be placed on the screen at one time is critical: not more than two-thirds of the screen area should be covered by the sample after it has been leveled out by shaking. If the screen is loaded too heavily, the material will not be able to flip over and allow the fines to reach the screen. The screen must be shaken in such a manner that the material is flipped over a sufficient number of times to assure passage of fines.

When using a mechanical screen, the amount of material that can be fed to the machine in a single batch operation is limited by the quantity of material that accumulates on the most heavily loaded screen, so that this screen will not become overloaded, lose efficiency and produce false data.

Figure 6-8 shows typical PSDs for raw MSW plotted on Rosin-Rammler graph paper, which shows size distribution for many materials as a straight line. Also shown is a histogram showing the percentage of total material that would be retained by each of a series of screens having holes one-half the size of the next larger screen. The ASTM screen series contains square root and fourth root screen series to offer wide flexibility in choosing a set of screens and to avoid overloading individual screens. To obtain a complete picture of a material, especially a mixture, the series must be broad enough to collect the entire sample, with not more than 10% retained on the top screen or passing through the smallest screen to the pan. For many materials, a series of six screens and the bottom pan will retain the entire sample.

Consider the RDF: the mean, or "characteristic," particle size is at the 63% passing point, about 20 mm. Between the 16- and 32-mm screen, 35% of the material has been retained. This screen would be the limiting screen for capacity. ASTM standards (see Appendix) call for inserting another screen in the series when one retains more than 25% of the sample to assure consistent accuracy in the measurements.

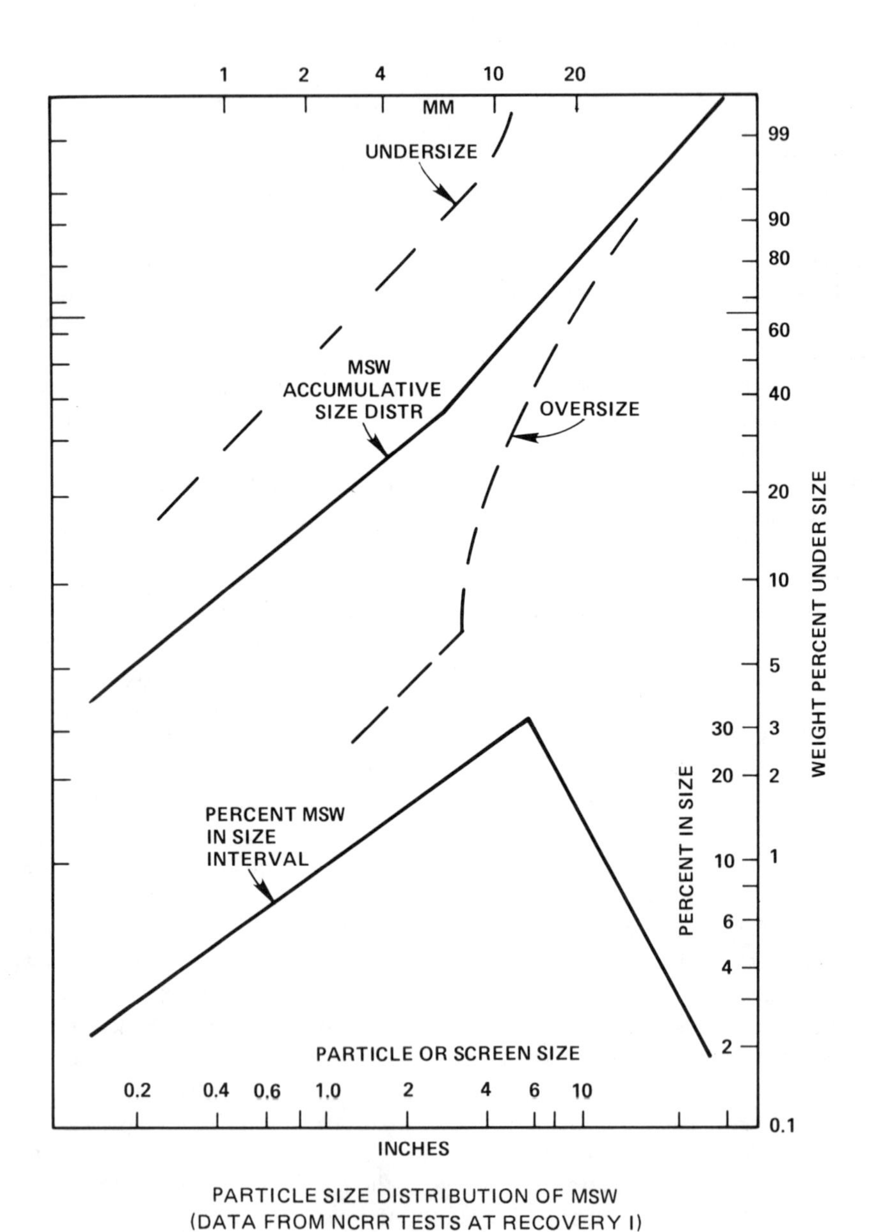

Figure 6-8. MSW particle size distribution plotted on Rosin-Rammler graph paper.

Influence of Half Size and Oversize Material on Capacity

The efficiency and capacity of a screen depends on the quantity of material above the screen and the amount that is near size or half size, relative to the screen size. Figure 6-9 shows the test performance of a screen with zero, 25% and 50% oversize material over the screen. The three curves show the same slope in the declining range, but the 25% and 50% curves exhibit substantial time lags before this slope is established, reflecting screen-limited discharge rates. Phenomena such as this are not unusual with materials with mixed sizes. Only actual tests will show the true screening characteristics.

A large amount of oversize material on the screen tends to block the screen and reduce its effective area. Substantial amounts of particles near to screen size, that is "difficult grains," and the fraction of particles that are "half size," that is half the size of the screen aperture, influence the screen capacity significantly. As these particles have a poor chance of getting through the holes, they also tend to block the fine undersize material from passing. The mechanical action of the screen can minimize screen blocking by mixing the material and assisting the fines in reaching the screen.

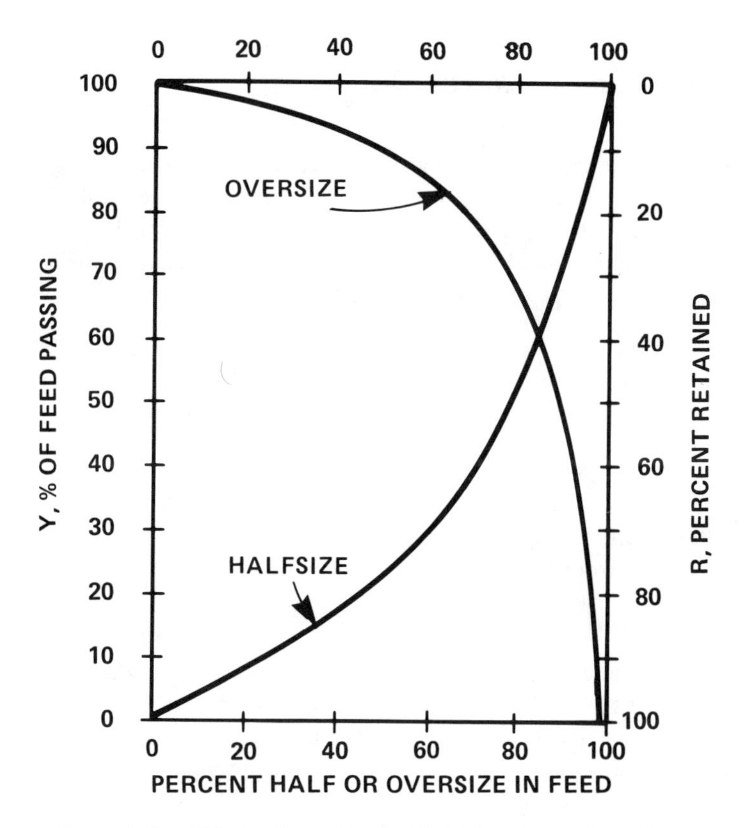

Figure 6-9. Effect of oversize and half-size material on flow through a screen.

The influences of these types of particles can best be understood by plotting curves of rate of flow through the screen. These can be obtained by integrating the curve of screen yield or efficiency versus time, as was done to obtain the typical curves in Figure 6-6. With a small amount of oversize material, the flowrate reduces at a relatively steady rate as the undersize material above the screen becomes depleted.

Figure 6-9, derived from screen sizing data cited by Taggert [1], shows how the percentage of the feed that is oversize or half size influences the capacity of a screen as compared with its nominal capacity without oversize. This kind of empirical data is useful in estimating screen capacity and especially in extrapolating from known data to other conditions.

To use this graph, screen the sample to obtain the undersize fraction. Next, run that fraction through the screen to determine the flowrate for this material, without oversize interference, for the retention time characteristic of the actual machine. This flowrate, expressed as TPH/ft^2, is then corrected by the factors shown in Figure 6-9, based on measurements of the oversize and half size percentages obtained by screening the entire sample.

Empirical Estimates of Screen Capacity

Analysis of data presented by Taggert giving screen capacities as a function of screen aperture size for gravel, stone and coal, for 95% screen efficiency and 25% oversize in the feed, reveals that the capacity of the screen is proportional to the two-third power of the aperture, and directly to the density of the material being handled. On this basis, the following approximation was derived:

$$\text{Nominal screen capacity, } Q_{nom} = \frac{d \times (D)^{2/3}}{K} \qquad (6\text{-}3)$$

where Q_{nom} = TPH/ft^2 of open area
d = bulk density (lb/ft^3)
D = aperture width or diameter (in.)
K = a constant

While this equation agrees reasonably well with some test data when corrected for oversize and half size, it should not be used without confirmation using the specific material and machine. It can be used in extrapolating test data, in which case the constant would be determined from the data, using appropriate densities, hole sizes, and oversize and half size corrections.

APPLICATION OF BASIC PRINCIPLES TO SCREEN DESIGN AND OPERATION

So far we have considered particle behavior, capacity and screen efficiency and, in effect, tried to decide the size of screen necessary to handle a given capacity for given material and screening size. Now let us consider the performance of screens under different operating conditions.

Flat-Deck Screens

Flat-deck screens come in many shapes and configurations, including gyrating (with either horizontal or both horizontal and vertical motion), vibrating or humming, shallow-inclined and steep-inclined, single deck and multideck, and with a variety of screen sizes and shapes.

An unfavorable characteristic of flat-deck screens is their sensitivity to loss of capacity with flaky materials that do not readily turn over to discharge smaller particles and especially fines. They are also subject to blinding, especially between decks, when light, flaky material is processed, and snagging, when wires are present in the feed, as with shredded refuse. Otherwise, they are compact and efficient, allowing many screening operations to be performed within one mechanical unit.

The influence of difficult grains, especially under overloading conditions, is illustrated in Figure 6-10. The upper curve shows the efficiency of a screen handling 2-inch material with 20% difficult grains, achieving 95% efficiency at 25 TPH. At 100 TPH, the efficiency falls to 80%. When the same screen handles 1/2- to 1/8-inch material with 55% difficult grains (the lower curve), the peak efficiency falls to 90% at 10 TPH, and to below 60% at 30 TPH. This illustrates the importance of testing the actual material and its variability to make certain that such severe losses of screening efficiency do not occur due to variations in the amounts of difficult grains.

Disc Screens

Disc screens overcome some of the problems experienced with flaky materials. Although they have limited ability to tumble material, they give more positive velocity to the oversize material than gyrating or vibrating screens. The size material for which they are suited is limited by mechanical considerations.

The disc screen illustrated in Figure 6-11 has a series of discs attached to rotating shaft assemblies arranged perpendicular to the flow of material, which is fed from one end above the assembly. Discs of two diameters and widths, alternately assembled on the rotating cylinders, create gaps through which particles of lesser size

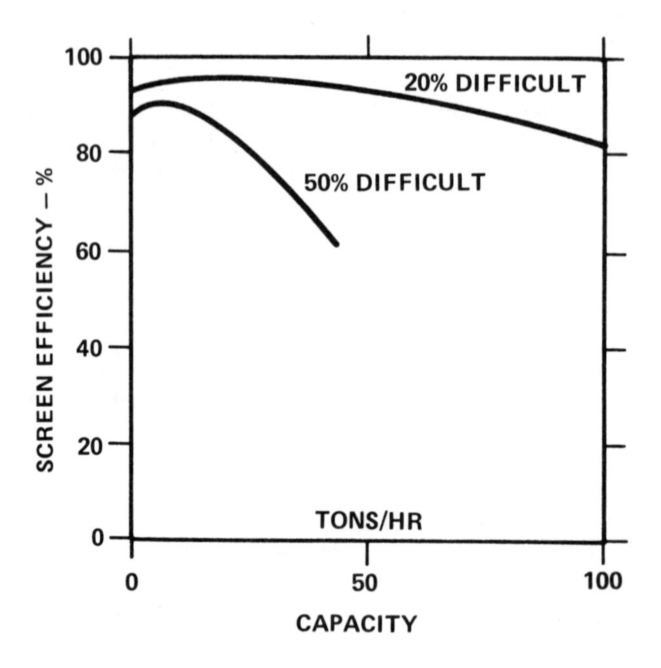

Figure 6–10. Performance of flat screens with different amounts of different grains.

can pass. The rotary motion and scalloped profile of the discs serve to progress the material, which does not pass to the exit opposite to the entry point. The width of the cylindrical spacers determines the gap through which the fines can fall. The oversize material is subjected to a bouncing action, which assists in separating out the fine materials from the oversize. The configuration of the discs can be designed to suit the materials being processed, within the mechanical limits of these machines.

Disc screens have long been used in the wood industry to separate bark and tramp material from sawdust, to remove fines from coarse wood, and in other size separation operations that are appropriate to the characteristics of this device. The Rader Disc Screen (RDS) has been applied to refuse processing at the Americology plant in Milwaukee [16] and at the City of Ames Solid Waste Recovery System, where its performance has been reported in some detail [15].

At Ames, two RDS screens were installed in 1978. They served to reduce the load on the second-stage shredder, which used almost twice the electrical energy as the first-stage shredder. They also reduced the erosion of the pneumatic conveying lines and the RDF storage bin floor, as well as the ash content of the RDF, which cut down boiler slagging and bottom ash accumulation.

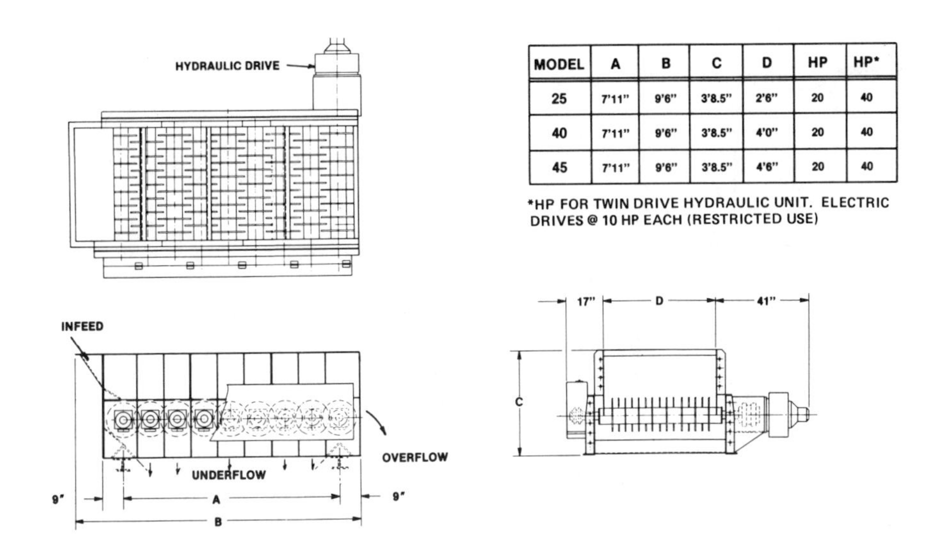

MODEL	A	B	C	D	HP	HP*
25	7'11"	9'6"	3'8.5"	2'6"	20	40
40	7'11"	9'6"	3'8.5"	4'0"	20	40
45	7'11"	9'6"	3'8.5"	4'6"	20	40

*HP FOR TWIN DRIVE HYDRAULIC UNIT. ELECTRIC DRIVES @ 10 HP EACH (RESTRICTED USE)

Figure 6-11. Disc screen (courtesy Rader Company).

In the Ames installation, a primary disc screen was installed directly after the first-stage shredder and magnetic separator. This screen was equipped with 1.5-in (38-mm) clearances between the discs. The overscreen material was fed to the second-stage shredder and the underscreen material to a second disc screen, designed to remove material smaller than 3/8-in. (9.5 mm), which is rejected by the plant. The overscreen material is combined with the product of the secondary shredder and fed to the air classifier.

This arrangement, shown in Figure 6-12, provided two-stage separation of the gritty materials from the combustible fraction, and also served to bypass a considerable amount of material around the secondary shredder. As the top deck of the screen is acting as a conveyor, the capacity of the screen is determined by the velocity at which the material enters the screen, the width of the screen and the depth of burden that can be carried. Figure 6-11 shows the dimensions of several screen models. At the rating of 20,000 ft^3/hr (100 "units" of wood), the Model #55 with a burden of 18 in. (0.5 m) would convey at a velocity of about 40 ft/min (0.25 m/sec). This machine is provided with 13 rotary shaft assemblies.

Multiplying the volumetric capacity by a typical density of 5 lb/ft^3 for shredded MSW, we would estimate the capacity of this screen to be 20,000 $\times$ 5/2000 = 50 TPH of MSW. However, actual bulk density of the overscreen material was found to be 2.5 lb/ft^3 (40 kg/m^3), resulting in an actual capacity of 25 TPH at the 18-in. (0.5-m) burden. This illustrates the importance of bulk density data in design of volumetric machines.

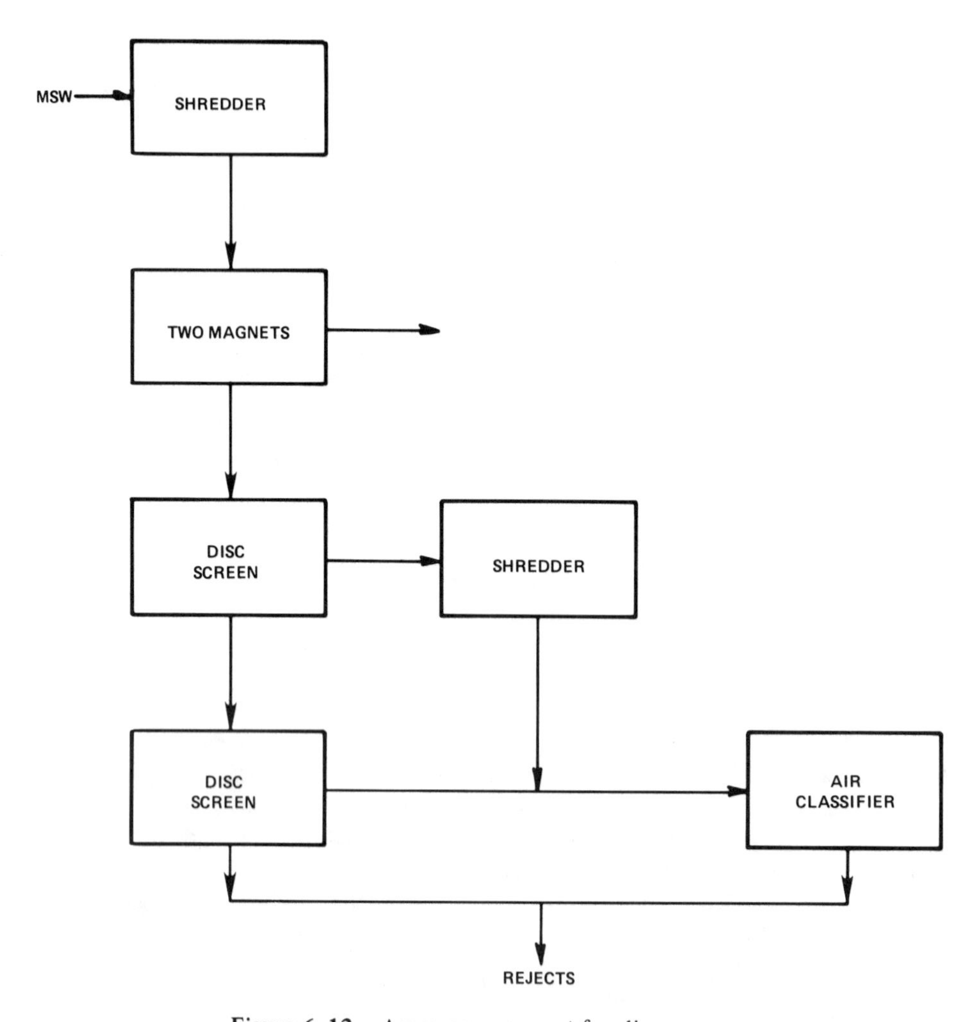

Figure 6–12. Ames arrangement for disc screens.

The Ames primary screen was driven by a 30-kW hydraulic drive. As the Ames system was operating at about 25 TPH during the test period, the burden of material averaged about 18 in (0.5 m). A heavier burden would result in a loss of efficiency as compared with the reported results. The second screen was equipped with a 5.6-kW drive and eight rotary shaft assemblies.

In the Ames installation, the objectives of reduced load on the secondary shredder and reduced ash in the RDF were achieved. The RDF heating value was increased and the bulk density reduced by 50% from 5.3 to 2.6 lb/ft^3 (84 to 42 kg/m^3). This reduction in bulk density improved air classification and conveying properties and greatly affected storage and retrieval properties.

The particle size distributions of the second-stage shredded material and RDF product were substantially changed as a result of the disc screen installation, as shown in Table 6-2. Figure 6-13a shows the comparative histograms of these streams. Figure 6-13b shows an estimate of the combustible and noncombustible distributions, illustrating clearly the separation of mainly noncombustible fines, which accounts for the improvement in the RDF.

Rotary Screens (Trommels)

While rotary screens have been largely replaced by flat screens in the minerals industry, their application to resource recovery has been proven to be advantageous due to the need for tumbling of light flaky materials to assure progression down the screen surface and inversion of flakes to release small or fine particles. In addition, the large-diameter trommels have been found to be effective in assisting breakage of garbage bags, substantial, but not excessive, breakage of bottles, and the removal of cans and heavy materials not susceptible to air classification. Trommels have been found to be effective in removing sand and glass from shredded RDF, either the lights or air-classified heavies.

Rotary screens, commonly called trommels or trommel screens, have long been used for sizing crushed rock and coal. Mechanically, they consist of drums with length/diameter ratios up to about four, supported either by axial shafts, external rings running on wheels (trunnions) or one of each. Axial bearings can be used only if the feed material does not tangle on the support spiders used to connect the drum to the shaft.

Table 6-2. Shredded MSW and RDF Particle Sizes

Size Range	*Percentage in Size Range*			
	Second Stage Shredder Infeed		*RDF Product*	
(mm)	*Before*	*After*	*Before*	*After*
minus 0.3	3	5	3	2
0.3–0.6	2	10	3	1.5
0.6–1.2	3	18	6	1.5
1.2–2.4	5	25[a]	10	3
2.4–4.8	7	20	12	4
4.8–9.5	10	14	17	10
9.5–19	13	3	20[a]	24
19–38	14[a]	1.5	16	34[a]
38–64	13	1.5	10	14
64–128	12	1.0	5	5
Over 128	8			

[a]Peak value

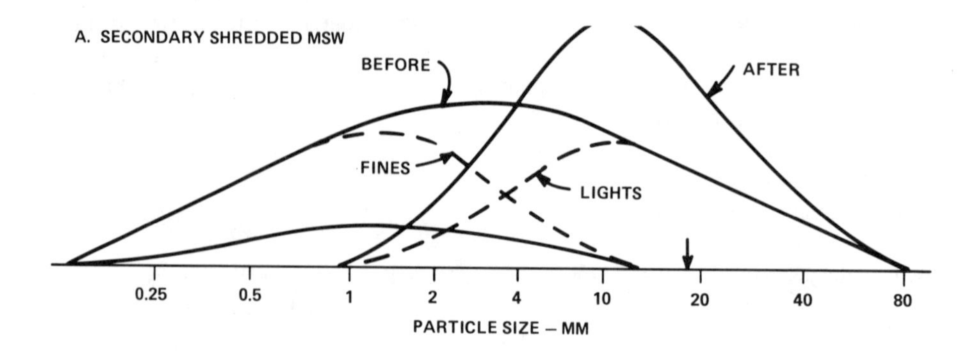

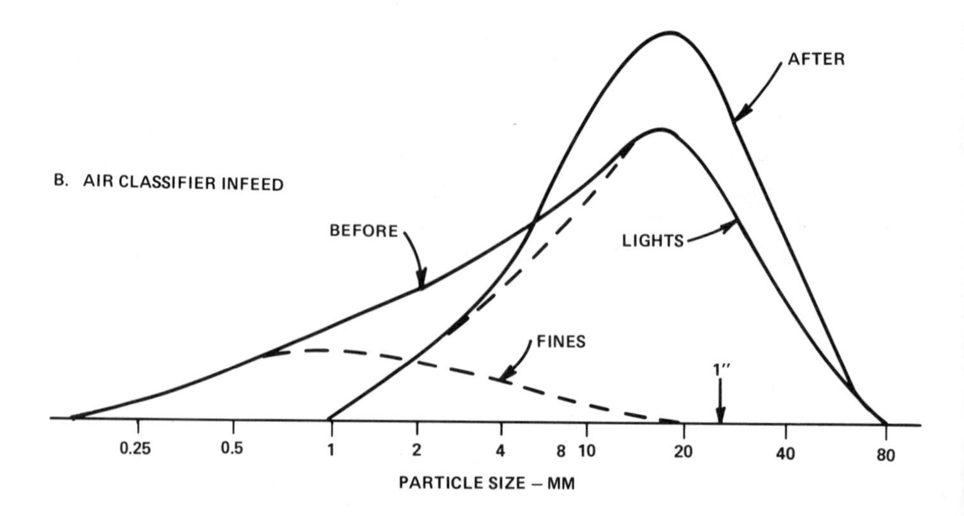

Figure 6–13. Effect of screening at Ames.

For processing raw MSW, a large-diameter trommel is necessary to handle the large items in the feed. External drive rings are necessary to have the benefit of a large opening at both ends. Shredded materials can be processed in trommels of smaller diameter with an axial bearing (trunnion) at the discharge end. The trommel operates as follows:

- Feed material is accepted through the feed end, which generally is supported by an external support ring riding on rollers.
- The feed is moved axially toward the exit end, which may be supported through a spider to a central bearing, or by an external support ring riding on rollers.
- Undersize material falls through apertures to hoppers and one or more conveyor belts below.

- Oversize material discharges at the exit through the spider supports or off the drum to a conveyor belt below.
- The feed material is lifted and dropped many times, assisted by lifters of appropriate shape, which may be set to progress the material backwards, forwards or radially.
- The trommel may be inclined so that the material progresses axially each time it falls. Lifters can provide the advancing action when the trommel is horizontal.

A satisfactory trommel must progress the required amount of material through the interior of the cylinder without the level of material building up to a level high enough to interfere with progression of the feed or with the tumbling action within the trommel. The level and the axial velocity of the feed determine the feed capacity of the trommel.

In addition, the separating efficiency of the trommel must be high for the materials and sizes of concern at the maximum anticipated feedrate and reasonably tolerant of variations in flow and material quality. Criteria that determine quality may include particle size, moisture, relative content of various constituents, such as paper, cardboard, plastics, metals and glass.

The machine must be able to start, stop and restart under fully loaded and overloaded conditions, without motor overload or drive slippage. The drive system must be protected from contamination by material.

Feed capacity (FC) equals overs capacity (OC) plus unders capacity (UC), except during transient conditions such as surging, stoppage of feed or plugging of outlets. The feed capacity is limited by the fillage (F) that can be tolerated without adverse effects. Fillage is defined as the ratio of cross-sectional areas of the material and the inside of the cylinder. This can be measured at any given time by stopping the trommel and measuring the height of the bed of material in the trommel. Fillage will not necessarily be the same throughout the trommel length, depending on the slope of the trommel, the rate of feed and thrus, and the influence and design of lifters. The entrance of the trommel is most prone to overloading due to the admission of feed, and to the time required for it to acquire axial motion after falling in. Special lifters may be provided at this point to accelerate the flow of material along the trommel, to reduce the burden in the front section.

Figure 6-14 shows the primary trommel at Haverhill, MA. Small holes are used at the feed entrance. Large holes pass only the accepted material. Beyond the entrance region, the progression of material is controlled by the slope of the trommel (S), the ratio of elevation to length, the diameter (D) of the drum and the speed of rotation (N). The angle of repose of the material is a factor that has somewhat less influence and is fairly consistent for a given type of material. The aerodynamic properties of the materials in MSW have an important influence on the capacity of trommels because the trajectories of the materials differ due to air resistance and any axial movement of air that may exist or be deliberately provided.

Figure 6-14. Lifters used in trommel screens at Haverhill, MA (courtesy the Heil Company).

Behavior of Particles in a Rotary Cylinder

Particles in a rotating cylinder have three basic forms of motion: (1) rotation about their own axes; (2) cascading (tumbling or rolling down the surface of the load; and (3) cataracting (parabolic free fall above the mass).

At low rotational velocities only cascading takes place. Under these conditions, the material lying on the shell is carried, at the shell velocity, to the top of the "kidney," where it falls over on top of the downmoving cascade of material, whereby it returns to the shell. As Figure 6-15 makes clear, under these conditions there is a bed of material that is rising, a bed that is falling and an interface between these two beds that has, on average, no net velocity. The friction caused by this interface causes a velocity gradient to take place through both the rising and falling beds: this causes tumbling to take place in rough materials and rolling in smoother shapes.

The region in which the falling bed reaches the shell is called the toe and is violently agitated, having a perpendicular shear force applied to the falling material, plus the energy of the materials that have fallen from above. As the speed of the

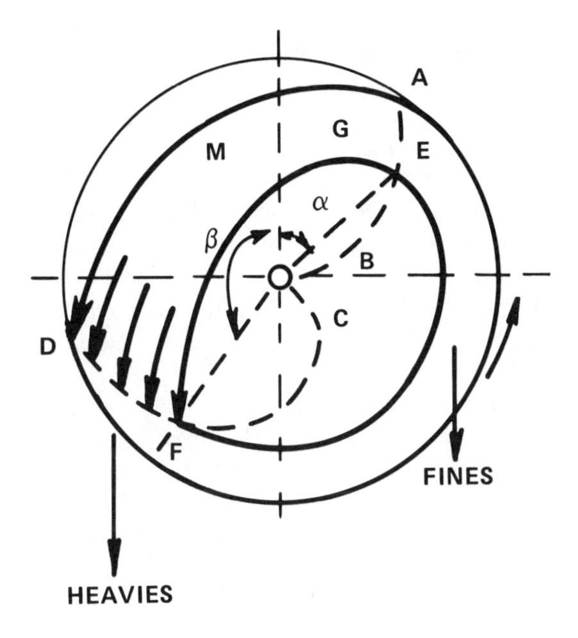

Figure 6-15. Path of particles in a rotating cylinder.

cylinder is increased toward the critical velocity, the trajectory of the particles increases to the limit of the cylinder, which defines the critical speed. Above the critical speed the material adheres to the shell and no falling trajectory takes place unless the material slips relative to the shell or other forces are applied.

The critical speed is

$$N_c = \frac{76.6}{(D)}$$

The ratio of rotary speed to critical speed is

$$C_n = \frac{N}{N_c}$$

where N_c = critical speed (rpm)
D = diameter
N = rotating speed (rpm)

At two-thirds to three-fourths of the critical speed, the length of fall of the lifted material is maximum and approximately the same as the diameter of the drum. In this range of velocities, cataracting is taking place rather than cascading, and the falling stream has become separated from the rising bed. The isolation of these two streams reduces the tumbling action.

The point of separation of the two streams is similar to the point of planing of a boat in that the velocities increase substantially at this point and fall back rapidly below it. After planing, speeds can be cut substantially before bogging down occurs. It is evident that the performance of a trommel would be very unstable if kept near this transition range: higher velocities are preferable for screening refuse.

If it were not for the turbulence at the toe, the kidney of material could just roll, resulting in the inner layers never seeing the screen: this turbulence and mixing is important because small particles and fines must be given an equal, if not preferential, chance to reach the apertures.

Vorstman and Tels [17] have shown that the fines tend to segregate from the flaky material during the vertical rise of the material, discharging from the screen as shown in Figure 6-15.

The progression of material down the axis of the trommel takes place mainly because of the inclination of the axis and the length of fall due to gravity (unless air movement is provided or lifters cause the displacement). The higher the drop and the steeper the slope, the greater the progression per turn.

As the speed of the trommel is limited to a large fraction of the critical speed and the drop cannot exceed the diameter, the progression rate is determined by the slope and the lifter angle. The size and shape of lifters has an influence on capacity and screening: the optimum can best be found by test.

Trommels normally are operated at angles from 3 to 8°, although some rotary screens are operated at angles as high as 16°. The advance per turn is then

$$dL = H \times \tan = H \times S \qquad (6\text{-}4)$$

where L is a unit of axial length.

The height, H, approaches the diameter, D, so we will substitute D for H. Multiplying by the number of turns per minute, N, gives the axial movement per minute:

$$\text{Axial velocity} = D \times S \times N \text{ (units of length per minute)}$$

The feed capacity of the trommel is then the cross-sectional volume of material, A, which is the fillage, F, times the cross-sectional area of the cylinder:

$$A = D^2 \times \frac{\pi}{4}$$

Therefore, the volumetric capacity is Q, volume units per minute:

$$Q = N \times S \times F \times \frac{\pi}{4} \times (D)^3 \qquad (6\text{-}5)$$

or

$$Q = 76.6 \times C_n \times S \times F \times \pi \times (D) \qquad 6\text{-}6)$$

To get the capacity on a weight basis, the volumetric capacity must be multiplied by the density in weight per unit volume. Tests show that the volumetric capacity of a given trommel will increase at a rate that is approximately linear with N, until a speed where it suddenly breaks off. This speed is close to the point where the catapulting is weakened by contact with the outer shell, hence is likely to be unstable. Failure to transport material at this point would result in plugging of the trommel. Therefore, it is prudent to operate safely below this point once it has been determined for a given material and configuration.

The behavior of particles in a rotary cylinder has been described mathematically by Alter et al. [18] and Glaub et al. [7], based on kiln studies by Davis and Gaudin [19, 20]. Mathematical analysis of particle movement in a rotating trommel is presented in the Appendix, from which we can develop theoretical performance curves such as those shown in Figure 6-16.

As the rotational speed increases, the particles adhere to the shell at higher levels before they fall back to the bottom in a trajectory. At the higher speeds, approaching the critical speed at which the material would adhere to the shell by centrifugal force, the particles are thrown to the other side of the drum, where they contact the maximum screen area, thus achieving maximum screening efficiency. Above the optimum speed, the efficiency drops off due to contact of the material with the top of the shell.

Table 6-3 shows the calculated performance of a 10-foot-diameter trommel operating at various speeds, including the number of drops that would take place in a 40-foot-long trommel. It is apparent that a speed somewhat above 50% of critical gives the maximum feed capacity, based on an arbitrary 6-inch depth of material in the trommel. The influence of bed depth has not been evaluated mathematically, but actual tests show this to be a significant factor in screening efficiency because a thick bed inhibits tumbling and access of particles to the screen holes.

Figure 6-16 shows the calculated performance parameters of a trommel as a function of critical speed. As speed increases, the length of drop increases to the diameter of the trommel, maximizing the axial movement of the material and, hence, the trommel capacity for a given bed thickness or fillage. At the same time, the number of drops decreases, thereby reducing the screening efficiency, which then peaks at 50-60% of critical speed and results in the largest number of drops per unit of time as well as the greatest axial advance.

The multitude of factors controlling trommel performance while screening a heterogeneous material such as shredded MSW require extensive testing with real materials to determine and confirm the applicability of theoretical approaches. The work of Glaub et al. has presented a summary of trommel theory and the results of extensive testing performed for the U.S. Department of Energy (DOE). Some of these data have been used for the analysis that follows.

How Probability Determines Screening Efficiency

The tumbling action of a trommel provides a means for presenting the feed material to the screen surface, while allowing flaky particles to turn over and give fines access

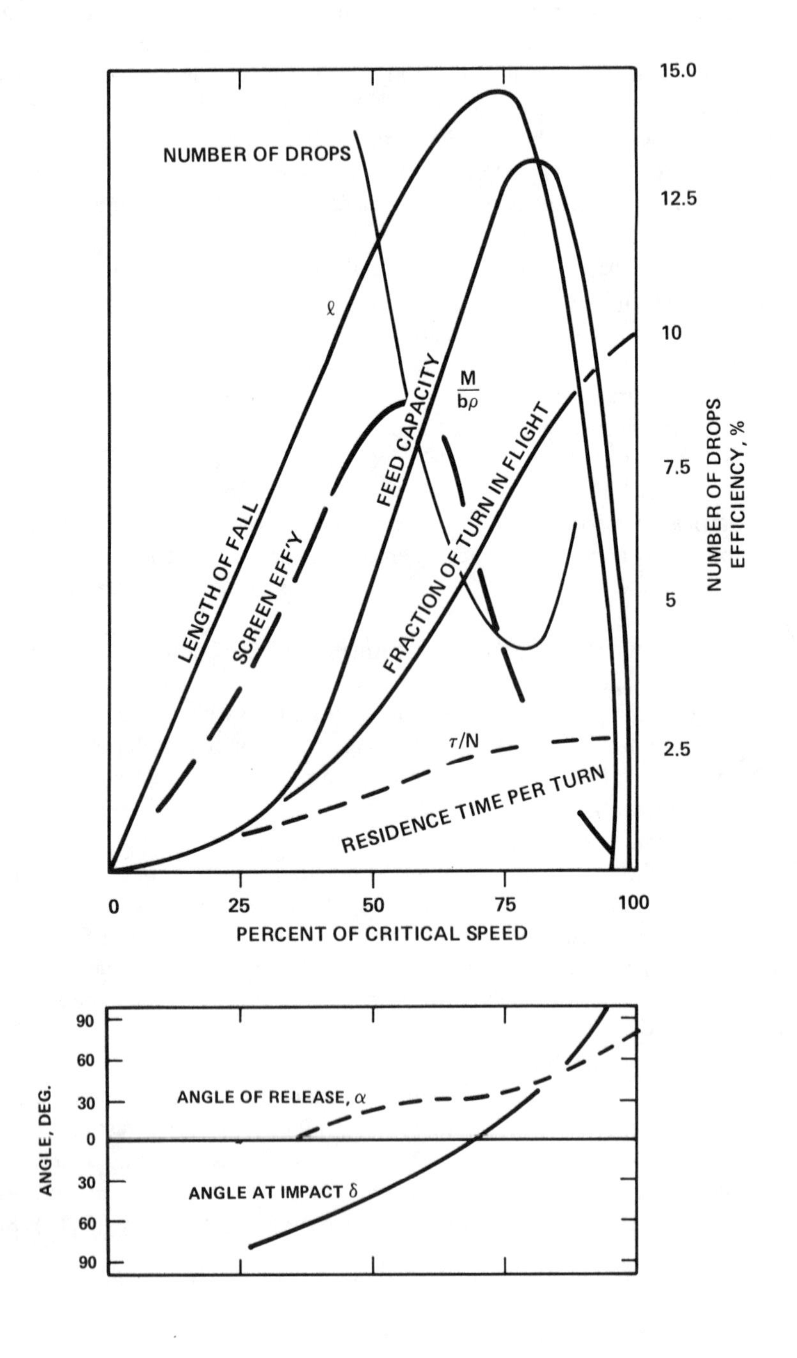

Figure 6-16. Calculated performance parameters of trommel vs percentage of critical speed.

Table 6–3. Performance Parameters of a 10-Foot-Diameter Trommel

Percent Critical Speed:	*50*	*60*	*70*	*75*	*80*	*85*	*90*
Rpm	12	15	16	18	19	20	22
Angle of Separation, a	14	21	30	35	40	50	55
Angle at Hit Point, d	35	20	0	10	30	55	80
Advance per Turn, 1	1.1	1.3	1.5	1.5	1.4	1.1	0.8
Drops per Turn, 1/f	0.30	0.45	0.61	0.68	0.76	0.77	0.78
Drops per TPH	5	2	1	0.8	0.7	0.9	1.4
Number of Drops in 45′	133	75	50	45	43	53	72
Retention Time, seconds	200	143	105	104	103	130	180
Capacity, TPH (based on 6–inch bed)	25	38	52	58	60	58	50

to the screen. In addition, the fines have been found to slide off of flakes while they rise up the drum, as evidenced by the discharge of fines from this portion of the screen [17].

The probability of a particle passing through a hole in a screen once it has been presented to the surface depends on the size and shape of the particle and the hole, and the ratio of hole area to total screen area (open area ratio). The probability of passing through the screen depends on the number of tries and is equal to the probability of one try raised to the nth power, where n is the number of tries. These probabilities can be calculated and are presented in Table 6–4 as a function of particle size, open area and number of tries for a 4.75-inch (210-mm) hole, such as was used in New Orleans and other primary trommels.

For each interval of particle size there is an average probability of passage, which increases with the number of tries. The PSD of the feed material can be divided into similar intervals, as shown in Table 6–5, listing the fraction of the total weight of material in each size interval. By multiplying the weight in each interval by the probability of passing, the actual weight passed is calculated. Adding the weights passed in all size intervals, we obtain the total weight passed by the screen. Dividing this weight by the feedweight gives the screen efficiency, which increases with the number of tries, as shown.

The screen efficiency therefore varies with the ratio of particle size to screen hole size and the number of tries, as shown in Figure 6–17, showing both actual particle size and particle/hole size ratio. This is a universal curve. It has a slightly different form, depending on whether the holes are square or round, whether bounding takes place and for different ratios of hole to shell area (percentage open area).

By superimposing the probability curve over the histogram of the PSD of a material to be screened, we can visualize the behavior of that material when it is given various numbers of tries to penetrate the screen, as shown in Figure 6–18. The validity of this procedure can be confirmed by comparing theoretical predictions with actual test data.

Table 6-4. Probability of a Particle Passing on a Single Try

Particle Size		*Open Area Ratio, Q (Hole Area/Screen Area)*			
(in.)	*(mm)*	*Q = 1.0*	*0.75*	*0.50*	*0.25*
0.05	1.5	0.98	0.74	0.49	0.245
0.1	3	0.96	0.72	0.48	0.24
0.25	6	0.90	0.70	0.45	0.25
0.5	12	0.81	0.61	0.41	0.20
1.0	25	0.64	0.48	0.32	0.16
2.0	50	0.36	0.27	0.18	0.09
2.85		0.14	0.10	0.07	0.04
3.8	25	0.04	0.03	0.02	0.01
4.28	75	0.01	0.008	0.005	0.0025
4.75	100	0	0	0	0

Table 6-5. Calculating Number Passing vs Tries

For a = 4.75 in.		*Number of Tries or Impingements*			
x	*d/a*	*1*	*10*	*50*	*100*
0.5	0.105	0.400	0.994	1.000	1.000
1.0	0.211	0.312	0.976	1.000	1.000
1.5	0.316	0.234	0.930	1.000	1.000
2.0	0.421	0.168	0.840	0.999	1.000
2.5	0.526	0.112	0.696	0.997	1.000
3.0	0.631	0.068	0.505	0.970	0.999
3.5	0.737	0.035	0.297	0.828	0.971
4.0	0.842	0.013	0.118	0.466	0.715
4.5	0.947	0.201	0.014	0.067	0.129

Figure 6-19 shows histograms plotted from actual test data presented by Cambell [11] that were obtained by NCRR tests on the Recovery I trommel at various feedrates from 50 to 150% of the design capacity of 65 TPH. Close inspection of the curves shows that the fraction of undersize material that remained in the material leaving the top side of the screen is significantly greater at 150% than at 50% capacity, as would be expected. The areas under the curves represent the relative weights of the oversize, undersize, and undersize in the oversize materials. These results can be compared with the predicted screening rates in Table 6-6.

Analysis by this procedure can be used not only for the total amount of material passing through the screen, but also for individual species of materials such as glass, metals and combustibles, ash and moisture, when the size distribution of each species or property is known.

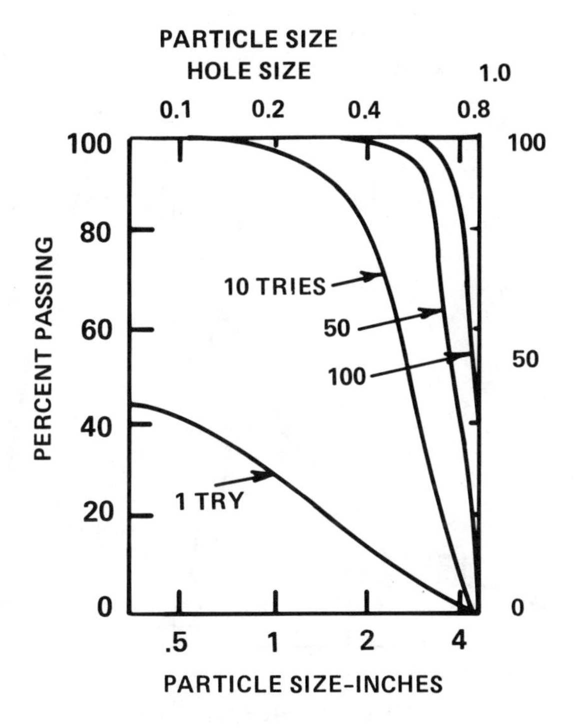

Figure 6–17. Percentage of material passing holes vs number of tries and ratio of particle size to hole diameter.

As the glass contained in MSW has a smaller size distribution than the combustibles, it would be expected that the glass will be screened with a higher efficiency than the combustibles, which is indeed the case.

ACTUAL PERFORMANCE OF SCREENS IN RESOURCE RECOVERY

Pilot-Plant Tests in Minneapolis

A major test program was undertaken in Minneapolis by Menon et al. [8] using packer truck refuse. The raw MSW was processed in batches by a 10-foot-diameter pilot trommel equipped with a weight scale to give a continuous record of the thrus. Samples were taken periodically during runs for analysis of moisture, inerts, glass, metals and combustibles. The major factors studied were as follows:

- type of refuse
- bag breakage
- trommel speed
- trommel "fillage"

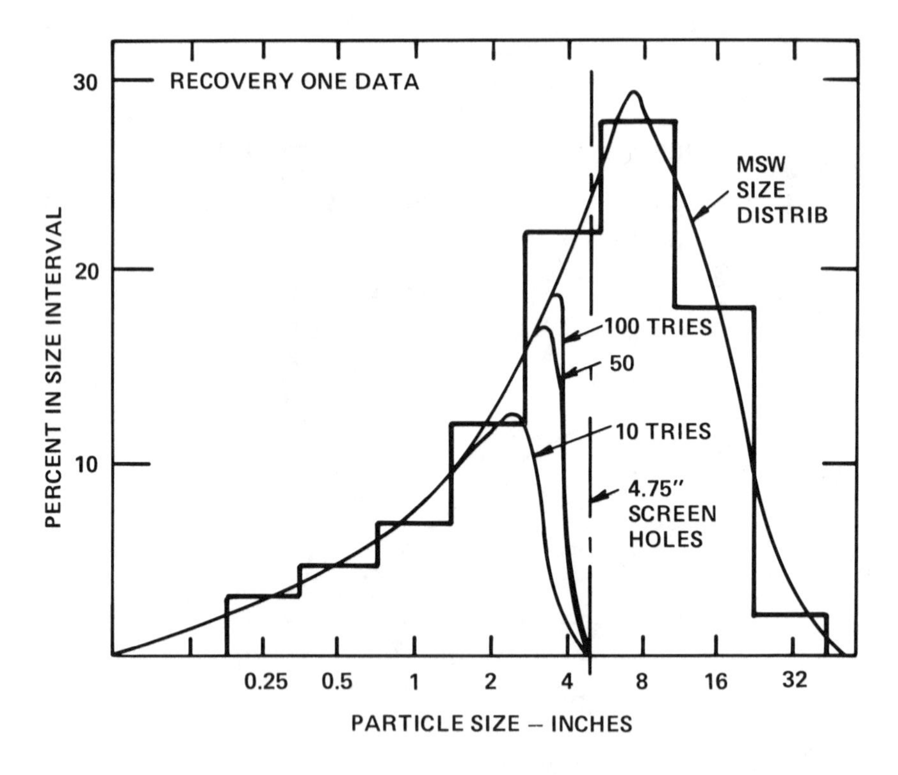

Figure 6-18. Probability of particle passing superimposed over histogram of particle size of MSW.

- lifter design
- capacity versus efficiency
- effectiveness of glass removal
- glass removal versus combustibles loss to the unders
- effect of glass breakage prior to trommeling
- glass breakage within the trommel
- effect of various hole sizes
- moisture, ash and combustible content of overs and unders

Tests were conducted on eight-inch, 4-3/4-inch and 2-inch holes to determine the quantity of thrus, effectiveness of glass removal and combustibles loss. Preliminary tests were run at the conventional 50% of critical speed (12 rpm) and at other speeds to determine the optimum trommel rotational velocity in terms of screening efficiency and capacity. Most of the subsequent testing was done at 17 rpm (70% of critical speed), which appeared to be optimum for a 10-foot (3-m) trommel diameter.

In the tests, the trommel was loaded with 250 lb of refuse, representing a design capacity of 75 TPH for a 40-foot length (ten times the pilot unit length) and

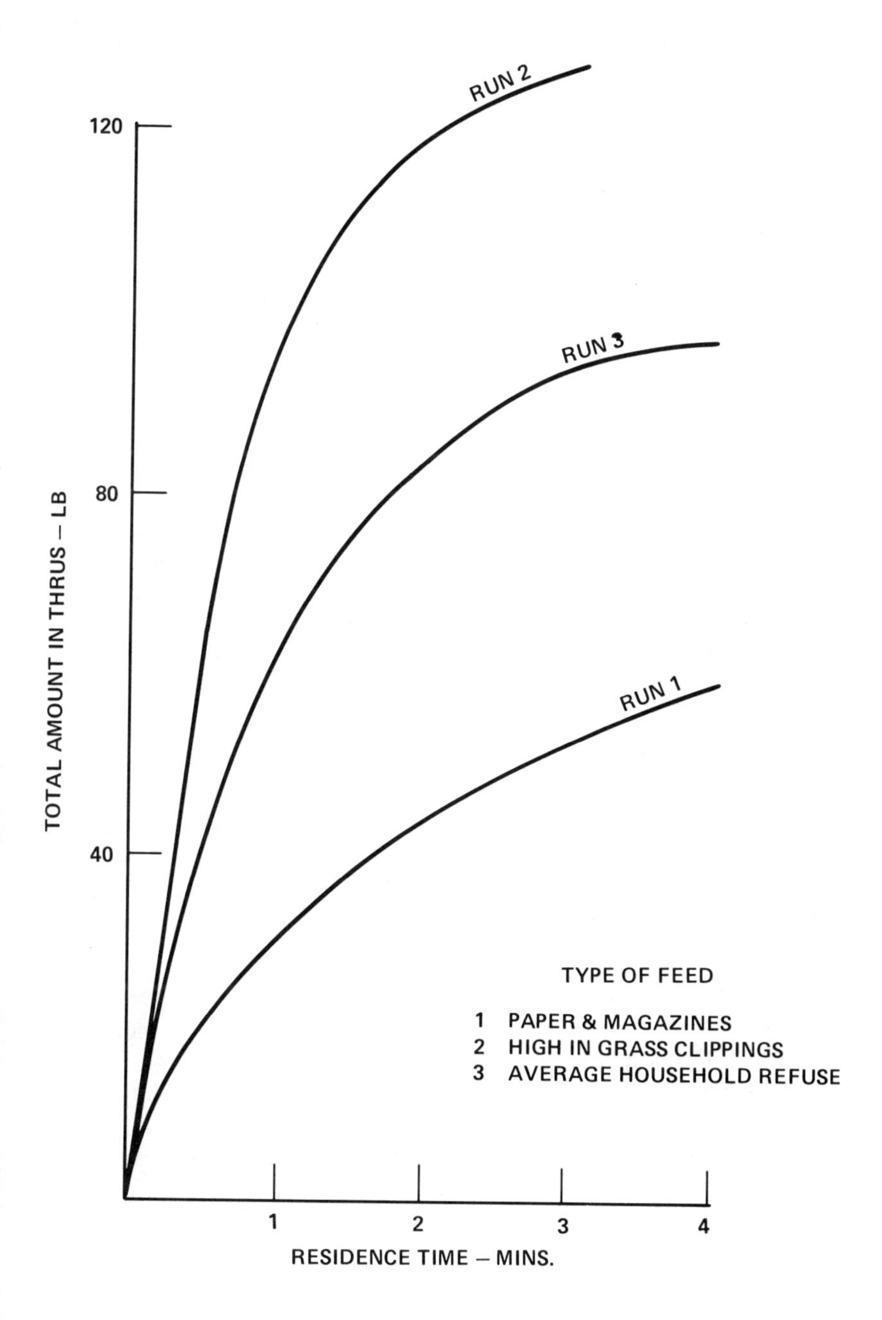

Figure 6–19. Effect of feed type on screen performance. Trommel screen, 54.5–ft^2 open area, 4.75-inch holes.

Table 6-6. Calculating Number Passing vs Number of Tries

For a = 4.75 in.,		*Q = 0.5*	*Number of Tries or Impingements*			
x	*f(x)*	*df(x)*	*1*	*10*	*50*	*100*
0.5	0.03	1.9	0.76	1.9	1.9	1.9
1.0	1.93	8.5	2.65	8.3	8.5	8.5
1.5	10.4	15.4	3.6	14.3	15.4	15.4
2.0	25.0	16.1	2.7	13.5	16.1	16.1
2.5	41.1	13.7	1.5	9.5	13.3	13.3
3.0	54.8	9.6	0.7	4.8	8.0	8.0
3.5	64.4	5.3	0.2	1.6	4.3	5.1
4.0	69.7	1.5	0.02	0.2	0.7	1.0
4.5	71.2	28.8	0.03	0.4	1.9	3.7
Efficiency, %:			12.2	54.5	70.0	73.0

rotated until the thrus rate was insignificant. This is an uncertain point because the trommel keeps grinding the material, especially if it has high moisture content. The "infinite" weight of thrus was considered to be the available thrus and used to calculate the screening efficiency. Samples were taken at regular intervals (such as 0.25, 0.5, 1, 2 and 4 minutes) to monitor the composition of the thrus during the test.

Type of Refuse

The influence of the type of refuse on the thrus capacity of the trommel is significant. Figure 6-19 shows that the fraction of the feed that passed through 4.75-inch holes after 2 minutes of screening ranged from 15% for packer truck waste heavy in paper and magazines to 50% for waste high in grass clippings; about 35% of average household waste passed the screen. Dividing the total thrus by the "ultimate" thrus gives the screening efficiency for the various feed materials, as shown in Figure 6-20.

Bag and Glass Breakage

The quantity of glass collected under the screen varies substantially, according to the extent to which the bags and the glass are broken before the material is fed to the trommel. Figure 6-21 shows the screening efficiency when the glass is unbroken (as received from trucks without compaction), as compared with material fed with the bags cut open and the bottles crudely broken. While the efficiency after 1.5 minutes appears to be about the same, the prebroken glass was screened out more rapidly at first.

Trommel Rotational Speed

The thrus capacity of the trommel is greatly influenced by rotational speed in the range from conventional speeds around 40-50% to the most effective speed of 67%. Figure 6-22 shows the relative screening capacity at various speeds. After 2 minutes

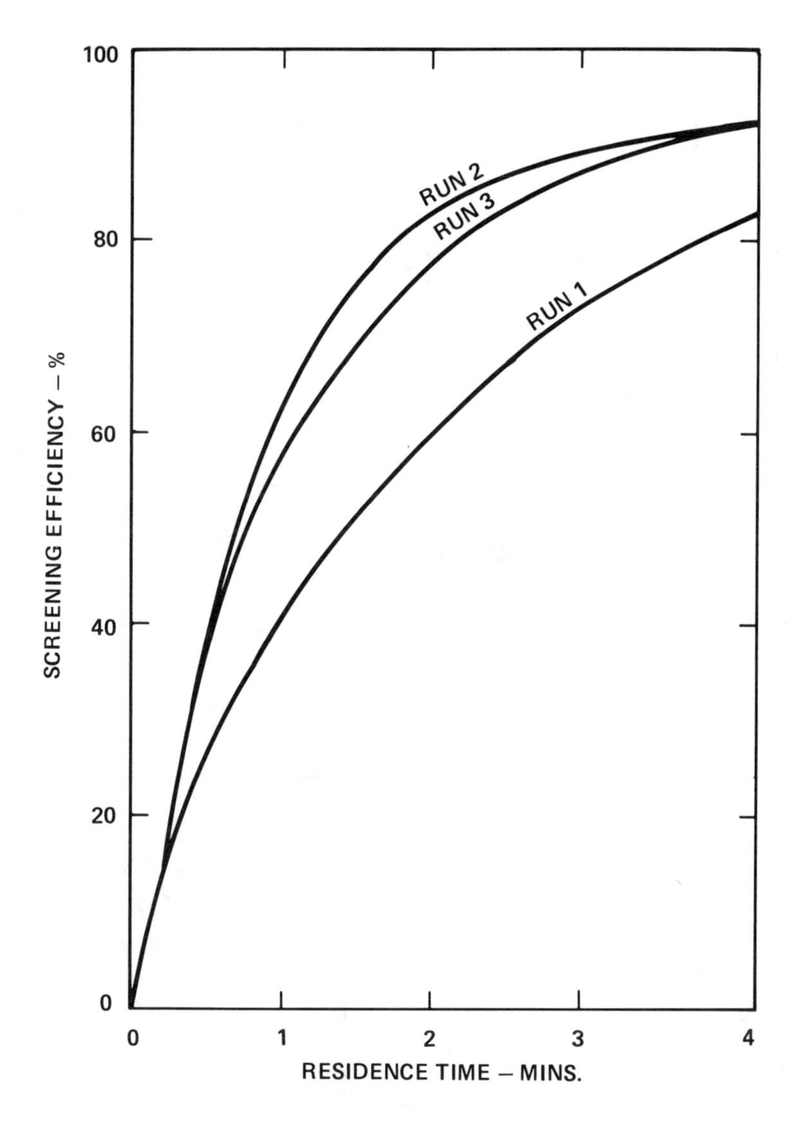

Figure 6-20. Screening efficiency – data for Figure 6-16.

of screening at 50% of critical speed, only 85% of the capacity at 67% is achieved. The severe drop-off of capacity above the optimum speed would create unstable operating conditions; hence, trommels should be operated safely below this point.

Material Trajectories

Material trajectories, observed and photographed, are essentially as shown in Figure 6-23. At low speeds the material tends to roll, whereas at the higher velocities it cataracts free of the rising mass, impinging over a wide section of the screen.

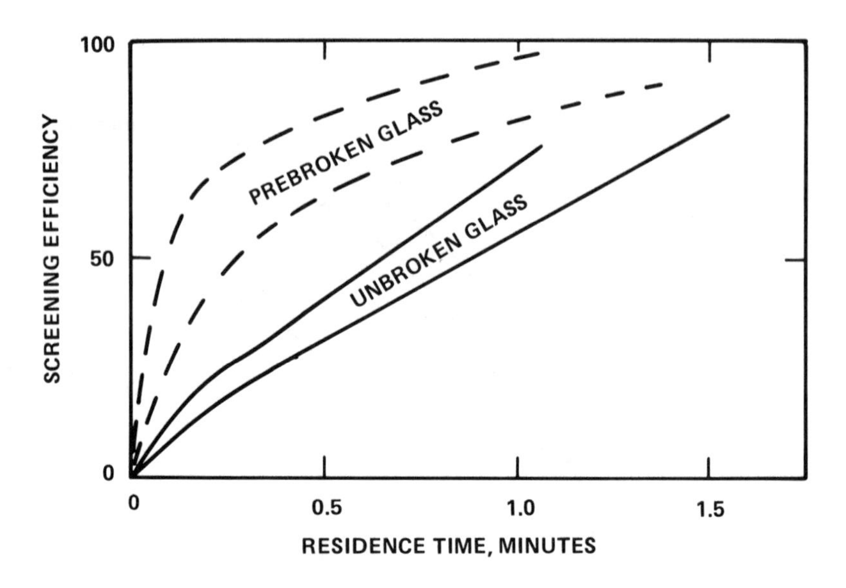

Figure 6-21. Screening efficiency of trommel when glass is broken or unbroken.

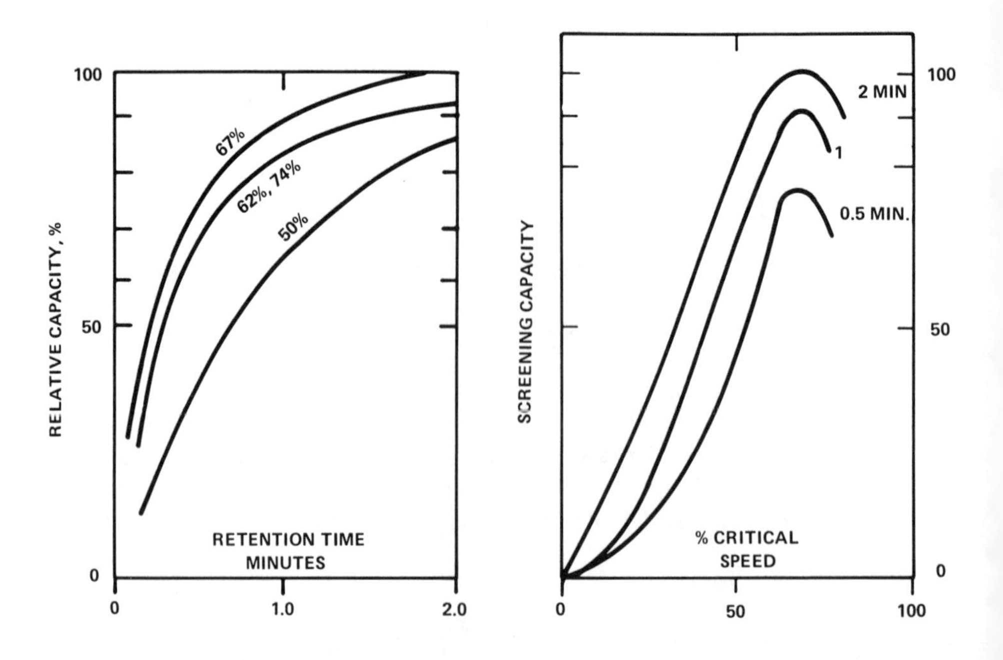

Figure 6-22. Effect of rotational speed on trommel performance.

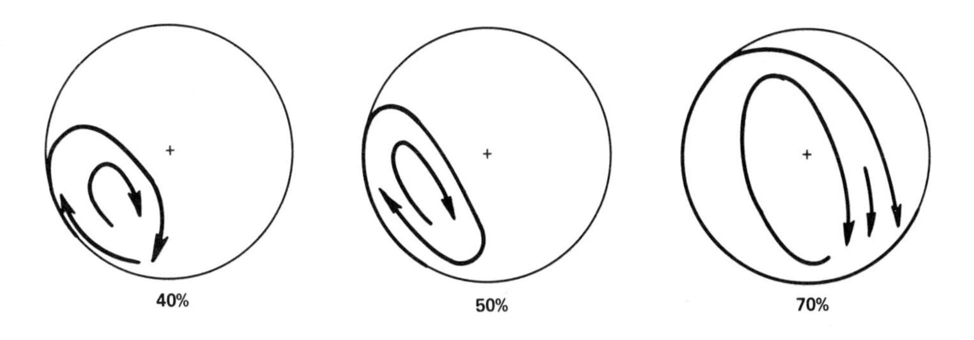

Figure 6-23. Material trajectories in a trommel at three different percentages of critical speed.

Lifters

Lifters increased the capacity by 40% above a bare screen; 6-inch lifters appeared to be optimum.

Trommel Fillage

The rate at which material passed through the holes with 15-32% fillage seemed to be directly proportional to fillage for the same screen area and screening time. The screening efficiency falls in direct proportion to fillage for the same screening time. If the fillage is doubled, the same screening efficiency could be achieved by doubling the screening time. This linearity would not be expected at less than 15% fillage at the point at which undersize material was reduced to the probability level. At some point (not determined) above 32% fillage, performance would be expected to deteriorate.

Screen Capacity versus Efficiency

The capacity and screening efficiency of different components and qualities of the refuse varied but showed remarkably similar trends. The efficiency of screening glass and combustibles is shown in Figure 6-7. Most of the data are plotted on straight lines on a semi-log plot.

Screen Efficiency versus Hole Size

The difference in screening efficiency versus time is seen in Figure 6-6: the larger 4.75-inch holes discharged the glass more rapidly, but the 2-inch holes point to 100% efficiency in two minutes of screening. The efficiency of screening combustibles has a reverse trend: the smaller screen achieves a somewhat higher efficiency than the larger one, although the total quantity passed is much less.

Loss of Combustibles to Thrus

For normal MSW, the loss of combustibles through the screen is below 15% and is half as much with 2-inch holes as with 4.75-inch holes, as shown in Figure 6-7.

Tests of Primary Trommel at Recovery I, New Orleans

The New Orleans trommel was the first full-scale machine used for raw refuse. Early operating data confirmed it to be effective in removing glass from raw MSW; however, as it was in an operating facility that was not designed for analytical tests, testing was limited to analysis of the over and under flows during normal running conditions. These analyses were valuable but left unanswered the questions about the relationship between screening and capacity. Finally, in 1980, a series of tests, reported by Campbell [11], was run under the supervision of the NCRR to fill this gap.

Figure 6-24 shows the overall split of feed to thrus of the New Orleans trommel, having 4-3/4-inch holes, a 9-foot diameter and a 40-foot length. The data points lie on a band ±5% wide through the range from 50% to 150% of the rated capacity (30-109 TPH). The rate of unders flow decreases linearly with capacity, indicating a potential of 80% at no load.

The potential thrus can be determined from the zero intercept and called the "available thrus," serving as a measure of the trommel's effectiveness. The graph shows that the effectiveness or efficiency is 62.5% at the rated load, based on total refuse feed.

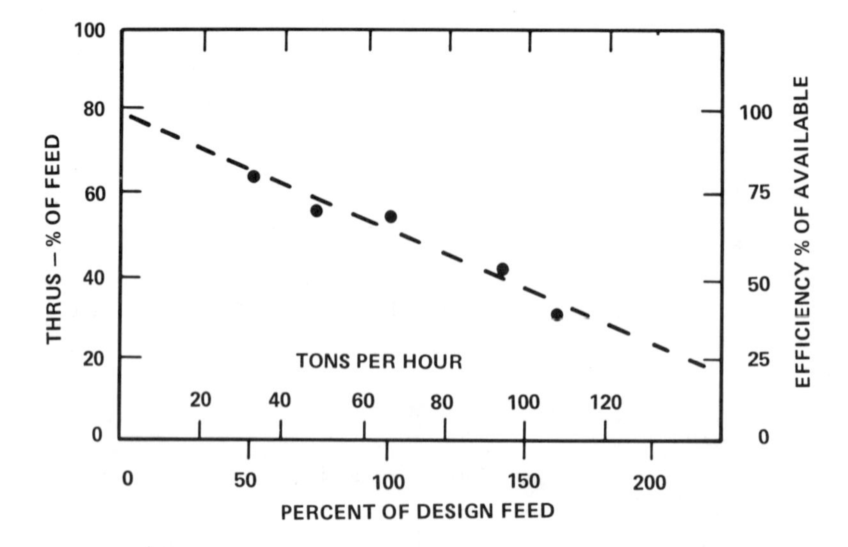

Figure 6-24. Capacity and screen efficiency of the New Orleans trommel [11].

The machine can routinely be fed at as high as 150% capacity, but the screening efficiency for the overall flow and the various constituents decreases substantially with increased capacity. Figure 6-25 shows the trommel efficiency for the target components of refuse, determined from samples taken from the overs and unders and verified by sampling from the tipping floor.

The objective of removing glass is achieved admirably, although it falls off steadily as loading increases, passing 90% at the rating point. The minus-1/2-inch fines follow the same line. A serious drop-off occurs above the 100-TPH loading. The behavior of paper, aluminum and ferrous metal is about the same, showing a rapid drop-off with loading. As paper is undesirable in the thrus, this is a favorable development. Because ferrous may be removed more effectively from the unders, it is unfortunate that load increase drives it into the overs. The overall efficiency and the aluminum efficiency follow the same line, showing the tendency of the seeded aluminum cans to follow the material wherever it goes. Plastic flakes (4 inch X 4 inch), seeded into the feed, showed a very rapid drop-off in the thrus, indicating that plastics will be found mainly in the overs.

The NCRR tests included extensive, detailed data on particle size distribution as well as weight distribution for the various species. Figure 6-26 shows the size distribution of the feed material determined during the 36- and 96-TPH tests. As loading is increased, the amount of undersize material in the overs increases, reducing the screening efficiency. The presence of combustible lights, combustible heavies and inerts such as sand and glass is indicated in the histograms of the feed. The influence of hole size can be understood from these histograms. It is clear that the trommel hole size determines the upper end of the unders distribution because no unders should be larger than the hole size. On the other hand, considerable amounts of

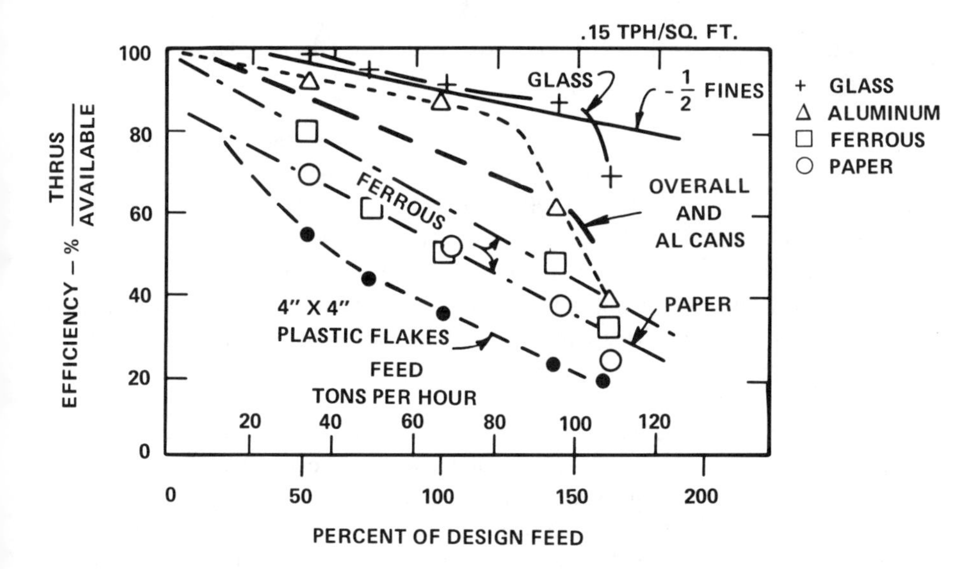

Figure 6-25. Screen efficiency of New Orleans trommel for various components [11].

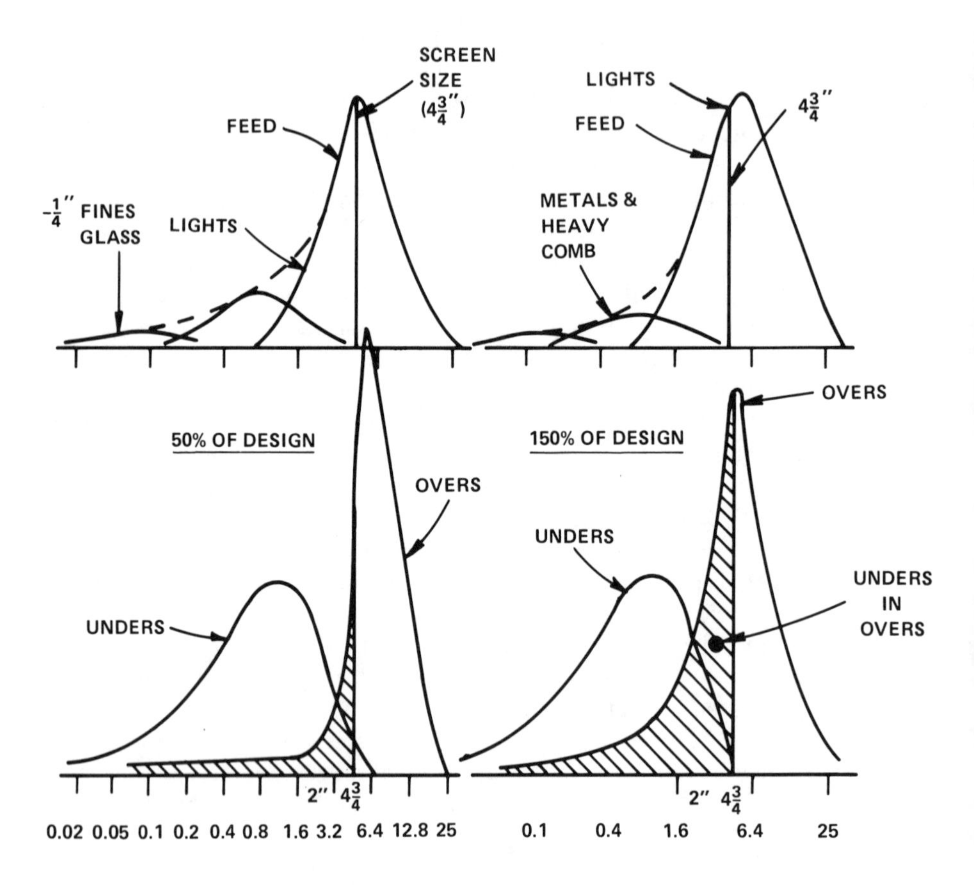

Figure 6-26. Particle size distribution of feed, unders and overs at 50% and 150% of design feed.

material that are truly undersize will remain in the overs if the material does not have a chance to pass through. At 50% of rating (36 TPH) very little of the unders remained in the overs, but at 150% (94 TPH) almost half the overs are unders.

The data and the histograms show that the PSD of the unders did not change during the entire range of testing and that the quantity of unders did not change much as the load was increased. This explains why the unders in the overs increased: the screen capacity increased slightly when more material was available but not in proportion to the increased load.

The NCRR data are based entirely on 4-3/4-inch holes, which were selected to pass a maximum amount of ferrous and aluminum cans, while coincidentally passing about 50% of the paper and concentrating the glass in the unders fraction. This arrangement produces a low-ash overs product, which can be shredded to make a good fuel. However, if half the paper is in the unders or thrus, it must be recovered as fuel or taken to landfill.

Another stage of processing can be used to recover the combustibles: a trommel, disc screen, or air knife can be used to remove the oversize combustibles, while dropping the glass. This has been done using 1-1/2- to 2-inch holes, forcing the metals into the "middling" fraction.

With 4-3/4-inch holes, the unders combustibles may be too large for transport and combustion, so smaller holes may be desired. Depending on the economics of the facility, the unders may be either processed or thrown away. To consider the options, trommel performance with other hole sizes must be evaluated.

Trommel Tests at East Bridgewater

The trommel at East Bridgewater, Massachusetts, was installed as a "scalper" to return oversize material to the flail mill and pass the underflow fed to the process via an air classifier and secondary trommel (to remove glass fines). In this arrangement the trommel holes were the orifice to the process. It was found that the 9-foot × 30-foot trommel could not consistently pass more than 12 TPH with 5-inch square mesh because it blinded rapidly with textiles. When larger holes (up to 9 inches) were tried, blinding continued, and the downstream process was plagued with rags. By cutting the holes back to 4-3/4 inches, the large rags were screened from the process, but blinding reduced capacity. Finally, it was confirmed that having a ligament equal to the hole size stopped blinding and gave maximum capacity of about 12 TPH, with 250 ft^2 of holes (0.05 TPH/ft^2, or 20 ft^2/TPH).

A secondary trommel, installed after the air classifier and having 5/8-inch holes, reduced fuel ash to 15-25%, depending on its moisture content. When the air classifier was operated as a drier, the lower ash level could be maintained.

Bridgeport Trommel Tests

The Bridgeport Resource Recovery Facility (BRRF) was provided with a primary trommel prior to shredding so that the shredder would get only oversize material and not the glass fraction. A secondary trommel was used after the air classifier to screen out the glassy fines that were lifted. Hole sizes from 1/4 inch to 8 inches were tested with these trommels [9]. The primary trommel was started up with 4-3/4-inch holes. To obtain quantitative evidence for evaluating process options, 3/4-inch and 2-1/2-inch screens were tested. The 3/4-inch screens were installed in the first two sections to prevent long objects from falling through the holes and jamming the trommel, as well as to remove the glass. The 2-1/2-inch holes were tested as a means of reducing the size of textiles in the combustible fraction.

Figure 6-27 shows the distribution of glass, fines and minus-1/4-inch combustibles dropped along the length of the trommel with the 3/4-inch × 2-1/2-inch combination. About 10% of the feed passed through the screen. Most of the thrus were glass, which discharged heavily for about half of the trommel, then fell off for lack of supply. The fines, consisting of sand and grass, were removed almost entirely by the 3/4-inch

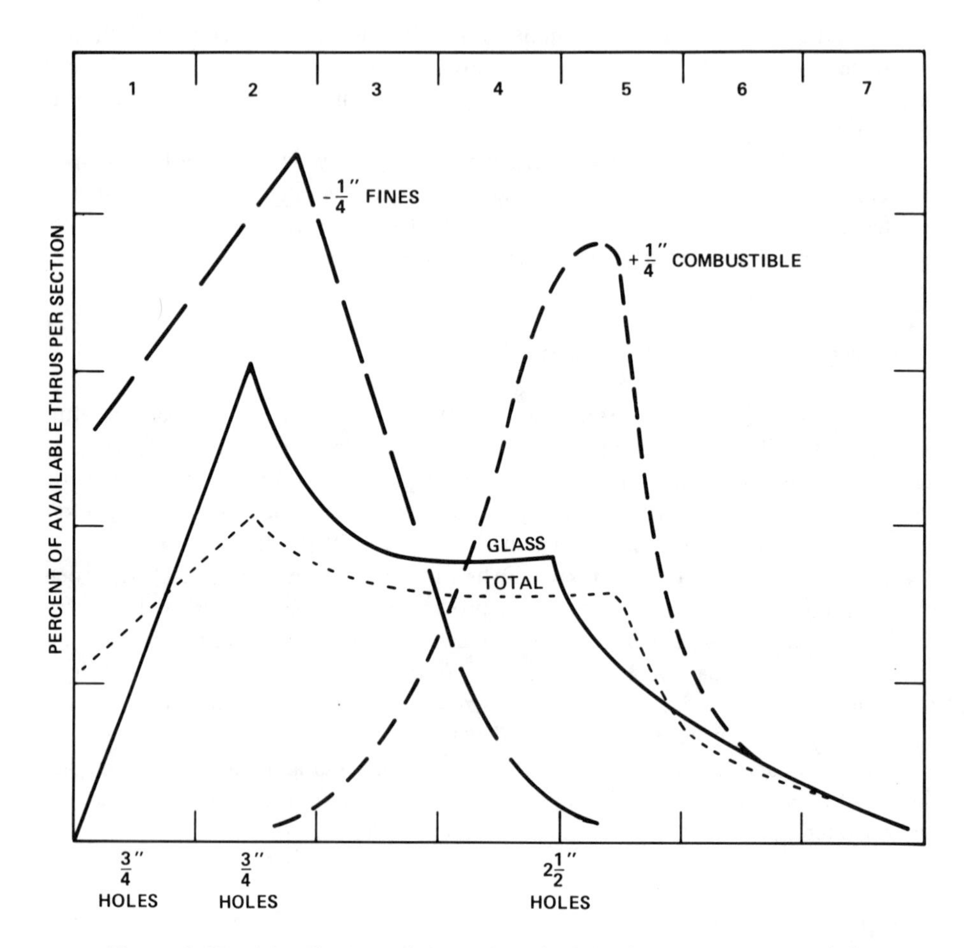

Figure 6–27. Distribution of thrus along length of primary trommel [9].

section. The plus-1/4-inch combustibles, mostly paper and plastics, started to come through in the 2-1/2-inch section, peaked and were depleted before the end of the trommel.

Tests of Trommels by Cal Recovery

Extensive studies by Cal Recovery, funded by the DOE [7], have addressed many of the questions on trommel theory and performance. The tests were limited to pretrommeling of MSW and posttrommeling of air-classified light fraction (ACLF).

Cal Recovery tested a 7.5-foot (2.36-m)-diameter by 20-foot (6.1-m)-long trommel processing raw MSW as well as ACLF. Various speeds and loading rates were used, and samples were taken to determine overall screening efficiency and efficiency for specific components such as glass and combustibles. Screen inclinations of 2°, 4°, 6° and 8° also were tested.

Analysis of the tests at different inclinations shows that the range from 2° at 40% of critical speed to 6° at 60% speed constitutes the range of optimum screening efficiency, wherein rotational speed can compensate for angle. As capacity increases with both speed and inclination and it is easier to feed with the higher angle, the higher speed would seem to be the optimum.

The set of curves shown in Figure 6-28 has been constructed from the Cal Recovery data by mathematical analysis, showing how the specific capacity and specific area of a trommel operated at a 6° slope vary with the percentage of critical speed, while screening raw MSW with 120- and 89-mm apertures. It is apparent that at low specific areas, such as 5 ft^2/TPH, the efficiency of the trommel is extremely sensitive to loading when operated at speeds outside of a narrow band centered at 55% of critical speed. At high specific areas, such as 20 ft^2/TPH, the speed makes relatively little difference. At the optimum speed, the screening efficiency does not fall much with increased loading. These curves help explain why the performance of trommel screens has been so difficult to correlate.

The curves for ACLF (Figure 6-29) [21], using 12-mm apertures, have a strikingly similar shape but substantially different specific areas due to the large amount of material of near-screen size. Similar curves were produced from test data on a screen inclined at 15°, with the exception that at light loadings higher speeds were required to attain optimum efficiency.

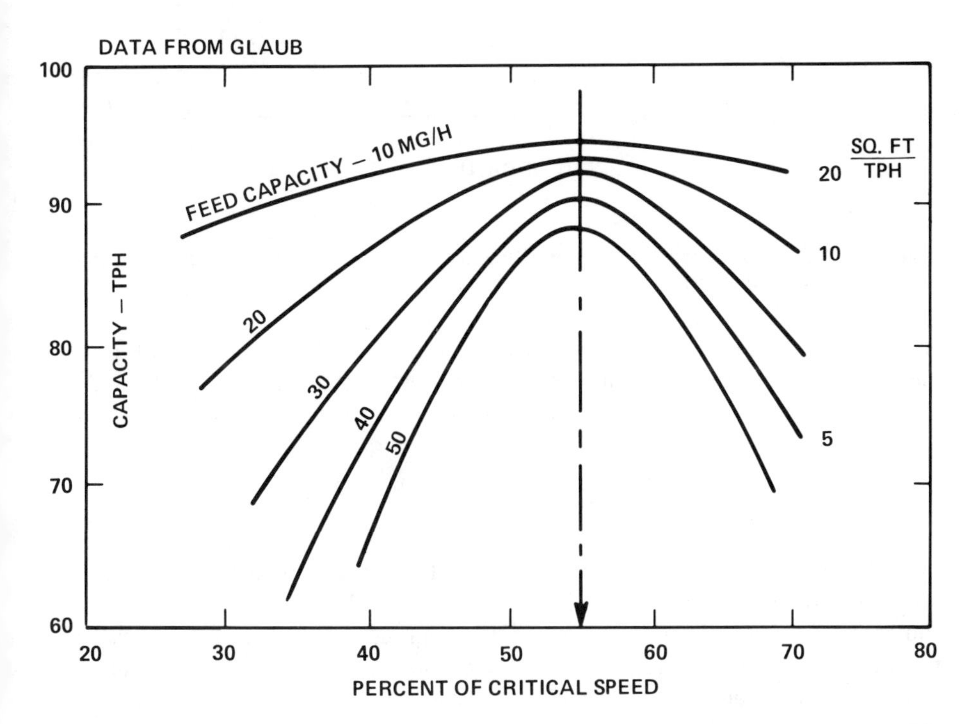

Figure 6-28. Variations of screening efficiency with capacity and percentage of critical speed – MSW (data from Glaub et al. [7]).

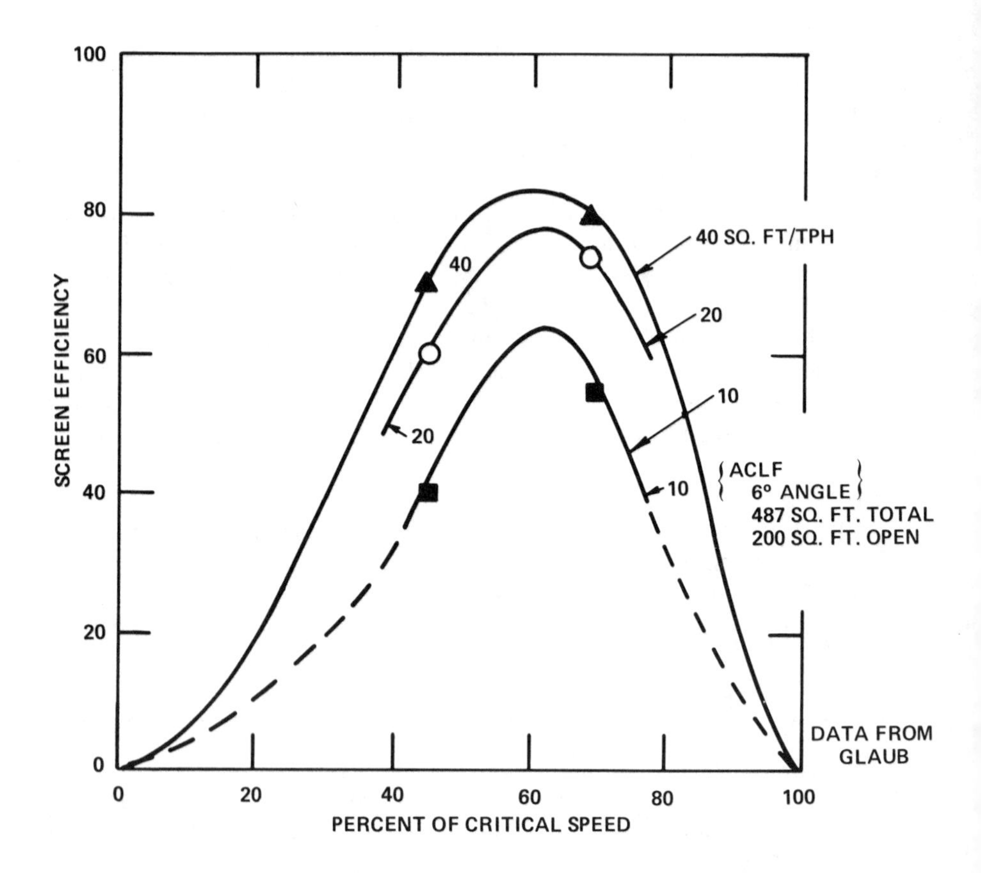

Figure 6–29. Variations of screening efficiency with capacity and percentage of critical speed Air-classified Light Fraction (ACLF) (data from Glaub et al. [7]).

Efforts to determine the relationship between recovery of energy with product quality had these results: beyond a certain point additional screening results in excessive energy losses, which are not justified by improved product quality.

Tests of Disc Screens by SPM Group in Denver

The testing performed by the SPM Group in Denver, Colorado for the DOE [22] produced data on the screening efficiency, product quality and yield of combustibles and energy when MSW was processed on a line consisting of a bag-breaking flail mill, three disc screens and a winnowing device. As the objective was to produce a low-ash RDF without losing excessive amounts of combustible material, all streams were weighed and analyzed after optimum operating conditions had been determined and established. The resulting data are presented in Figure 6–30, showing ash removal

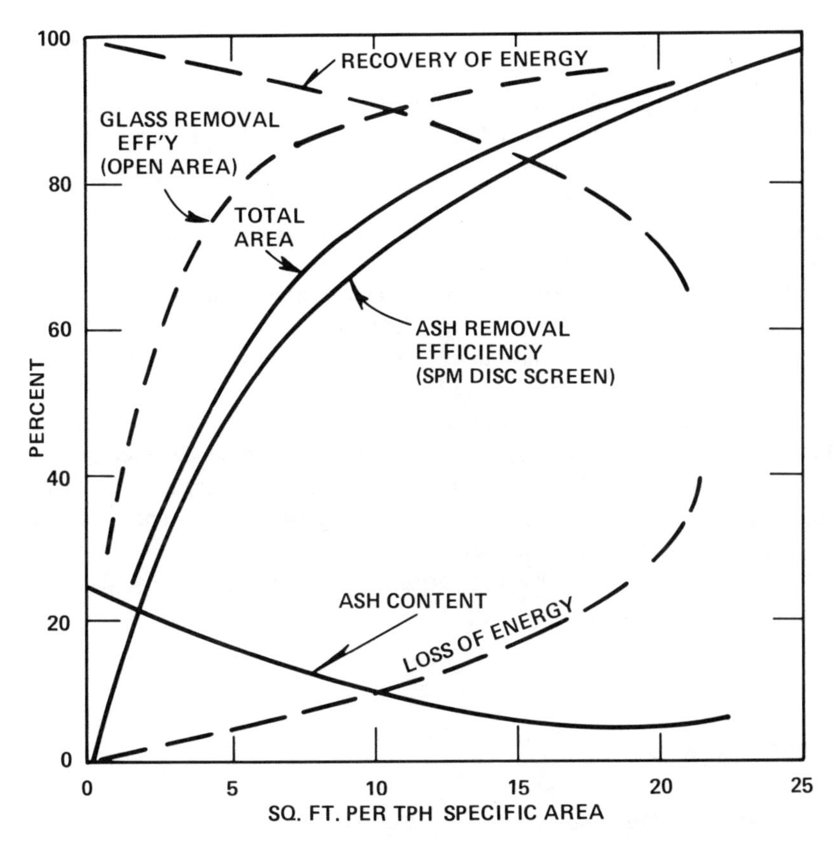

Figure 6-30. Percentage removal of glass and ash and loss and recovery of energy versus specific area of SPM disc screen (data from SPM Group [22]).

efficiency, ash content and energy losses plotted against disc screen plan area. The three screen stages define points of increased efficiency with increased screen area.

Comparable data from the Cal Recovery tests, based on total screen area, plotted for comparison, show approximately the same screening efficiency for glass removal. This would indicate that a single trommel, 7.5 feet wide × 30 feet long, having 470 ft^2 of surface, might perform approximately as well as a set of three disc screens 5 feet wide × 15 feet long each, totaling 225 ft^2 of surface. This is not surprising when one considers that something less than half the trommel surface is active.

The curves showing loss of energy due to screening out combustibles along with high-ash materials indicate that while the first two stages of screening resulted in a rapid reduction in ash content with relatively small losses of energy, the third stage paid a heavy energy penalty to achieve a slight improvement in ash content.

SUMMARY

Size separation by screening has been found to be highly effective as a means of removing glass and other inert materials from raw MSW so that they are not shredded into fines that cannot readily be removed from the combustible fraction.

In addition, screens are effective both in removing glassy grit from the light fraction from air classifiers and in recovering combustibles from primary trommel unders and from air classifier heavy fractions.

Separations are most effective when the material to be passed through the screen has a size distribution considerably less than the screen apertures. This assures a high probability of passage and, hence, a high screen efficiency, thus reducing the surface requirements.

Flaky materials that approximate hole size have low screening efficiencies, making it possible to recover these materials as screen overs while filtering out fines.

The conditions that control screen feed capacity, thrus capacity and efficiency for the total stream and specific components are sensitive to the speed of trommels and disc screen rotors and to the rate of feeding.

It is important to optimize the performance of screens so that the objectives are efficiently met, such as recovery of maximum energy while producing high-quality products.

REFERENCES

1. Taggert, A.F. *Handbook of Mineral Dressing* (New York: John Wiley & Sons, Inc., 1945).
2. Sullivan, P.M., M.H. Stanczyk and M.J. Spendlove. "Resource Recovery from Raw Urban Refuse," Bu Mines RI 7760 (1973).
3. Douglas, E., and P.R. Birch. "Recovery of Potentially Re-useable Materials from Domestic Refuse by Physical Sorting," *Resource Recovery Cons.* 1:319–344 (1976).
4. NCRR. "Evaluation of Trommels for Waste-to-Energy Plants, Phase 2. Report on the Warren Springs Laboratory Test Series," final subcontractor report by Warren Springs Laboratory under US DOE Contract No. ACO3–80CS–24317, Task 8, Stevenage, U.K. (1981).
5. Woodruff, K.L., and E.P. Bales. "Preprocessing of MSW for Resource Recovery with a Trommel," paper presented at the 1978 ASME National Waste Processing Conference, Chicago, 1978.
6. Savage, G., and G.J. Trezek. "Screening Shredded Municipal Solid Waste," *Compost Sci.* 17(1):7 (1976).
7. Glaub, J.C., D.B. Jones, J.U. Tleimat and G.M. Savage. "Trommel Screen Research and Development for Applications in Resource Recovery," DOE/CS/20490–1, U.S. Dept. of Energy, Washington, DC.
8. Menon, K., W. Wellan, D. Bergstrom and F. Hasselriis. "Primary Trommeling of MSW in Eco-Fuel and other RDF Processes," CEA corporate R&D (unpublished report), 1978.

9. Hasselriis, F., and N.V.P. Menon. "Pilot and Full-Scale Tests of Primary Trommels used to Improve Quality of Refuse-Derived Fuel," paper presented at a conference entitled Resource Recovery from Municipal, Hazardous and Coal Solid Wastes, Miami, FL, May 1982.
10. Bagalman, P., J. Bernheisel and W. Parker. "Trommel Performance at Nominal Design Conditions," EPA Contract 68-01-4423, NCRR, Washington, DC (1978).
11. Campbell, J. "New Orleans Full Scale Trommel Evaluation, Final Report," DE-ACO3-80CS-24315, U.S. Dept. of Energy, Washington, DC (1981).
12. Parker, W.S. "Application of Trommeling to Prepared Fuels," in *Proc. Int. Conf. on Prepared Fuels and Resource Recovery Technology*, U.S. Department of Energy, Argonne National Laboratory, Chicago, IL (1981).
13. Hasselriis, F., H. Master, C. Konhein and H. Betzig. "Eco-Fuel II: The Third Generation," in *Proc. Int. Conf. on Prepared Fuels and Resource Recovery Technology*, Nashville, TN, Argonne National Laboratory, Argonne, IL (1981).
14. Morey, B., and A. Gupta. "A Review of Resource Recovery Technology," in *Energy and Resource Recovery from Industrial and Municipal Solid Wastes, AIChE Symp. Ser. No. Ser. No. 162* 73 (1977).
15. Fiscus, D.E., A.W. Joensen, A.O. Chantland and R.A. Olexsey. "Evaluation of the Performance of the Disc Screens Installed at the City of Ames, Iowa Resource Recovery Facility," paper presented at the ASME 1980 National Waste Processing Conference, Washington, DC.
16. Malan, G.M. "The Americology Resource Recovery System," in *Proc. Int. Conf. on Prepared Fuels and Resource Recovery Technology,* U.S. Department of Energy, Argonne National Laboratory, Chicago, IL (1981).
17. Vorstman, M.A., and M. Tels. "Segregation in Trommel-Sieving," in *Recycling Berlin '79*, Vol. 2, K.J. Thome-Koziesky, Ed. (Berlin: Springer-Verlag, 1979), pp. 1026-1033.
18. Alter, H., J. Gavis and M. Renard. "Design Models of Trommels for Resource Recovery Processing," in *Proc. Ninth Biennial ASME Nat. Waste Processing Conf.* New York: American Society of Mechanical Engineers, 1980.
19. Davis, E.W. "Fine Crushing in Ball Mills," *AIMME* 61:250-296 (1919).
20. Gaudin, A.M. *Principles of Mineral Dressing* (New York: McGraw-Hill Book Co., 1939).
21. Hasselriis, F. "Optimization of Solid Waste Separating Systems," in *Fourth Annual Oak Ridge National Laboratory Life Sciences Symposium* (Ann Arbor, MI: Ann Arbor Science Publishers, 1981).
22. SPM Group, Inc. "Refuse Derived Fuel Pre-separator Process Technology Evaluation," DOE/CS/20523-1, U.S. Department of Energy (1982).

CHAPTER 7

Selecting Unit Operations and Optimizing the Process

The total process objectives and the nature of the MSW feed to the plant determine the type and arrangement of the selected unit operations. Also, available equipment determines what objectives are reasonable and possible with the actual MSW. The matching of equipment to MSW to objectives is an optimization process that must be based on extensive experimental data.

The intimate relationship between the inputs and outputs of the various streams in the process creates a major problem in the development of resource recovery processes. What enters a given operation leaves in a different state; it then becomes the input to one or more other devices, affecting the outputs of these units. We must know at the outset how each machine operates with the outputs of upstream machines under various capacities. Lacking data obtained from a full-scale plant, pilot operations or plants have been used to obtain preliminary data. For reasons that are not always apparent, large plants scaled up from pilot plants do not necessarily produce the same results. While the operations may not seem to change much, the actual machinery needed may be drastically changed, and performance is highly load-sensitive.

The process by which resource recovery plants were developed will illustrate the difficulties that have been encountered in obtaining information about processing refuse and applying this to larger plants.

HISTORY OF PROCESS SELECTION IN RESOURCE RECOVERY

Resource recovery was practiced extensively in Dickens' day but was mechanized only recently. The trommel screen was promoted in the United Kingdom in 1924. The present concepts of processing refuse have developed rapidly in the last two decades, when it was suddenly realized that landfills were beginning to fill up rapidly. And incinerators, which had been an alternate method of waste disposal, were suddenly condemned as polluters. Finally, the energy crisis directed our attention to this waste of energy and recoverable materials. Tidiness was an objective that was supported by economic benefits, at least in the eyes of the beholders.

The Bureau of Mines was mandated to study recovery of urban ore, including MSW and incinerator residues. The solid waste research program at Salt Lake City, reported by Dean [1] in 1971, produced some of the first data on recovery of a low-ash shredded material that showed potential as a source of heat and power. The material was from shredded refuse from the city of Madison, which was testing reduction of waste volume by shredding in 1972. Boettcher [2] and others did further testing of air classifiers and screens and recommended that a pilot plant be built for further investigation. Sullivan and Stanczyk [3] reported on the potential of recycling metals, minerals and fuels from urban refuse, and on the operation of a College Park pilot plant. The National Center for Resource Recovery [4] prepared an Engineering Feasibility Study in 1972 and did extensive development and demonstration of methods for separation of metals, glass and RDF from MSW. The NCRR proceeded to evaluate the basic unit operations associated with resource recovery, culminating in the design, construction and testing of Recovery I in New Orleans [5].

The Franklin, Ohio Solid Waste Recycling plant, using wet processing, was operating in 1972, and its wet RDF product was tested in a stoker-fired boiler in 1973 [6], as the emphasis moved from recovery of metals and glass toward recovering the energy component of MSW and toward recovery before burning, rather than from the burned residue of municipal incinerators.

Over the last decade there has been no dearth of schemes to convert refuse into metals, glass and fuel. With the full support of the Bureau of Mines, EPA, DOE and private enterprise, a number of demonstration plants were designed and built. The most significant of these are the St. Louis project, which shredded and air classified the refuse for combustion in Union Electric's coal-fired boilers, and the more complex Ames, Iowa system, which included a trommel with various hole sizes to remove different materials from the air-classified heavy fraction. Other projects demonstrating pyrolysis have been less successful than direct combustion.

The College Park pilot plant of the Bureau of Mines became operational, demonstrating the composition and characteristics of urban refuse, as well as methods by which the various components can be sorted out and reclaimed for useful purposes [7].

American Can, Continental Can, General Electric, Garrett Research, Combustion Engineering and CEA developed proprietary processes for resource recovery, which ultimately led to construction of Americology in Milwaukee, Wisconsin, the Eco-Fuel process in Bridgeport, Connecticut, Hempstead, New York and Dade County, Florida and the new facility in Madison, Wisconsin.

CEA designed and built a full-scale facility (15–25 TPH) plant to process the refuse of Brockton, Massachusetts, modeled after the College Park pilot plant: a flail mill was installed with a built-in air classifier, followed by a secondary shredder. The fluff RDF product was stored and then fed through a pneumatic system into a Double-Vortex® burner for disposal by combustion. The shortcomings of this system were quickly demonstrated. It was revamped several times in the evolution of process, which produced a dried, pulverized-RDF Eco-Fuel II. The story of this highly developmental period is worth reviewing because it illustrates the problems in scaling up from pilot to full scale.

The College Park and Warren Springs pilot plants demonstrated the basic methods for separating the metal, glass and food heavies from the light, paper and plastics fraction, yielding a combustible fraction, and for refining the ferrous, nonferrous and aluminum metal and glass fractions into valuable products.

A major shortcoming of the data generated by these pilot plants was that the plants were operated at very light loadings, which would not be practical when scaled up to full-sized plants. When heavier loadings are used, the efficiency of these operations deteriorates substantially, resulting in large amounts of material reporting to the wrong fraction and contaminating the products, as well as loss of good product.

The initiative of Dr. Harvey Alter of the NCRR to develop the ASTM Committee E38 on Resource Recovery led to an industry-wide activity to write specifications for the by-products of resource recovery for the recovered metals, glass and energy products. Participation by producers, consumers, laboratories to carry out the testing of these materials, designers and operators of refuse-processing plants and utility users led to publication of a series of product specifications, with accompanying methods for sample collection, preparation and testing of the products.

In many instances, these specifications prepared for metals, glass and fuel products have been difficult to meet by the producers on the first try. Plant modifications subsequently have been made, with varying degrees of success. The process of testing the plant products led to identification of the problems within unit operations, the optimization of some of them, the addition of others and elimination of yet others.

A major obstacle in the resource recovery industry lies in development of markets for the recovered products needed to produce the anticipated revenues. The steady and occasionally drastic increases in the cost of energy have increased the value of fuel products. The markets for ferrous metals, aluminum and especially glass have been disappointing by comparison with the enthusiastic predictions.

With this historical background, we can look now at the unit operations that have become a part of resource recovery plants, starting with the objective products, the data that describe the interaction between the machines and the process streams, the matrix of operations that describes the flow of materials through such plants and, finally, the effectiveness of the individual operations themselves and their relationship to the overall effectiveness of the process.

Pilot-Plant Test Data

The College Park pilot plant (Figure 7-1) is a good illustration of the complexity of plant required to meet the objective of recovery and recycling of all values in the MSW feed. Also, it is the prototype for the Monroe County plant, which is designed to recover metals and glass, as well as processed RDF for the utility boilers of Rochester Gas and Electric. It may be compared with simpler plants, such as the St. Louis demonstration, the more complex Ames, Iowa plant, and the equally complex Americology and Recovery I facilities.

The published reports on the College Park plant did not give the flowrates of the streams, but concentrated on the materials balances of the refuse separations and the quality of the recovered products. No provision was made in the pilot plant to measure the streams in the course of their flow through the plant. This problem persists even in full-scale plants because it is difficult to weigh the materials in transit

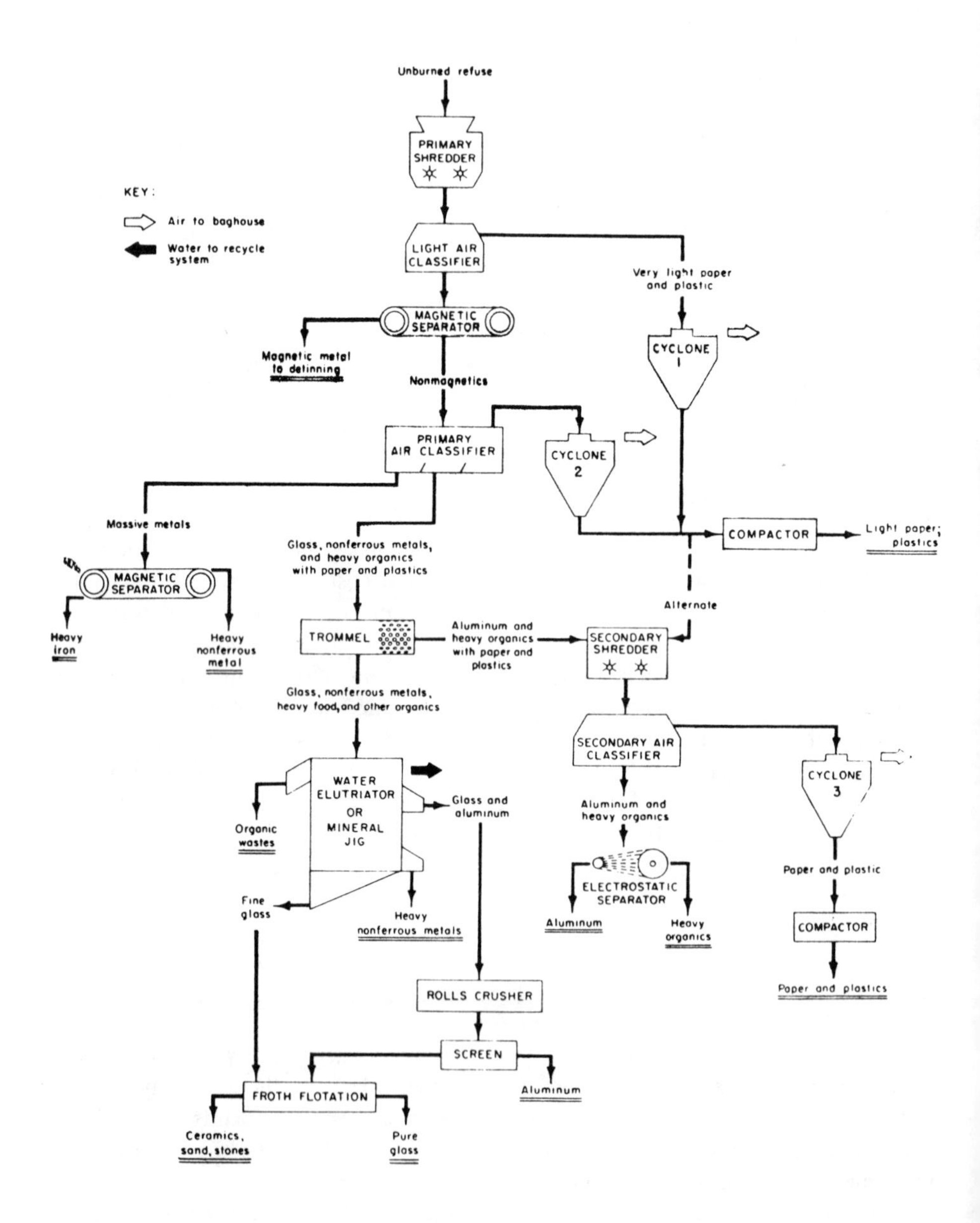

Figure 7–1. College Park pilot plant, Bureau of Mines.

and to stop the plant to make measurements. It is possible, however, to weigh the output streams resulting from measured inputs and to analyze the output streams to report their quality. These data can be reassembled to obtain the properties of the internal streams. The difficulty in measuring instantaneous flowrates of both internal and exiting streams leaves a gap in the information about the variations in load to which the machines were subjected during generation of the averaged data. These variations influence the effectiveness of the individual pieces of equipment. The averaged data obtained from the output streams are conveniently smoothed, however.

The final report published on the work done at the College Park facility covers seven years of testing refuse from many localities. It analyzes in detail the nature of the refuse supplied and the variation in samples tested [7]. These data detail the properties of the product streams and the degree of recovery possible, along with the amounts recovered per ton of refuse and the theoretical value of the recovered products based on their quality. Unfortunately, little information has been provided that could be used to select machinery that could produce products of this quality on a full-scale basis.

The same approach was used at Warren Springs, where many methods were demonstrated for separating materials, without much quantification of the flowrates as they relate to the quality of the separations. The data showing the quality of product cannot be scaled up to design full-sized plants.

The early designers of resource recovery plants generally went to vendors who, they hoped, would know how to build and guarantee equipment to perform the unit operations. The available vendors had background in somewhat related fields but little, if any, experience with refuse. Shredders and air classifiers were familiar equipment, long used for paper, wood, grain and other uniform products. A great effort was made to test these machines with refuse; however, in general it was necessary to build a full-scale facility to permit any real testing or to use existing facilities designed for other purposes. Americology and Continental Can built pilot plants to test air classifiers and other equipment. CEA built a 15-25 TPH commercial-scale plant at Brockton and, later, a pilot plant to optimize the design of the Bridgeport process.

Shredders were demonstrated in a number of plants where the objective was to reduce the bulk of refuse before landfilling, to avoid the requirement for cover. These demonstrations led to the widespread use of shredders in resource recovery plants, even though other equipment might have been more suitable.

The trommels tested in Warren Springs attracted the interest of the NCRR, as well as the Bureau of Mines, and led to the application of this device in many places within the process.

Pneumatic conveying and boiler feeding equipment had long been used for handling bagasse and wood products and seemed suitable for shredded refuse.

Whatever equipment is selected and whatever sequence of operations is used, the result will be that the MSW feedstream will be altered in its characteristics and relative amounts of the various species contained in it.

Size separation equipment will split the stream by size, and size reduction equipment will change the size of different species in different ways, permitting further size separations. Air classification devices will make aerodynamic separations of the light materials, including paper, plastics, fine dirt and sand, from glass, wood and other heavy materials.

MATRIX APPROACH TO PROCESS SELECTION

The complex collection of machinery that constitutes the refuse processing plant is not unlike that assembled in a chemical plant. It is not surprising that the mathematical approaches developed to cope with chemical plant design would find their application in resource recovery.

In the matrix approach, each machine is a node, or focal point, with entering and leaving streams. These streams may consist of a number of species, which may be treated separately or as a group, as they flow through the process. The mathematical approach takes into account the manner in which the streams are split, diverted, recycled or recirculated as required or desired by the process. Each node is a unit operation, which performs divisions both of the stream flow of each of the species and of the entire flow. A set of equations may be developed for each of the operations that determine the splits or division that will take place under the conditions of operation.

This systematic approach can be used to design a flowsheet, provided there are data on the input materials and on the "splits" that are characteristic of each unit operation. To obtain these data, it is necessary to run tests on at least fairly typical material, using appropriate machines, loading the machines to realistic levels so that the splits obtained from analysis of the products are representative of full-scale plant conditions.

Example: Mass Balance of a Primary Trommel

The mass balance of a primary trommel will serve as an example. Tables 7-1 and 7-2 show data collected on a wet and dry basis respectively, from the operation of the primary trommel installed at Recovery I, in New Orleans. The trommel was fed in a normal manner. Samples of the overs and unders were taken to determine the relative flowrates and compositional analysis of these streams. The feed was calculated by adding the overs and unders compositions and corrected for the measured flowrates.

The "split" of each constituent species to unders was calculated as follows:

$$\text{Split to unders} = (\text{unders flow})/(\text{unders} + \text{overs})$$

Total unders were 39% of the feed on both wet and dry bases. The split fractions ranged from 26% for paper to about 100% for glass, stone, metals and fines, including yard waste.

Table 7-3 shows the screening efficiencies obtained from two tests to point out the variation in performance that can be observed between two successive runs. Note that the heavies (food, rubber, leather, textiles and wood) have the highest variation

Table 7-1. Mass Balance of a Primary Trommel (Recovery I)

Species	*Feed (wet)*	*Total*	*Split (under)*	*Overs*	*Total*	*Unders*	*Total*
Paper	26.86		0.260	19.88		6.98	
Plastics	3.08		0.393	1.87		1.21	
Food							
Rubber	23.24	62.97	0.267	17.03	39.88	6.21	23.09
Textiles							
Wood							
Yard	9.79		0.888	1.10		8.69	
Ferrous	3.51		0.614	1.35		2.16	
Aluminum	0.28	3.84	0.923	0.02	2.53	0.26	3.50
Nonferrous	0.05		1.000	0.00		0.05	
Glass	14.26	14.26	0.971	18.06	18.06	14.09	14.09
Stone	3.32	3.32	1.000	0.00		3.32	3.23
Fines	15.61	15.61	0.976	0.37	0.37	15.24	15.24
	100.00			60.87		39.13	

Table 7-2. Mass Balance of a Primary Trommel (Recovery I)

Species	*Feed (dry)*	*Total*	*Split (under)*	*Overs*	*Total*	*Unders*	*Total*
Paper	23.59		0.201	18.85		4.74	
Plastics	3.62		0.389	2.21		1.41	
Food	1.96		0.265	1.44		0.52	
Rubber	0.38	56.8	0.695	0.12	40.59	0.26	16.21
Textiles	15.92		0.201	12.72		3.20	
Wood	5.69		0.221	4.43		1.26	
Yard	5.64		0.892	0.82		4.82	
Ferrous	4.53		0.612	1.76		2.77	
Aluminum	0.35	5.29	0.920	0.03	2.53	0.32	3.50
Nonferrous	0.41		1.000	0.00		0.41	
Glass	18.47		0.988	18.06		0.41	
Stone	3.94	22.41	1.000	0.00	18.06	3.94	22.00
Fines	15.84	15.84	0.973	0.43	18.06	15.41	22.00
	100.00			60.87		39.13	

between runs, and that yard waste is a near second. The main species, paper and plastics, do not vary as much, nor do glass and fines, which are far smaller than screen size. The operation can be evaluated in terms of these parameters:

$$\text{Yield of combustibles} = \frac{15.3}{15.3 + 3.2} = 82.7\%$$

Table 7-3. Screening Efficiency[a] of Primary Trommel

Species	*Run #1 (%)*	*Run #2 (%)*	*Percentage Variation (%)*
Paper	73.6	61.8	2.6
Plastics	78.2	70.5	10.0
Heavies (FRLTW)	51.9	80.8	41.0
Yard Waste	92.3	64.1	35.0
Ferrous	84.8	79.8	19.0
Aluminum	92.3	89.8	2.7
Glass	98.8	99.1	0.3
Stone	100.0	100.0	0.0
Minus-1/4-inch	97.6	97.5	1.0

[a]Screening efficiency = $\dfrac{\text{\% of component to unders}}{\text{\% of component under 4 inches in feed}}$

Recovery:

$$\text{Ferrous} = \frac{1.72}{1.72 + 0.03} = 98.3\%$$

$$\text{Glass} = \frac{2.01}{2.01 + 0.49} = 80.0\%$$

$$\text{Aluminum} = \frac{0.21}{0.21 + 0.04} = 84.0\%$$

Contamination of combustibles:

$$\text{C (ferrous)} = \frac{0.03}{15.3} = 0.2\%$$

$$\text{C (inerts)} = \frac{0.49 + 1.07}{15.3} = 9.8\%$$

$$\text{C (aluminum)} = \frac{0.04}{15.3} = 0.26\%$$

Noncombustible content of extracts:

$$\text{Extraneous inerts} = \frac{1.56 + 0.08}{16.94} = 9.68\%$$

It should be noted that the total ash of the extracted fuel product includes not only the extraneous inerts, but also the inherent ash, which ranges from 3 to 9% and averages about 6% for the combustible content of MSW. Thus, the total ash, including 9.68% extraneous inerts, would be about 16%.

Example: Splits of an Air Classifier

Now let us take another example, showing the performance of an air classifier following a shredding operation. The feed material is listed as a series of species, paper, food, yard waste and textiles (combustibles); ferrous, nonferrous and aluminum (metals); and glass, dirt and rock (inerts), with the percentages of each in the feed.

The total feedrate is multiplied by the percentage to get the flowrate of each species, which is, in turn, multiplied by the split or fraction lifted by the air classifier to obtain the flowrate of the light fraction. The difference between the feed and the light fraction is the heavy fraction, of course. Subtotals of the combustible, metal and inert fractions may be taken.

The mass balance for the air classifier is shown in Table 7-4. Based on the information in this table, we can make the following calculations:

$$\text{Extraneous ash} = \frac{1.56 + 0.08 + 15.3 \times 0.06}{16.94} = 15.68\%$$

$$\text{Noncombustible content of rejects} = \frac{2.06 + 2.81}{8.07} = 60\%$$

Again, the total ash probably would be about 22% when the inherent ash of the light fraction combustibles is taken into account.

Example: Flowsheet of a Complete Pilot Plant

To further illustrate the use of this approach, we may make use of the extensive data generated by the Bureau of Mines at the College Park pilot plant. One can refer to the flowsheet shown in Figure 7-1 to follow the process and to understand which sample points were used to define the composition of the flowstreams and the products. Figure 7-2 shows the flowsheet in a simplified form.

The products of this plant are shown in Table 7-5, expressed as a percentage of the incoming refuse. Two samples are shown for comparison, one with a larger percentage of corrugated cardboard, which results in a shift from Cyclones 1 & 2 to Cyclone 3. Notice that the total amount is quite close, however. This illustrates the effect of variability of the MSW feed on the equipment and the need to design for the range of variation, rather than for spot data.

Table 7-6 lists the plant flows (TPH). The logic in this table becomes apparent when it is read in combination with Figure 7-2. The TPH of each component are traced through each unit operation.

The sum of all of the sampled output streams gives the total input to the plant, except for the moisture that is lost in processing. (Note that most data obtained

Table 7-4. Mass Balance of an Air Classifier

Species	*Percentage*	*Feed 25 TPH*	*Total*	*Split (%)*	*Lights (TPH)*	*Total C/M/I*[a]	*Heavies (TPH)*	*Total C/M/I*
Paper	43	10.75		0.924	9.93		0.82	
Food	12	3.00	18.5	0.660	1.98	15.3	1.02	3.20
Yard	14	3.50		0.826	2.89		0.61	
Textiles	5	1.25		0.400	0.50		0.75	
Ferrous	7	1.75		0.017	0.03		1.72	
Nonferrous	0.5	0.13	2.13	0.077	0.01	0.08	0.13	2.06
Aluminum	1.0	0.25		0.160	0.04		0.21	
Glass	10	2.50		0.198	0.49		2.01	
Sand	7.5	1.87	4.37	0.572	1.07	1.56	0.80	2.81
Totals:	100	25.00	25.00	1.000	16.94	16.94	8.07	8.07

[a]C = Combustibles
M = Metals
I = Inerts

from pilot plants or full-scale plants by summing the outputs suffers from this problem.) It has been estimated that the MSW feed may contain as much as 10% more moisture than is reported. The work done in shredding, plus the drying done by the fan and cyclone systems, removes a substantial amount of moisture, especially if the throughput of the plant is relatively small. When the plants are scaled up from pilot plant data, the throughputs will be relatively large and the moisture removed relatively less.

The economic viability of resource recovery facilities depends heavily on the quality of the products because this will determine the price that can be obtained and the market for these products. If ferrous metals are to be sold, they must meet the requirements of the facility that will further process them into usable metal. The degree of processing directly affects the market value: clean metal without contamination gets top dollar, while dirty, mixed metal may have to be taken to landfill. Likewise, the fuel product may not be economical to burn if it has excessively high ash and moisture content. The standard that must be met has to be worked out between producer and user and a balance struck between what the user would like and what the producer can deliver for the price.

Before the agreement can be reached, someone must provide sufficient sample quantities of the product, which the user can evaluate to make the balance between the cost of production and the value to the user.

The data generated at College Park can be used to illustrate the processing steps that produce valuable products from MSW, as well as the manner in which the unit operations affect the qualities of the streams within the plant and its products. However, these data do not reveal the effect of loading on the effectiveness of separation processes.

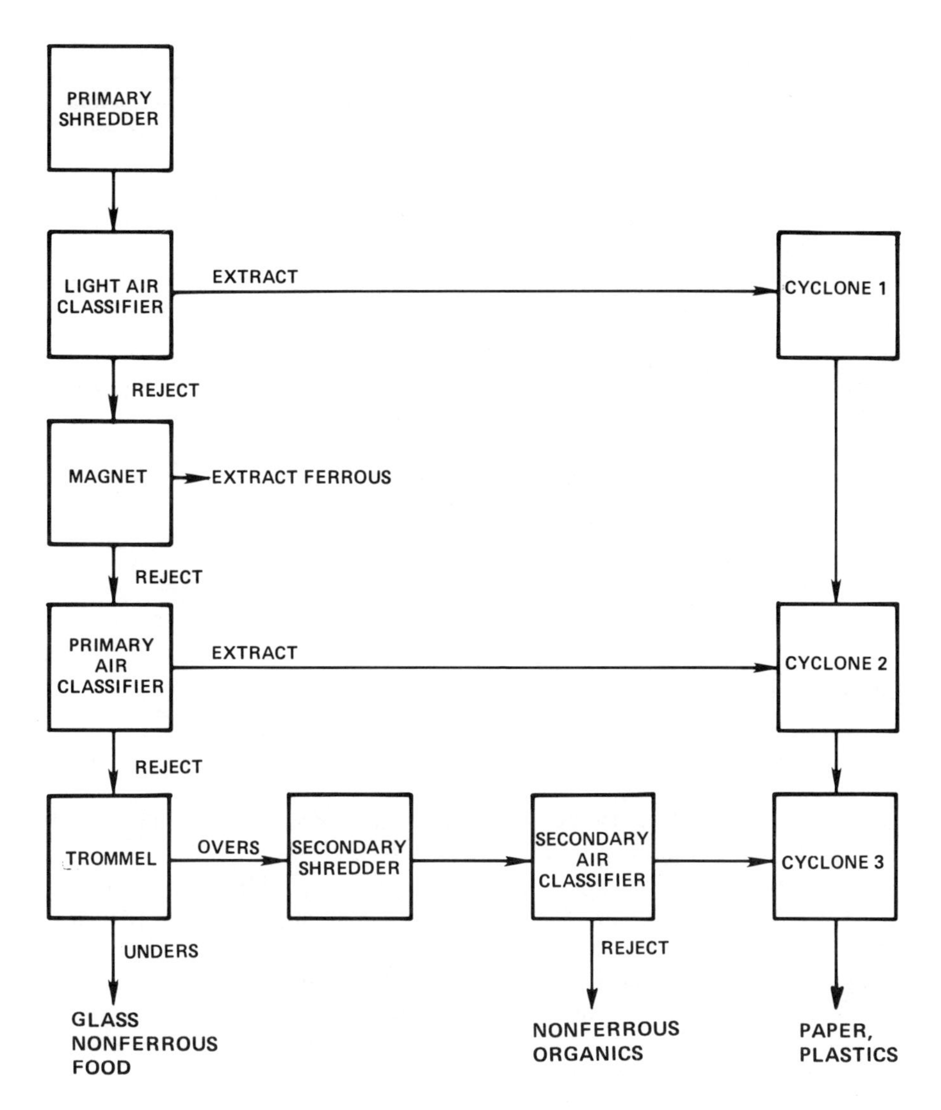

Figure 7-2. Simplified diagram of the College Park pilot plant.

MATRIX ANALYSIS OF MATERIALS FLOW

The data presented in Table 7-6 can be used to write a matrix in which each number in a line, representing a specific material, is multiplied by a split representing how the successive operations treat the material. This number we may call the fraction (f) of that material that continues down the line toward final product. Then 1-f is the amount that goes elsewhere at this operation.

Table 7-5. Percentage of Input Refuse in Process Streams

	Sample A	*Sample B*
Products and Streams		
Magnetic belt	7.1	6.8
Primary A-C Heavies	1.4	1.2
Trommel unders	22.6	25.0
Aspirator Heavies	3.2	6.9
Cyclones 1 and 2	54.6	36.1
Cyclone 3	10.9	23.9
Cyclones 1, 2 and 3	65.5	64.0

Taking paper, for example, we can write the line showing the weights at each point in the series of operations as follows:

$$\text{Paper: } 1686 \times 0.24 = 402 \times 0.98 = 394 \times 0.964 = 380 \times 0.755$$
$$= 287 \times 0.975 = 280$$

This is the same as a matrix expressed this way:

$$\text{Paper: } 1686 \times (0.24\ 0.98\ 0.964\ 0.755\ 0.975) = 402\ 394\ 380\ 287\ 280$$

The entire flowchart then can be expressed as a matrix and then calculated by hand or by computer.

The matrix of fractions, f, for a given process at a given flowrate (or near it) can be determined from test data, then used to process input materials other than those used as feed during the test. This assumes that each type of material has properties that do not change substantially when the relative amount of the material is changed.

The general form of the matrix can be expressed this way:

$$\begin{matrix} w1 \\ w2 \\ \cdot\cdot \\ \cdot\cdot \\ wi \end{matrix} \times \begin{matrix} (f11\ f12\ f13\ f14\ f15\ f16\ f17\ f18) \\ (f21\ f22\ f23\ f24\ \ldots\ldots\ldots\ f28) \\ \\ \\ (fil\ \ldots\ldots\ldots\ldots\ldots\ldots\ fij) \end{matrix} = D$$

The convenience of this approach is that once the determinant, "F," has been found from the test, the vertical columns represent individual unit operations, which can be shifted around to see what would happen if the machines were arranged differently. This is a treacherous procedure, however, as a change in operating order

Table 7-6. Process Materials Flow Analysis – College Park[a]

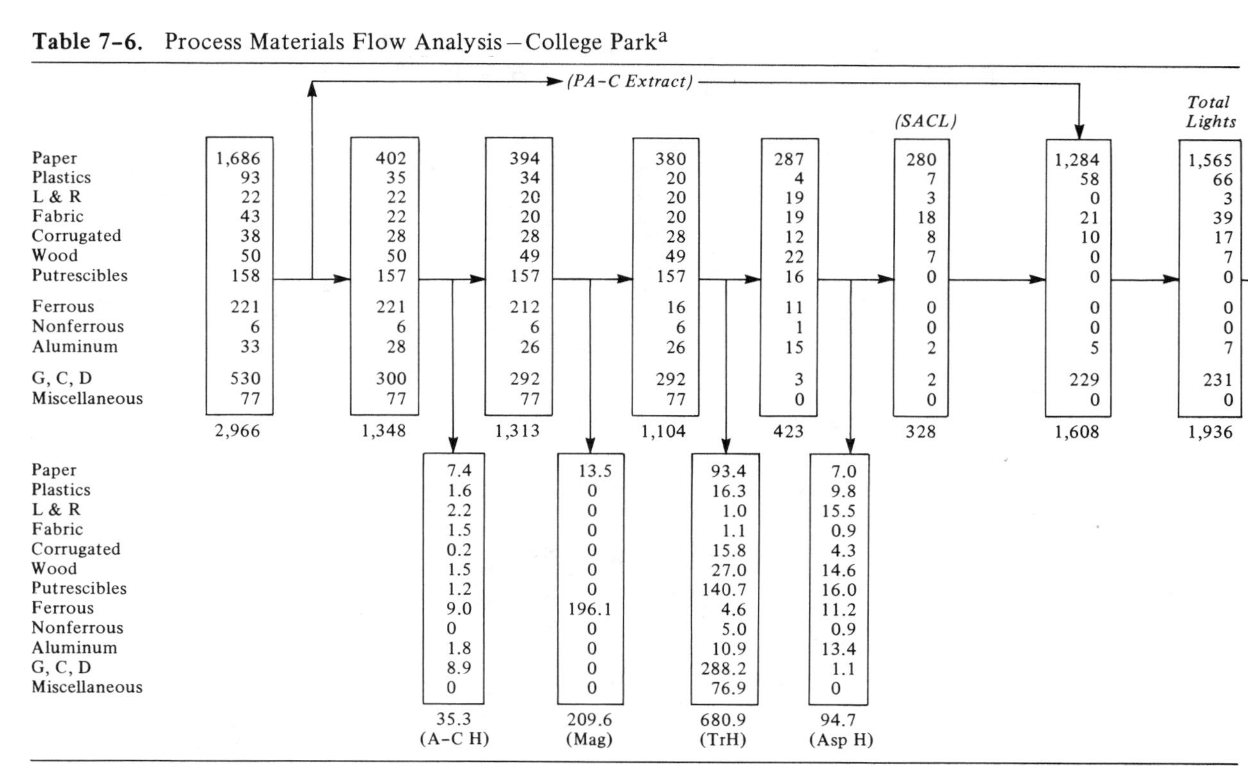

	Feed	Stream 2	Stream 3	Stream 4	Stream 5	(SACL)	(with PA-C Extract)	Total Lights
Paper	1,686	402	394	380	287	280	1,284	1,565
Plastics	93	35	34	20	4	7	58	66
L & R	22	22	20	20	19	3	0	3
Fabric	43	22	20	20	19	18	21	39
Corrugated	38	28	28	28	12	8	10	17
Wood	50	50	49	49	22	7	0	7
Putrescibles	158	157	157	157	16	0	0	0
Ferrous	221	221	212	16	11	0	0	0
Nonferrous	6	6	6	6	1	0	0	0
Aluminum	33	28	26	26	15	2	5	7
G, C, D	530	300	292	292	3	2	229	231
Miscellaneous	77	77	77	77	0	0	0	0
Total	2,966	1,348	1,313	1,104	423	328	1,608	1,936

	(A-C H)	(Mag)	(TrH)	(Asp H)
Paper	7.4	13.5	93.4	7.0
Plastics	1.6	0	16.3	9.8
L & R	2.2	0	1.0	15.5
Fabric	1.5	0	1.1	0.9
Corrugated	0.2	0	15.8	4.3
Wood	1.5	0	27.0	14.6
Putrescibles	1.2	0	140.7	16.0
Ferrous	9.0	196.1	4.6	11.2
Nonferrous	0	0	5.0	0.9
Aluminum	1.8	0	10.9	13.4
G, C, D	8.9	0	288.2	1.1
Miscellaneous	0	0	76.9	0
Total	35.3	209.6	680.9	94.7

[a]PA-C = primary air classifier
SA-C = secondary air classifier
L & R = leather and rubber
G, C, D = glass, ceramics, dirt
A-C H = air classifier heavies
TrH = trommel heavies
Asp H = aspirator (air classifier) heavies

[b]Pounds per batch processed, numbers rounded off.

probably would change the splits or F-factors because the machines operate differently on different materials. The procedure must be checked against test data before it can be used safely.

COMPUTER-AIDED ANALYSIS OF REFUSE PROCESSING

The conventional approach to stream analysis in MSW processing has been to obtain the concentrations of major components by dissecting samples and weighing the components. This is valuable in tracing the splits of metals, glass and combustibles, but does not show the splits of moisture, extraneous and inherent ash and combustible material, which are the essential components of the fuel product.

To trace moisture, inerts and combustible matter, the major components can be characterized in this respect and the mass flow of moisture, inerts and combustibles traced throughout the process by using the mass flow of each component and calculating the contribution to moisture, inerts and combustibles from each component. This procedure lends itself readily to computer analysis.

The flowsheet of any refuse-processing plant can be set up in matrix form and analyzed by matrix mathematics in a manner similar to that used for computer simulation of chemical processes. As illustrated in the previous example, the flowsheet is arranged as a series of nodes through which a multispecies material flows. Each node performs operations on each of the species, diverting or splitting them to either of two outlets. The outlets can exit the system or be directed to another node or device. Recirculation of streams back to an upstream point is permitted.

Computer programs were developed by the author for the Bridgeport and Newark processes to permit analysis of the entire process while varying the composition of the MSW input and the configuration and functions of the processing operations. They were written in APL language, which permits interactive programming and facilitates the use of matrix mathematics.

In the memory of the program is a species table, which contains a list of the components of refuse and their generic composition tabulated as combustibles, moisture, inerts, ferrous, nonferrous and glass. The term "combustibles" refers to the dry material, including inherent ash. The term combustible refers to dry material without inherent ash, in which case the ash is tabulated under "inerts."

Table 7-7 shows a typical species table, including combustibles with inherent ash included (inerts are zero) and another listing (i.e., "paperincl"), with the inherent ash listed under inerts. "Wetpaper" is included to reflect the influence of rainy weather on the moisture content of the refuse. The entry for wood can be used for food and yard wastes as both tend to have approximately the same high moisture and low ash content.

The processing system is set up in the form of nodes, as shown in Table 7-8. The feedrates to the system are listed, with their entering points. Each node is named and given a number. The nodes can have more than one input from the outside or from other nodes but can have only two exits, either to the outside or to another node.

Table 7-7. Table to Convert Species to Generic Composition

Species	*Combustible*	*Moist*	*Inert*	*Ferrous*	*Nonferrous*	*Glass*
Paper	0.920	0.080				
Cardboard	0.940	0.060				
Plastics	0.970	0.030				
Textiles	0.970	0.030				
Food	0.500	0.500				
Wood	0.460	0.540				
Yard Waste	0.460	0.540				
Wet Paper	0.550	0.450				
Combustibles	1.000					
Moisture		1.000				
Dirt			1.000			
Ferrous				1.000		
Nonferrous					1.000	
Glass						1.000
Paperincl Ash	0.860	0.060	0.080			
Cardincl	0.890	0.060	0.050			
Plasticincl	0.910	0.030	0.060			
Woodincl	0.440	0.540	0.020			
Textilincl	0.910	0.030	0.060			
Wetpaperincl	0.500	0.450	0.050			

Table 7-8. Feed Specifications and System Schematic

Species Name	*Entry Point*	*Feed-rate*
Paper	1	62.7
Plastics	1	1.4
Food/Yard Waste	1	6.0
Ferrous	1	7.6
Nonferrous	1	.8
Glass	1	9.0
Inerts	1	3.2
Moisture	1	16.1

Schematic

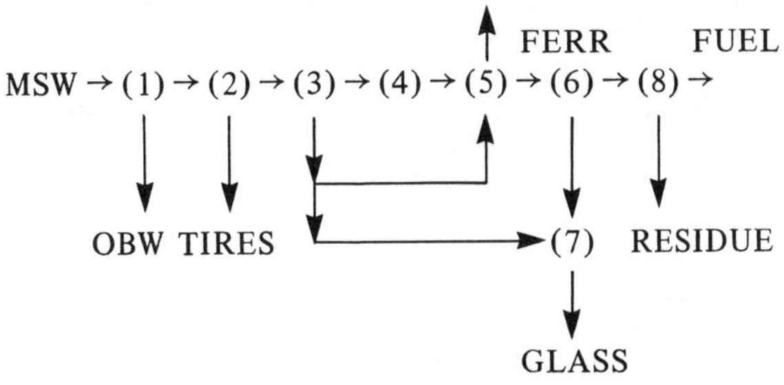

FERR FUEL

MSW → (1) → (2) → (3) → (4) → (5) → (6) → (8) →

OBW TIRES (7) RESIDUE

GLASS

Table 7-9 lists the nodes, or operations, and their destinations, along with the diversions or fractional splits that each device performs on combustibles, moisture, inerts, ferrous, nonferrous and glass. These data were obtained from pilot plant or full-scale tests and can be altered readily to see the sensitivity of the process to the splits performed by specific devices.

After the configuration and feed tables have been set up, the computer prints out Table 7-10, showing the mass flows in the internal streams and the outlets and giving the weight compositions in terms of the generic properties. When the feed is entered as paper, plastics, textiles, food and so forth, the computer consults the species table and converts the species composition to generic composition, as shown in the first line of the table. Table 7-11 shows the percentage composition of the feed and the internal and external streams, for reference.

Benefits of Computer Matrix Analysis

The illustration above is not presented to show the actual performance of the plant, although it is not far from reality, but rather to illustrate the use of computer matrix analysis as a tool for studying the process. With the process installed in the computer memory, these operations can be performed:

- The MSW feed can be varied to see what happens to the internal, product and residue streams. The moisture, paper and inert contents of the feed can be varied to match the changes that take place in the waste stream, seasonally and daily, as the refuse source fluctuates.
- The assumed fractional splits of the operations can be varied to discern the effects on the final products and the levels of performance required by the operations to meet product specifications or needs.

Table 7-9. Operation Configuration/Split Fraction Table

				Fractions Going to First Destination					
No.	*Operation Name*	*Destination (1st)*	*(2nd)*	*Combustibles*	*Moisture*	*Inert*	*Ferrous*	*Nonferrous*	*Glass*
1	OBW	2	0	0.97	1.00	1.00	0.98	0.99	1.00
2	Tires	3	0	0.98	1.00	1.00	1.00	1.00	1.00
3	Primary Trommel	5	4	0.35	0.35	0.50	0.80	0.90	0.80
4	Shredder	5	0	1.00	1.00	1.00	1.00	1.00	1.00
5	Magnet	6	0	0.95	0.99	1.00	0.08	0.98	1.00
6	ADS	8	7	0.90	1.00	0.85	0.85	0.90	0.13
7	Recovery	0	3	0.15	0.15	0.20	0.15	0.10	0.95
8	Secondary Trommel	0	0	0.06	0.06	0.65	1.00	0.00	0.65

Table 7-10. Summary of Flows in Each Branch, TPH

Stream	*Combustibles*	*Moisture*	*Inert*	*Ferrous*	*Nonferrous*	*Glass*	*Total*	*Product*
Entering Streams								
0→1	53.4	26.2	9.8	7.6	0.8	9.0	106.8	
Internal Streams								
1→2	51.8	26.2	9.8	7.4	0.8	9.0	105.0	
2→3	50.7	26.2	9.8	7.4	0.8	9.0	104.0	
3→5	19.3	9.2	5.0	6.0	0.8	7.5	47.8	
3→4	35.9	17.0	5.0	1.5	0.1	1.9	61.4	
4→5	35.9	17.0	5.0	1.5	0.1	1.9	61.4	
5→6	52.4	25.9	10.0	0.6	0.9	9.4	99.2	
6→8	47.2	25.9	8.5	0.6	0.8	1.2	84.1	
6→7	5.2	0.0	1.5	0.1	0.1	8.2	15.1	
7→3	4.5	0.0	0.1	0.1	0.1	0.4	5.2	
Exiting Streams								
1→0	1.6	0.0	0.0	0.2	0.0	0.0	1.8	OBW
2→0	1.0	0.0	0.0	0.0	0.0	0.0	1.0	Tires
5→0	2.8	0.3	0.0	6.9	0.0	0.0	9.9	Ferrous
7→0	0.8	0.0	1.3	0.0	0.0	7.8	9.9	Glass
8→0	2.8	1.6	5.5	0.5	0.0	0.8	11.2	Residue
8→0	44.4	24.4	3.0	0.0	0.8	0.4	72.9	Fuel

- Tests can be run in pilot or full-scale plants to obtain splits that can then be entered into the computer for study.
- Different arrangements of the processing equipment can be tried with the computer model to ascertain which produce optimum results.

The flowsheet analysis, converted from species to combustibles, moisture and noncombustibles, becomes a vehicle for analyzing the performance of the total process. As an example, the composition of the products and the rejects can be analyzed as follows:

- Determine the composition of the fuel product [8]:

Combustibles	– 60.9%		
Moisture	– 33.4%		
Inerts	– 4.1%	Combustibles	60.9%
Ferrous	– nil	Moisture:	33.4%
Nonferrous	– 1.1%	Inerts:	5.8%
Glass	– 0.6%		

If the combustibles contain 6% inherent ash, the true ash content of the product is 11.8 and true combustible is 54.9%.

Table 7-11. Composition of Streams and Products, %

Stream	*Combustibles*	*Moisture*	*Inert*	*Ferrous*	*Nonferrous*	*Glass*	*Total*	*Product*
Entering Streams								
0→1	50.0	24.5	9.2	7.1	0.7	8.4	100.0	
Internal Streams								
1→2	49.3	24.9	9.4	7.1	0.8	8.6	100.0	
2→3	48.8	25.2	9.5	7.2	0.8	8.7	100.0	
3→5	40.4	19.2	10.4	12.6	1.6	15.7	100.0	
3→4	58.5	27.7	8.1	2.5	0.1	3.1	100.0	
4→5	58.5	27.7	8.1	2.5	0.1	3.1	100.0	
5→6	52.6	26.1	10.1	0.6	0.9	9.5	100.0	
6→8	56.1	30.8	10.1	0.6	0.9	1.5	100.0	
6→7	34.7	0.0	9.9	0.6	0.6	54.2	100.0	
7→3	86.1	0.0	2.9	1.6	1.5	7.9	100.0	
Exiting Streams								
1→0	90.9	0.0	0.0	0.2	0.0	0.0	100.0	OBW
2→0	100.0	0.0	0.0	0.0	0.0	0.0	100.0	Tires
5→0	27.8	2.6	0.0	69.4	0.2	0.0	100.0	Ferrous
7→0	7.9	0.0	13.6	0.1	0.1	78.2	100.0	Glass
8→0	25.2	13.8	49.1	4.5	0.0	7.1	100.0	Residue
8→0	60.9	33.4	4.1	0.0	1.1	0.6	100.0	Fuel

- Determine losses of combustibles to reject streams:

OBW	–	1.6%
Tires	–	1.0%
With ferrous	–	2.8%
With glass	–	0.8%
In residue	–	2.8%
TOTAL:		9.0%

- Determine yield of combustibles:

$$\text{Yield} = \frac{53.4 - 9.0}{53.4} = 80\%$$

- Determine how reprocessing rejected streams would affect the yield of combustibles. By using an air knife to recover 1.7% of the combustibles from the ferrous stream and a trommel to recover 1.4% from the residue stream, the yield could be increased:

$$\text{Yield} = \frac{53.4 - 9 - 1.7 - 1.4}{53.4} = 90\%$$

The ash and moisture contents of the product would be increased by adding the recovered combustibles in the residue stream. This increase can be calculated from test data on the splits of the recovery operations.

OPTIMIZING OPERATIONS AND THE OVERALL PROCESS

Maximum RDF yield can be obtained by minimizing rejected streams or by processing them to recover the combustible fraction.

Maximim quality can be obtained by extracting the cleanest light fraction and rejecting the more contaminated material along with the metals, glass and other inerts. Optimizing yield and quality can be done by using more stages of separation to recover combustibles from rejects and also by improving the effectiveness of the separation processes themselves.

The overall process ultimately includes the energy use, including sale or use of steam and power. The quality of the product influences its value to the energy user. The yield of combustible influences the revenue obtained by the producer. The optimum balance between the complexity and effectiveness of processing thus must take into account the economic factors, including capital and operating and maintenance costs, which must not overlook the considerable cost of transportation and disposal of the residues resulting from processing for quality.

Figure 7-3 presents two mass balances, showing high and low recovery of fuel product in a process that includes a drying operation. The major differences result from the effectiveness of the screening operations, resulting in an increase in combustible recovery from 73 to 90%.

Figure 7-4 shows a quality/yield diagram, which serves to compare the performance of a number of processes in terms of the yield of fuel versus its ash content, as follows:

- **St. Louis**: This process had a minimum of rejected material and a maximum yield of RDF, with corresponding high ash.
- **Americology**: The performance before and after addition of disc screens is shown. The original process had a low yield due to an ineffective air classifier. The improved process using disc screens was not much more effective because the location and space available for the screens were not optimum.
- **Ames**: The initial performance was comparatively good; installation of disc screens reduced the RDF ash at the cost of somewhat reduced yield.
- **Madison**: The fuel ash is relatively low at the expense of low yield (about 60%). If there were a market for more RDF in this particular case, more RDF could be recovered if higher ash content could be tolerated.
- **Bridgeport**: The primary and secondary trommels used in this process result in fairly high yield and low ash.

```
MASS BALANCE - LOW RECOVERY

           MSW --> TROMMEL--MAGNET--DRYER--SCREEN--> PRODUCT

                                   Moisture
                                    = 16.4               RDF
                                       A
                                       :
Combustible: 55.6 -----o-------o-------:------o--------> 40.6
                       :       :       :      :
Moisture:    25.0 -----o-------o-------o---------------> 1.4
                       :       :              :
Inerts:      19.4 -----o-------o--------------o--------> 3.3
             -----     :       :              :          ----
   Total:   100.0      :       :              :          45.3
                       :       :              :
                       V       V              V
    Combustible:      5.4  +  1.0     +      8.6  = 15.0
    Moisture:         6.8  +  0.4     +       -      7.2
    Glass:            7.8  +   -      +       -      7.8
    Metals:           1.0  +  6.2     +      1.1     8.4
                     ----     ----          ----    ----
    Removed:         21.0  +  7.6     +      9.7  = 38.4

MASS BALANCE - HIGH RECOVERY

           MSW --> TROMMEL--MAGNET--DRYER--SCREEN--> PRODUCT

                                   Moisture
                                    = 20.6               RDF
                                       A
                                       :
Combustible: 55.6 -----o-------o-------:------o--------> 50.0
                       :       :       :      :
Moisture:    25.0 -----o-------o-------o---------------> 2.0
                       :       :              :
Inerts:      19.4 -----o-------o--------------o--------> 3.3
             ----      :       :              :          ----
   Totals:  100.0      :       :              :          55.3
                       :       :              :
                       V       V              V
    Combustible:      2.0  +  1.0     +      2.6  =  5.6
    Moisture:         2.0  +  0.4     +       -      2.4
    Glass:            7.8  +   -      +       -      7.8
    Metals:           1.0  +  6.2     +      1.1     8.4
                     ----     ----          ----    ----
    Totals           12.8  +  7.6     +      3.7  = 24.2
```

Figure 7-3. Mass balances with low and high recovery of combustibles.

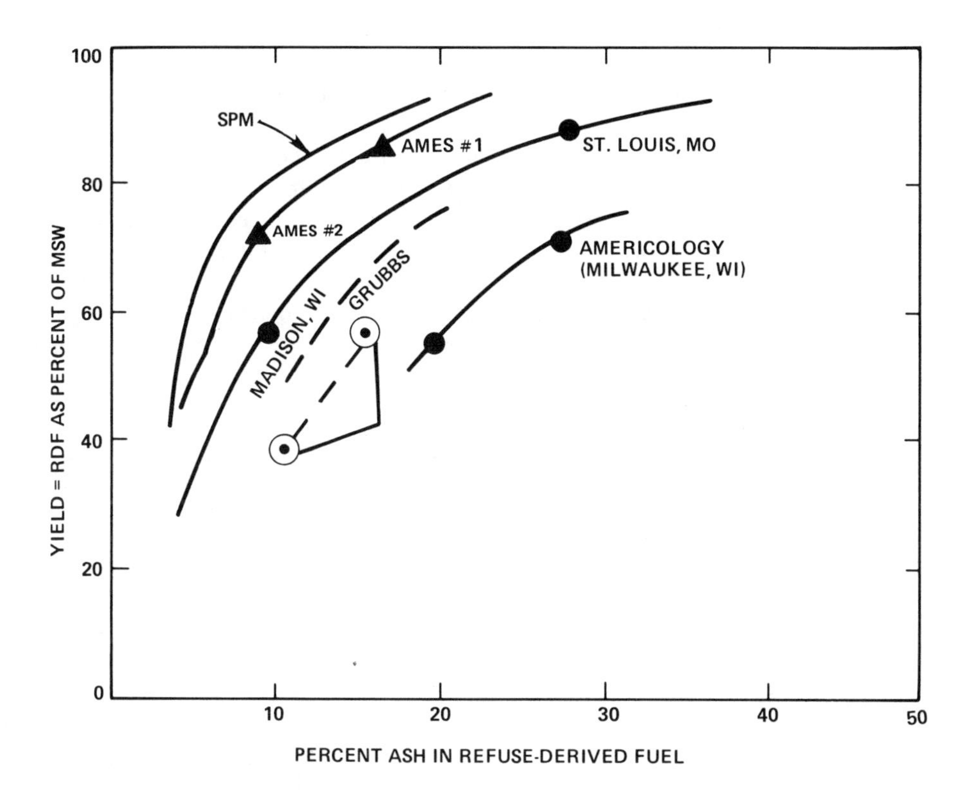

Figure 7–4. Yield of RDF product versus ash content of various processes.

- **SPM tests**: the high yield and low ash content of this MSW processing line were achieved by the use of three to four times the screening area used in some of the processes cited above, combined with the benefits of a low-moisture RDF collected in Denver, Colorado [8]. These points illustrate the degree of processing optimization that could be achieved.
- **Cal Recovery**: the two points shown indicate the comparative performance of the Cal Recovery laboratory process equipment arranged in two configurations: one having a trommel screen prior to shredding and one having the shredder first, followed by an air classifier and trommel screen. The equipment used in these tests determined the comparative effectiveness of the two alternates.

Similar graphs can be drawn showing the energy recovery efficiency, which is higher than the weight recovery because the rejected materials have higher ash and moisture content, and moisture content of the fuel is taken into account.

Optimizing Energy and Power Recovery

The combustible and energy yield of the processing plant determines the quantity of fuel delivered to the energy user. However, the quality of the fuel and the method of fuel use that can be applied influence the amount of energy and power recovery.

If the fuel moisture and ash content tend to limit the application of the fuel, the fuel revenue can be adversely affected. The value of the fuel displaced and the efficiency of the boiler and power plant using the RDF will determine the revenue generated.

Figure 7-5 shows how the percentage of combustible yield of the processing plant and the operating pressure of the boiler affect the amount of power produced per ton of MSW. An ideal process having a 90% yield of combustible, burned in a

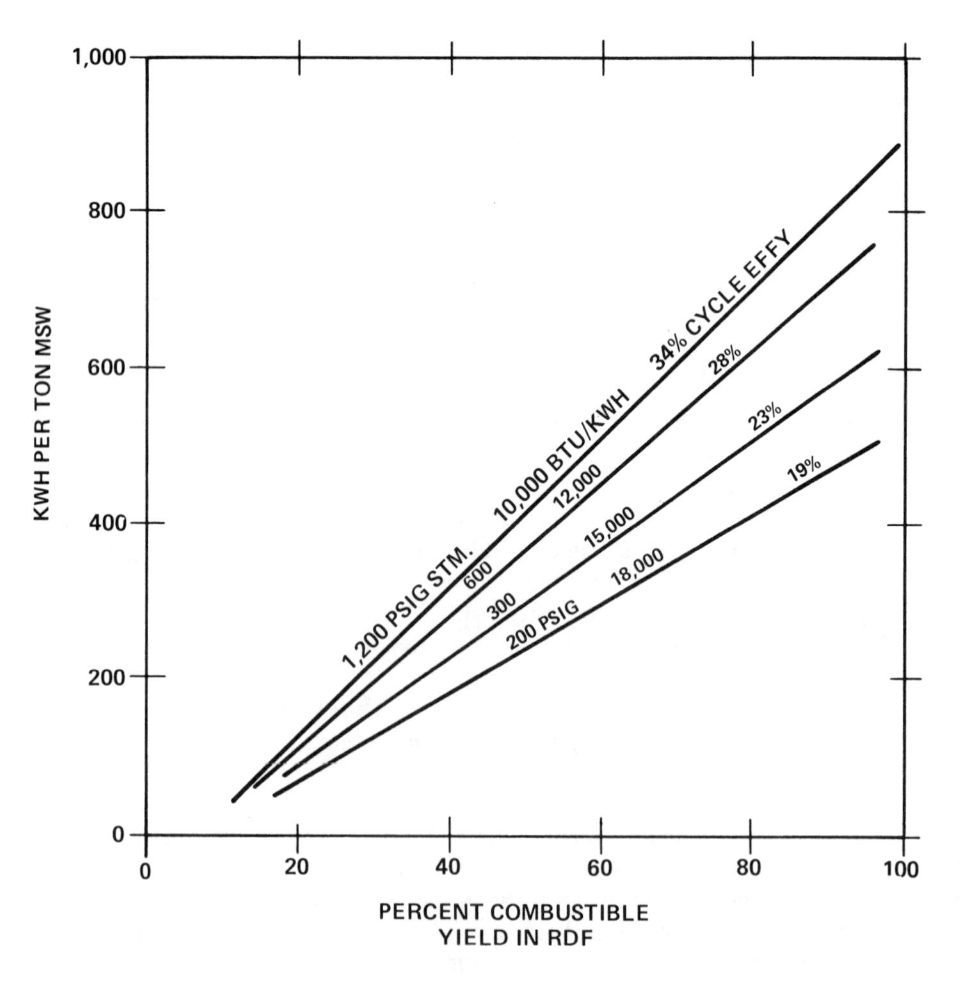

Figure 7-5. Power output per ton of MSW as a function of yield of RDF and power cycle heat to power efficiency.

boiler generating 1200-psig steam and having a steam cycle efficiency of 34%, can produce 800 kWh/ton of MSW, compared with a 200-psig boiler using fuel from a process with only 80% yield, which could produce only 400 kWh/ton.

SUMMARY

Optimizing the process of size reduction, separating and drying MSW to produce a prepared fuel adequately suited for generation of heat and power requires extensive analysis of data obtained from testing the performance of many unit operations that affect each other as the material flows through the plant. The design of new plants must be based on data obtained from laboratory, pilot or full-scale operations and processes having the same configuration. The entire process must be evaluated from the standpoint of effectiveness of the components, as well as of the conversion into energy and power, while taking into account capital, operating and maintenance costs per ton of MSW actually processed, revenue actually produced and the significant costs of residue disposal. This optimization process is complex, requiring numerous iterations and analysis of many options, taking into account the variability of the MSW and its influence on processing. Computer analysis can be helpful in carrying out optimization studies because models can be assembled including test data on the properties of MSW and equipment performance expressed in mathematical form, so that the performance of complex processes can be simulated. This capability greatly reduces the risks incurred in implementing processes for converting refuse to energy by economically viable methods.

REFERENCES

1. Dean, K.C., C.J. Chindgren and L. Peterson. "Preliminary Separation of Metals and Non-metals from Urban Refuse," Bureau of Mines Solid Waste Research Program, TPR–34 (1971).
2. Boettcher, R.A. "Air Classification of Solid Wastes," BuMines contract Ph 86–68–157, EPA publication SW–30c (1972).
3. Sullivan, P.M., M.L. Stanczyk and M.H. Spendlove. "Resource Recovery from Raw Urban Refuse," Report No. 7760, U.S. Dept. of Interior, Bureau of Mines (1973).
4. "Materials Recovery System: Engineering Feasibility Study," National Center for Resource Recovery (1972).
5. Kolhepp, D.H. "The Dynamics of Recycling," in *Sixth Annual Northeastern Antipollution Conf.* (1975).
6. "New Orleans Resource Recovery Facility Implementation Study," National Center for Resource Recovery (1977).
7. DeCesare, R.S., F.J. Palumbo and P.M. Sullivan. "Pilot-Scale Studies on the Composition and Characteristics of Urban Refuse," U.S. Bureau of Mines RI 8429 (1980).

8. SPM Group, Inc. "Refuse Derived Fuel Pre-separator Process Technology Evaluation," DOE/CS/20523-1 (1982).
9. Glaub, J.C., D.B. Jones, J.U. Tleimat and G.M. Savage. "Trommel Screen Research and Development for Applications in Resource Recovery," DOE/CS/20490-1 (1982).

APPENDIX A

Conversion Factors

Table A-1. English and Metric Units to SI Units

To Convert from	*to*	*Multiply by*
in	cm	2.540
in^2	cm^2	6.452
ft	m	0.3048
ft^2	m^2	0.09290
ft^3	m^3	0.02832
lb	kg	0.4536
lb/hr	Mg/s	0.1260
$lb/10^6$ Btu	ng/J	430.0
g/Mcal	ng/J	239.0
Btu	J	1054.0
Btu/lb	J/kg	2324.0
Btu/hr	W	0.2929
J/sec	W	1.000
J/hr	W	3600.0
Btu/ft/hr	W/m	0.9609
Btu/ft/hr	J/hr/m	3459.0
$Btu/ft^2/hr$	W/m^2	3.152
$Btu/ft^2/hr$	$J/hr/m^2$	11349.0
$Btu/ft^3/hr$	w/m^3	10.34
$Btu/ft^3/hr$	$J/hr/m^3$	37234.0
psia	Pa	6895.0
H_2O	Pa	249.1
Rankine	Celsius	C = 5/9R–273
Fahrenheit	Celsius	C = 5/9(F–32)
Celsius	Kelvin	K = C + 273
Rankine	Kelvin	K = 5/9R

Table A-2. SI Units to English and Metric Units

To Convert from	*to*	*Multiply by*
cm	in	0.3937
cm^2	in^2	0.1550
m	ft	3.281
m^2	ft^2	10.764
m^3	ft^3	35.315
kg	lb	2.205
Mg/s	lb/hr	7.937
ng/J	$lb/10^6$ Btu	0.00233
ng/J	g/Mcal	0.00418
J	Btu	0.000948
J/kg	Btu/lb	0.000430
J/hr/m	Btu/ft/hr	0.000289
$J/hr/m^2$	$Btu/ft^2/hr$	0.0000881
$J/hr/m^3$	$Btu/ft^3/hr$	0.0000269
W	Btu/hr	3.414
W	J/hr	0.000278
W/m	Btu/ft/hr	1.041
W/m^2	$Btu/ft^2/hr$	0.317
W/m^3	$Btu/ft^3/hr$	0.0967
Pa	psia	0.000145
Pa	H_2O	0.004014
Kelvin	Fahrenheit	F = 1.8K – 460
Celsius	Fahrenheit	F = 1.8C + 32
Fahrenheit	Rankine	R = F + 460
Kelvin	Rankine	R = 1.8K

Table A-3. Commonly Used Conversion Factors

atm × 101.325 = kPa	hp × 745.6999 = W
barrel × 0.15899 = m^3	in × 25.4 = mm
Btu × 1.055056 = kJ	in Hg × 3.37685 = kPa
Btu/lb × 2.326 = kJ/kg	in H_2O × 0.24884 = kPa
F-32/1.8 = C	mile × 1.609344 = km
ft × 0.3048 = m	$mile^2$ × 2.589988 = km^2
ft^3 × 0.02831685 = m^3	mm Hg × 133.3224 = Pa
ft^2 × 0.09290304 = m^2	lb × 0.4535924 = kg
ft/min × 0.00508 = m/s	psi × 6.894757 = kPa
gal × 0.003785412 = m^3	ton × 0.907184 = t
gal × 3.785412 = L	yd^3 × 0.7645549 = m^3

APPENDIX B

ASTM Standards for Resource Recovery

COMMITTEE E-38.06.01: WASTE GLASS

E688 Testing waste glass as a raw material for manufacture of glass containers.

E708 Waste glass as a raw material for the manufacture of glass containers.

COMMITTEE E-38-02: FERROUS SCRAP

E701 Testing municipal ferrous scrap.

E702 Specification for municipal ferrous scrap.

COMMITTEE E-38-03: NONFERROUS METALS

E753 Specification for municipal aluminum scrap.

E956 Classifications for municipal mixed nonferrous metals.

COMMITTEE E-38-01: ENERGY (REFUSE-DERIVED FUELS)

E711 Test for gross calorific value of RDF-3 by bomb calorimeter.

E775 Test for total sulfur in the analysis sample of RDF.

E776 Test for forms of chlorine in RDF.

E777 Test for carbon and hydrogen in the analysis sample of RDF.

E778 Test for nitrogen in the analysis sample of RDF.

E790 Test for residual moisture in RDF analysis sample.

E791 Calculating RDF analysis data from as-determined to different basis.

E828 Designating the size of RDF-3 from its sieve analysis.

E829 Preparing RDF-3 laboratory samples for analysis.

E830 Ash in the analysis sample of RDF-3.

E856 Definition of terms relating to the physical and chemical characteristics of RDF.

E885 Analysis of metals in RDF by atomic absorption spectography.

Prepared by the ASTM Committee E-38 on Resource Recovery.

E886 Dissolution of RDF-3 samples for analysis of metals.
E887 Test for silica in RDF-3 ash.
E897 Test for volatile matter in the analysis sample of RDF-3.
E926 Test for preparing RDF-3 samples for analysis of metals.
E949 Test method for total moisture in a RDF laboratory sample.
E953 Test method for fusibility of RDF-3 ash.
E954 Method for packing and shipping of laboratory samples for RDF-3.
E955 Test method for thermal characteristics of RDF-3 macrosamples.
EDS-18 Standard methods for collection of a gross sample of RDF.

COMMITTEE E-38.07: HEALTH AND SAFETY

E884 Practice for sampling airborne microorganisms at municipal solid-waste processing facilities.

COMMITTEE E-38.08: PROCESSING EQUIPMENT AND UNIT OPERATIONS

E868 Conducting performance tests on mechanical conveying equipment used in resource recovery systems.
E889 Test for composition or purity of a solid-waste materials steam.
E929 Test method for measuring electric energy requirements of processing equipment.
E929 Standard method for characterizing the performance of refuse size reduction equipment.
E929 Performance test methods for screening.
E929 Performance test methods for disc screens.
E929 Standard method of air drying RDF-5 (densified RDF) for further analysis.
E929 Standard method for measuring density of RDF-5.
E929 Standard methods for measuring particle size distribution of RDF-5.
E929 Standard method for measuring total moisture of RDF-5.
E929 Method of measuring throughput of resource recovery unit operations.
E929 Determination of the recovery of a product by a materials separation device.
E929 Method of determining the bulk density of solid waste fractions.
E929 Method of measuring the energy consumption of resource recovery equipment.

COMMITTEE E-38.09: SAMPLING

COMMITTEE E-38.10: HEAT RECOVERY INCINERATORS

APPENDIX C

Element Concentrations in RDF and RDF Ash

Table C-1 Average Element Concentrations in RDF Ash, $\mu g/g$

Element	*Range in Refuse (microgram/gram)*	*Concentration in Ash* %	*Concentration in Ash (microgram/g)*
Aluminum (Al)	5,500–21,000	32.0	15,500
Calcium (Ca)	1,500–17,000	29.0	14,000
Sodium (Na)	1,200–11,000	10.0	4,900
Potassium (K)	290–2,000	8.7	4,200
Titanium (Ti)	1,100–4,500	6.4	3,100
Iron (Fe)	710–5,100	5.6	2,700
Magnesium (Mg)	530–4,100	4.4	2,100
Zinc (Zn)	120–7,000	1.0	460
Barium (Ba)	20–1,400	0.8	400
Manganese (Mn)	40–480	0.7	330
Lead (Pb)	88–1,600	0.5	220
Copper (Cu)	21–1,500	0.3	140
Silver (Sb)	2–610	–	ND
Chromium (Cr)	13–240	0.2	92
Arsenic (As)	0.2–240	–	ND
Selenium (Sn)	20–80	0.07	35
Nickel (Ni)	5–47	0.05	27
Bismuth (Bi)	<30	0.03	15
Strontium (Sr)	10–62	0.02	12
Gold (Ag)	<3–20	0.02	10
Cobalt (Co)	<3–7	0.02	9
Cadmium (Cd)	<2–23	0.01	6
Lithium (Li)	<2–5	0.006	3
Berillium (Be)	<2	0.004	2
Mercury (Hg)	0.3–1.7	–	ND

Source: U.S. Bureau of Mines Publication RI 8426.

APPENDIX D

Mathematics

PARTICLE SIZE DISTRIBUTIONS

Weight distributions of particle size can be obtained by passing the sample through a series of screens and weighing the amount of material remaining on each screen. When particles are too fine to be measured by screens (less than 400 mesh or 40 micrometers), other methods can be used, such as the Bahco elutriator. Number distributions can be obtained by the Coulter Counter, microscopic counting, and so on.

The weights retained on each screen are added up and the individual weights are divided by the total to get the percentage retained by each screen size. This data may then be plotted on various types of graph paper.

Histograms

The data may also be plotted in histograms to show the fraction or percentage of the weight (or number) that lies between size increments. It is best that the size increments be equal so the shape of the curve may be compared with that from other data.

When the screen sizes have been selected in a series, such as square root of 2 and square root of 4 (ASTM standard series of screen sizes), the percentage in each size can be plotted directly as a histogram, as shown in Figure D-1.

When the screen sizes are not in a series, a line may be drawn through the plotted data points and divided into equal increments of size. The percentage passing each of these sizes can then be tabulated, as shown in Table D-1, and the differences taken for each size. These differences, the percentage in size, can then be plotted as a histogram.

Another procedure is used when the screen sizes are not in a series. In this case the screen sizes are tabulated with the percentage retained and the differential screen size tabulated. The percentage retained per unit screen size is then calculated and plotted versus the average screen size, on a graph such as the Rosin-Rammler graph (see Figure D-2).

Significance of the Graphs of Particle Size Distribution

The log-normal distribution, which is seen so frequently in shredded or broken materials, as well as in many natural materials, can be described mathematically, as in Eq. D-3. Log-normal distribution is plotted as an "accumulative" curve and as a

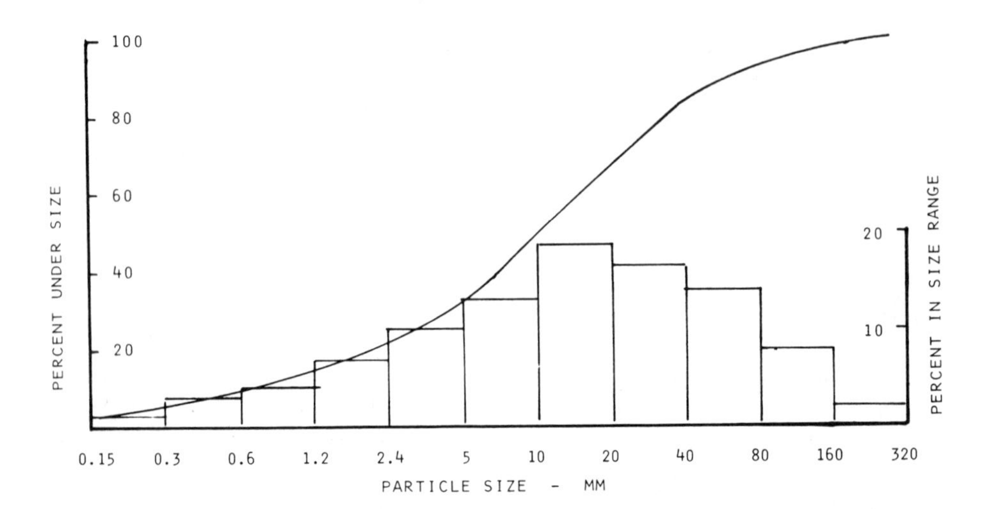

Figure D-1. Histogram of doubled-shredded RDF data plotted from Table D-1.

Table D-1. Tabular Form for Calculating Size Distribution from Screening Data.

Size Range from X_2 to X_1	*Weight Percentage in Size*	*Cumulative Percentage over Size*	*Cumulative Percentage under Size*
over 250	(0.6)*	(0.6)	(100.0)
250 to 125	(2.0)	(2.7)	(99.3)
125 to 63.5	(8.0)	(10.6)	(97.3)
63.5 to 38.1	14.1	24.7	89.3
38.1 to 19.1	16.4	41.1	75.2
19.1 to 9.5	18.7	59.8	58.8
9.5 to 4.8	13.4	73.2	40.1
4.8 to 2.4	10.1	83.4	26.1
2.4 to 1.2	(7.0)	(90.4)	(16.6)
1.2 to 0.6	(3.3)	(93.7)	(9.6)
0.6 to 0.3	(2.8)	(96.5)	(6.3)
0.3 to 0.15	(1.2)	(97.7)	(3.5)
Less than 0.15	(2.3)	100.0	(2.3)

*Total 100

histogram in Figure D-3. The curve is symmetrical on log coordinates, and has 50% of the weight on each side.

The mean, X, is the total weight divided by the number of units.

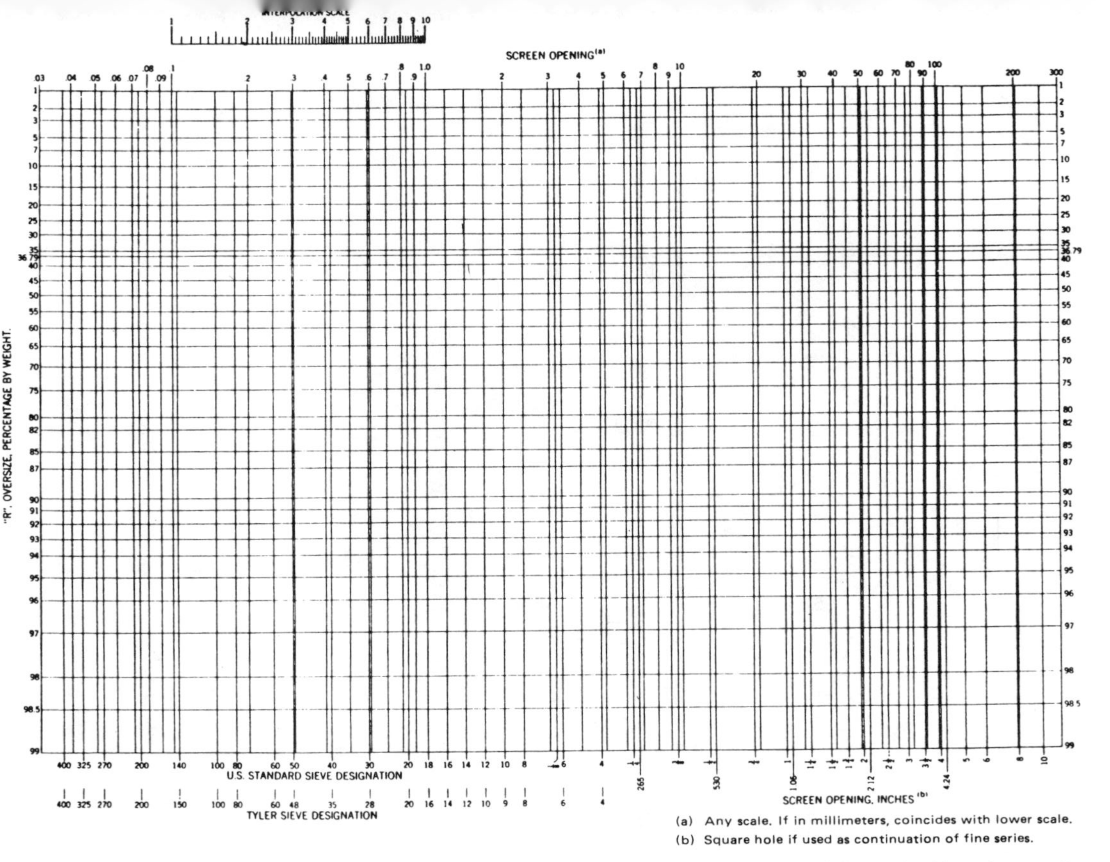

(a) Any scale. If in millimeters, coincides with lower scale.
(b) Square hole if used as continuation of fine series.

From Lenders, W.W., and W.T. Reid, "A Graphical Form for Applying the Rosin and Rammler Equation to the Size Distribution of Broken Coal," Bureau of Mines Inf. Circular 7346, 1946.

ASTM, 1916 Race St., Philadelphia, Pa. 19103. Designation: E 828, PCN 12-508280-41.

Figure D-2. Rosin-Rammler Graph.

The standard deviation, s, can be calculated by either of the following expressions:

$$s = \sqrt{\frac{\Sigma(X_i - X)^2}{n - 1}} \qquad \text{(D-1)}$$

$$s = \sqrt{\frac{\Sigma X_i^2 - (\Sigma X_i)^2/n}{n - 1}} \qquad \text{(D-2)}$$

One standard deviation, s, represents 34% of the area on each side. A range of one standard deviation represents 67.6% of the total population. A range of two standard deviations represents 95% of the population, and three represents 99.7%.

Mixtures

Figure D-3 shows the s-curve, which is characteristic of mixtures of two different types of material having intersecting particle size distributions. The separate distributions plot as straight lines as indicated by the dotted lines. The histograms plotted below reveal the relative amounts of the two materials.

The Log-Normal Distribution

The log-normal distribution is described by the following equation:

$$Y = \frac{1}{\sigma\sqrt{2\pi}} \exp\left(-\frac{(X - \mu)^2}{2\sigma^2}\right) \qquad \text{(D-3)}$$

The Rosin-Rammler Distribution

The Rosin-Rammler distribution is described by the following equation:

$$\text{Weight retained,} \quad Y = 100\left(1 - \exp\left[-\left(\frac{D}{a}\right)^n\right]\right) \qquad \text{(D-4)}$$

The geometric mean or characteristic size is a, which is the size at Y′ = 36.79% retained on the screen, and D is the size corresponding to Y.

$$a = D\left(-\ln\frac{Y'}{100}\right)^n \qquad \text{(D-5)}$$

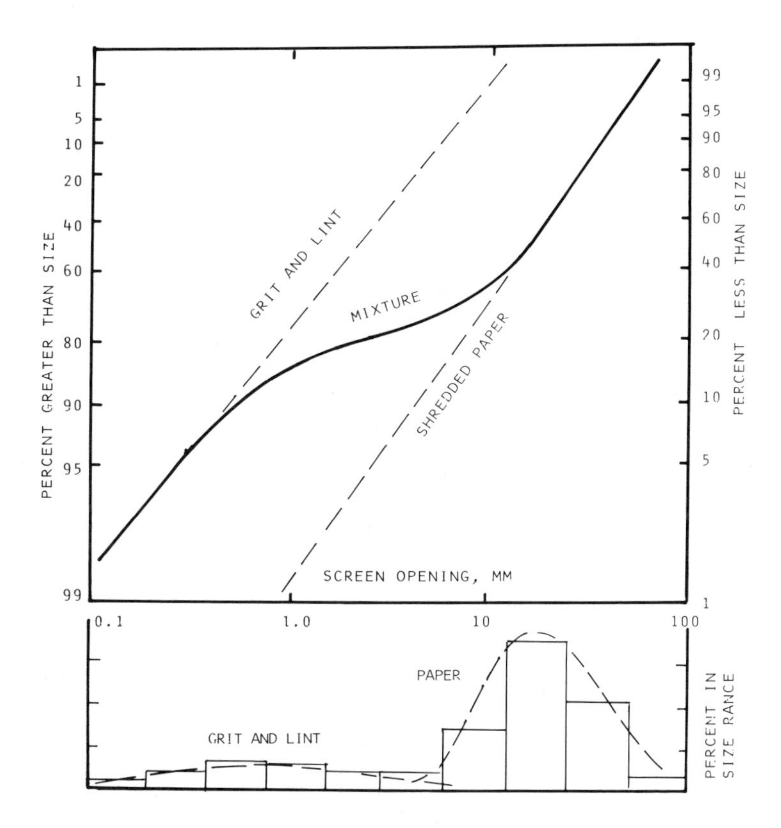

Figure D–3. Rosin-Rammler plot and histogram of RDF-3 showing intersecting distributions of two different types of particles.

The exponent, n, is the slope of the line on Rosin-Rammler paper. It can be found from the weight retained, Y, at two points, preferably located hear the mean:

$$n = \frac{\ln\left(\frac{Y_1}{100}\right) - \ln\left(\frac{Y_2}{100}\right)}{(\ln D_1 - \ln D_2)} \tag{D-6}$$

when D_X, Y_X and n are known, the size at the geometric mean, a can be found:

$$a = \frac{D_X}{\left(\ln \frac{100}{Y_X}\right)^{\frac{1}{n}}} \tag{D-7}$$

MATHEMATICAL RELATIONSHIPS

Conservation of Energy

Kinetic + Potential + Work = Kinetic + Potential + Heat + Friction

or

$$KE_1 + PE_1 + W = KE_2 + PE_2 + Q + F \tag{D-8}$$

Newton's First Law

Force = Mass × Acceleration

$$F = M \times a = \left(\frac{W}{g}\right) \times a \tag{D-9}$$

Aerodynamic Drag

$$\text{Drag Force} = F_d = C_d \times (\text{Frontal Area}) \times (\text{Gas Kinetic Energy})$$

$$F_d = C_d \times A \times \left(\frac{\rho V^2}{2Z}\right) \tag{D-10}$$

Reynold's Number

$$\text{Re} = \frac{(\text{Size}) \times (\text{Velocity}) \times (\text{Density})}{(\text{Viscosity})} = \frac{DV\rho}{\mu} \tag{D-11}$$

Drag Coefficient

$$C_d = \frac{(\text{Volume}) \times (\text{Density})}{(\text{Frontal Area}) \times (\text{Gas Kinetic Energy})}$$

$$C_d = \left(\frac{\text{Volume}}{\text{Face Area}}\right)\left(\frac{\rho_p}{\rho_g}\right)\left(\frac{g}{V^2}\right) \tag{D-12}$$

The drag coefficient, C_d, can be correlated by the Reynold's number, Re, which takes into account the relative velocity of the particle and the gas, the gas density and viscosity, and the particle dimension used to calculate C_d.

The following are coefficients of drag and sphericities for simple shapes, in the turbulent range:

Shape		*Coefficient of Drag*	*Coefficient of Resistance*	*Sphericity*
Sphere		0.45	0.45	1.000
Cube		1.10	1.40	0.806
Cylinder	3r		1.4	0.860
	10r		1.5	0.691
	20r	1.40	4.0	0.580
Disc	1/3r		5.0	0.594
	1/10r		6.0	0.323
	1/15r		8.0	0.220

Values of C_d in the viscous range are dependent mainly on the Re, and in the transition range on both shape and Re. Sphericity is the ratio of the surface of a sphere with the same volume as the particle to the actual surface of the particle. It is used in mineral dressing as an indication of the rate of settling of particles. The product of C_d and $(\text{sphericity})^2$ is roughly a constant.

VELOCITY, ACCELERATION, TIME, AND DISTANCE TRAVELED

$$V = \frac{dS}{dt} \tag{D-13}$$

$$a = \frac{dV}{dt} = \frac{d^2S}{dt^2} \tag{D-14}$$

$$V_2 - V_1 = \alpha dt = 2\frac{dS}{dt} \tag{D-15}$$

$$t_2 - t_1 = \frac{2dS}{V} = \frac{dV}{a} \tag{D-16}$$

FREE FALL UNDER THE FORCE OF GRAVITY (TURBULENT FLOW)

When a particle falls under the force of gravity, in a fluid environment that exerts a drag force proportional to the square of the differential velocity, the following equation applies:

$$M_p \frac{dV_p}{dt} = Mg - C_d A \frac{\rho_a V^2}{2g} \tag{D-17}$$

Dividing by M_p:

$$\frac{dV_p}{dt} = g - \frac{C_d\ A\ \rho_a\ V^2}{M_p 2g} \tag{D-18}$$

When the particle reaches the velocity, V_t, where the forces are in balance, $dV_p/dt = 0$, and $V = V_t$, then:

$$g - \frac{C_d\ A\ \rho_a\ V^2}{M_p 2g} = 0$$

Solving for C_d and V_t^2:

$$C_d = \frac{M_p g^2}{A\ \rho_a\ V_t^2}\ ,\quad V_t^2 = \frac{M_p\ 2\ g^2}{A\ \rho_a \bar{C}d} \tag{D-19}$$

Substituting back we get:

$$dV = g\left(1 - \frac{V^2}{V_t^2}\right) = g - \frac{(V_a - V_p)^2}{V_t^2}\ g \tag{D-20}$$

This derivation is universal, in that it did not have to consider the shape of the particle. By measuring V_t by simple tests (such as elutriation), the acceleration of this particle under gravity can be determined by Eq. D-15 without needing to know C_d. The value of V_t found experimentally can be used to calculate C_d, if the mass, frontal area, and density of the particle are known. Using the volume and frontal area of simple shapes, both C_d and V_t can be calculated:

For spheres

$$V_t = V\frac{4d(\rho_p - \rho_a)}{3\ Cd\rho_a} \tag{D-21}$$

For cubes

$$V_t = Vd \frac{(\rho_p - \rho_a)}{Cd\rho_a} \quad \text{(D-22)}$$

For plates

$$V_t = Vt \frac{(\rho_p - \rho_a)}{Cd\rho_a} \quad \text{(D-23)}$$

The differences in shape are apparent: 4d/3 for spheres becomes d for cubes and the thickness, t, for plates.

PARTICLE ACCELERATION UNDER AERODYNAMIC DRAG FORCES

The acceleration equation can be used to determine the distance a particle will move in a given time and the time to move a certain distance, by mathematical integration.

Vertical movement:

$$t = \frac{V_t}{2g}\left[\ln\left(\frac{B - V_p}{A - V_p}\right)\frac{A}{B}\right] \quad \text{(D-24)}$$

where

$$A = V_a - V_t$$

$$B = V_a + V_t$$

Distance traveled:

$$S = \frac{V_t}{2g}\left[A \ln\left(\frac{A}{A - V_p}\right) - B \ln\left(\frac{B}{B - V_p}\right)\right] \quad \text{(D-25)}$$

Horizontal motion (no friction):

$$t = \frac{V_t^2}{g}\left(\frac{1}{V_a - V_p} + \frac{1}{V_a}\right) \quad \text{(D-26)}$$

$$S = \frac{V_t^2}{g}\left[\frac{V_p}{V_a - B_p} - \ln\left(\frac{V_a}{V_a - V_p}\right)\right] \quad \text{(D-27)}$$

For movement at an angle θ, with coefficient of friction μ:

$$\frac{dV}{dt} = g\left[\left(\frac{V_a}{V_t} - \frac{V_p}{V_t}\right)^2 - \sin\theta - \mu\cos\theta\right] \tag{D-28}$$

Then

$$t = \frac{Vt}{2Kg}\left[\ln\left(\frac{B' - U\ A'}{A' - U\ B'}\right)\right] \tag{D-29}$$

$$S = \frac{{V_t}^2}{2Kg}\left[A'\ln\left(\frac{A'}{A' - U}\right) - B'\ln\left(\frac{B'}{B' - U}\right)\right] \tag{D-30}$$

where

$$A' = U_{at} - K$$

$$B' = U_{at} + K$$

$$U_{at} = \frac{V_a}{V_t}$$

$$U_{pt} = \frac{V_p}{V_t}$$

$$K = \sqrt{\sin\theta - \mu\cos\theta}$$

APPENDIX E

Calculation of Displacement of Particles in a Rotary Drum Classifier

THEORETICAL CONSIDERATIONS

The displacement of the particles in the drum during a single drop and the drop time of a given particle can be calculated from the following characteristics of the particles:

Fluidization velocity, V_f, cm/sec:

$$V_f = \sqrt{\frac{2mg}{C_d \rho A}} \qquad \text{(E-1)}$$

where m is the mass of the particle, g
g is the gravitational acceleration, cm/sec^{-2}
ρ is the air density, g/cm^{-3}
A is the particle cross section, cm^2
C_d is the drag coefficient

The drop time in seconds is:

$$t = \frac{V_f}{g \cos \theta^{1/2}} \log_e \left[e^{gD/V_f^2} + \sqrt{e^{2gD/V_f^2} - 1} \right] \qquad \text{(E-2)}$$

and the displacement during the drop, in cm, is:

$$\ell = V_a t - \frac{V_f^2}{g} \log_e \left[\frac{V_a}{V_f \sin \theta^{1/2}} \sin h \left(\frac{gt \sin \theta^{1/2}}{V_f} \right) + \cos h \left(\frac{gt \sin \theta^{1/2}}{V_f} \right) \right] \qquad \text{(E-3)}$$

Grubbs, M.R., M. Paterson, B.M. Fabuss, "Air Classification of Municipal Refuse," Fifth Annual Mineral Waste Symposium (Chicago, IL, 1976), pp. 169–174.

where θ is the drum inclination from horizontal in degrees
D is the drum inside diameter in cm
V_a is the air-stream velocity in cm/sec

From eqs. E-2 and E-3 the air-stream velocities (V_a) were calculated at which a particle of given fluidization velocity (V_f) shows zero displacement. At this air-stream velocity the particle does not move upward or downward in the rotary drum but stagnates in the same location.

APPENDIX F

Analysis of Behavior in a Rotary Cylinder

A particle within a rotating horizontal cylinder is acted upon by a centrifugal force, mV^2/R, acting radially outward, and by the force of gravity acting vertically downward. At very high speeds the force exerted by gravity is less than the centrifugal force and the particles will continue to adhere to the wall of the cylinder. As the speed of rotation is reduced, the critical speed is reached at which the centrifugal force and the force of gravity are just equal:

$$N_c = \frac{g}{(4\pi R)} = \frac{54}{\sqrt{R}} = \frac{76.6}{\sqrt{D}} \quad \text{(F-1)}$$

At velocities less than the critical velocity, N_c, the particle will leave the surface and follow a trajectory defined by the angle α at which it starts. It will then fly through this path until it lands on the cylinder at a point that is described by the angle β, as shown in Figure F-1.

For a thick bed of material, the material lies at different radii, resulting in different points for the trajectories, as shown in Figure 6-15. There is a locus of points, AE, along which trajectories start, rising to points similar to G, and following paths similar to M until they hit the cyclinder at points from D to F. If the bed thickness is increased from AE to AB, a point is reached where the rising and falling streams come into contact in the region BC, causing a high rate of shear and tumbling in this region. The thickness of the bed when the material falls (impinges) on the cylinder from D to F varies from nothing to the uniform thickness at F from which point the bed is carried up to the cascading points AE to begin another cycle.

The particles move at different velocities during this cycle: (1) a constant velocity related to shell rotation, which does not produce an axial flow component to progress the material down the cylinder; and (2) a changing velocity while falling under the acceleration produced by the force of gravity, during which time the material can advance axially to an extent determined by the inclination of the trommel, by lifters, or by air movement. By calculating the flying time of the material, the displacement per turn and the rate of flow of the material through the cylinder can be determined.

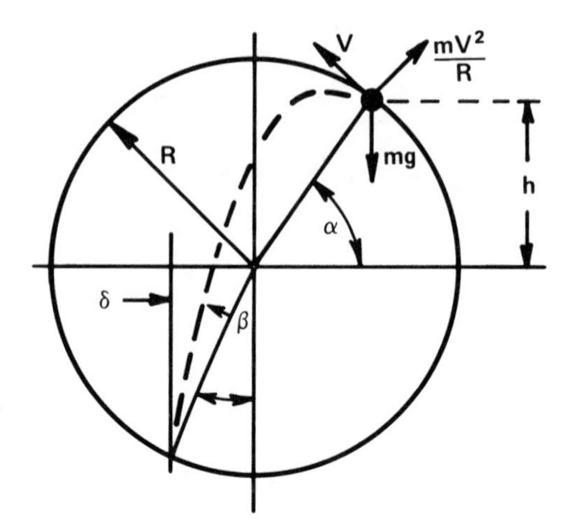

Figure F-1. Particle trajectories in a rotary drum.

PARTICLE TRAJECTORIES IN A ROTARY DRUM

Referring to Figure F-1, we note that the particle falls a net distance, h + R, where $h = R \sin \alpha$, to reach the bottom center of the drum. The velocity of the particle is the sum of the velocity resulting from the drum velocity at the time of separation, V_o or R_ω, and the velocity due to the acceleration of gravity, g_t. The sum of the vertical distances traveled in time t is:

$$R + h = R(1 + \sin \alpha) = \frac{1}{2}\, g_t - \omega R\, t \sin \alpha \qquad \text{(F-2)}$$

The time from the point of separation to reaching the drum is the flying time, t:

$$t = \frac{\omega R \sin \alpha}{g} + \sqrt{\frac{2R(1 + \sin \alpha)}{g} + \left(\frac{wR \sin \alpha}{g}\right)^2} \qquad \text{(F-3)}$$

For the general case where the material falls to a point defined by the angle δ, the following equation can be derived, based on the additional knowledge that $\cos \delta = \sin 3\alpha$:

$$\omega_t = \sin \alpha \cos \alpha + \sqrt{\sin^2 \alpha \cos^2 \alpha + 2\sin \alpha\, (\sin 3\alpha + \sin \alpha)} \qquad \text{(F-4)}$$

This equation can be solved in terms of the angle α, which was determined by the rotational velocity, giving the time in flight from which the distance advanced by the particle can be determined from the amount the drum has turned during that time.

If the drum is inclined at an angle β, then the particle while flying will move a distance along the cylinder that is defined by h the total distance dropped perpendicular to the axis of the cylinder and the tangent of the angle β. This can be visualized from Figure F-2, and expressed as follows:

$$\text{Total vertical drop} = R(\omega_t \cos\alpha + \cos\alpha + \cos\delta) \tag{F-5}$$

$$\text{Advance, } 1 = R \tan\beta\,(\omega_t \cos\alpha + \cos\alpha + \sin 3\alpha) \tag{F-6}$$

The number of drops made by the material while progressing through the trommel can now be determined:

$$\text{Number of drops} = n = \frac{L}{1} \tag{F-7}$$

where L = trommel length.

During the time the material is in flight, the angular displacement is ω_t. The remainder of the time is the sum of the angles encompassing the period when the material is in contact with the drum. The material remains in one axial location for the total of these two periods before being deposited into the next section in a spiraling fashion.

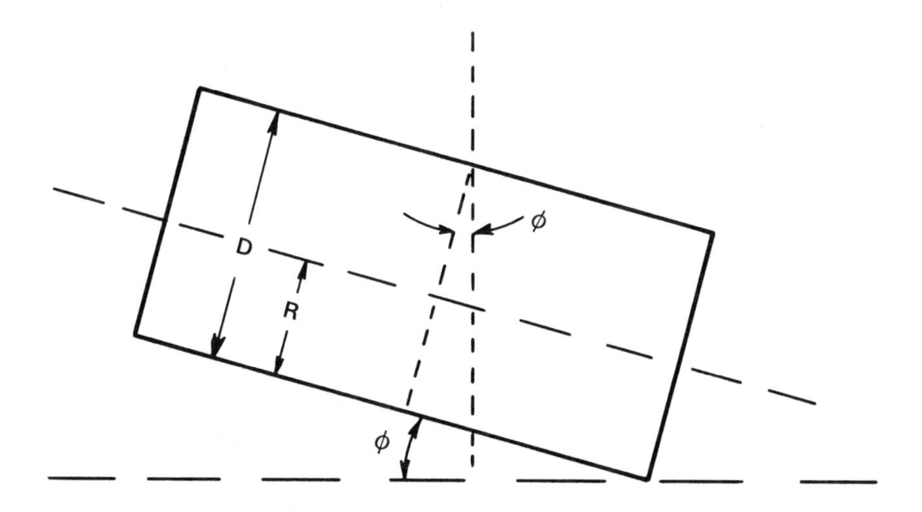

Figure F-2. Vertical drop of a particle.

The fraction of a revolution (2π) that is taken for transport from the point of impact to the point of separation plus the time of flight is:

$$\text{Fraction of a turn, f} = \frac{\omega_t + \alpha + \delta + \frac{\pi}{2}}{2\pi} \tag{F-8}$$

The reciprocal of f is the number of drops per revolution.

RESIDENCE TIME

The residence time of a particle in a trommel is determined by multiplying the number of turns made by the particle, nf, by the time required per turn. This is the time required to rotate the drum multiplied by f, the fraction of the time during which the material moves axially.

The residence time, τ, is:

$$\tau = \frac{2\pi\ \mathrm{nf}}{\omega} = 2\pi\ \mathrm{nf} \sqrt{\frac{R}{g \sin \alpha}} \tag{F-9}$$

FEED CAPACITY

The capacity of a trommel to pass material depends on the velocity of the material, the cross-section of the trommel through which the material is flowing (influenced by bed depth), and the density of the material volumetric rate of mass-flow.

The velocity of the material is the length, L, divided by the retention time, τ, or alternately, the length per drop, 1, divided by the retention time per drop, τ/n:

$$\text{Material velocity, V} = \frac{L}{\tau} = \frac{1}{\tau/n} \tag{F-10}$$

The cross-sectional area of the material, or fillage (a function of the bed depth), determines the rate at which the trommel can be fed.

How does the loading of the screen affect the capacity? How is capacity calculated? Given the velocity of the material along the screen and the fraction of the circumference covered by material, we only need the thickness of the material to determine the volume flow and, given the density, the mass flow. What then is the affect of thickness?

Alter et al. made a calculation of the capacity of the Recovery I trommel, based on the assumption that the thickness of the material layer was the same as the mean particle size (5-6 in) and obtained a result that agreed with the actual measured capacity. Since the hole size was 4.75 in, the assumption that the bed thickness should be assumed to be about equal to the hole size also agrees with actual capacity. The answer to this quandry is to run tests, measure the mass-flow, and check the height of material and its density when the trommel is stopped. From this information the thickness can be calculated.

Assuming the material carried by the screen has a thickness of β and knowing that this material lies along an angle $(\delta + \pi/2 + \alpha)$, we know that the cross-sectional area is $\beta R\ (\delta + \pi/2 + \alpha)$. This is the same material that recently was carried in flight during the flight time, t, which is $\beta R\omega t$. Since the material has a velocity of $(\omega/2\pi)1/f$, we can now calculate the volumetric feedrate and, by multiplying by the density of the layer, ρ, calculate the mass flow rate. Alter et al. offer the following expression for the mass-flow rate, M:

$$M = \chi g^{1/2}\ \beta\ \rho\ \tan\beta\ R^{3/2} \tag{F-11}$$

where

$$\chi = (\sin\alpha)^{1/2}\ (\omega t \cos\alpha \sin\alpha \sin 3\alpha) \tag{F-12}$$

INFLUENCE OF BED DEPTH

Increasing the bed depth increases the flow rate through the trommel, but decreases the chance for undersize material to reach and pass through the holes.

One limit of fillage is the point where the downward cascade interferes with the upward-moving material in the region BC of Figure 6-15. As we have noted, as soon as this happens the flow pattern changes drastically and with MSW tumbling becomes unstable. At some lesser fillage the optimum lies.

Due to the fact that the material moving at the radius point E is moving more slowly than that at A and the drop is correspondingly less, this material will advance more slowly through the trommel. This material also does not have much chance to see the holes and will therefore just roll around, taking space and energy, while at the same time compressing the material nearer the holes, reducing its freedom to change position, and accommodating itself to the holes. These factors are readily recognized, but not easily quantified except by test.

INFLUENCE OF SPEED ON THE PERFORMANCE OF A TROMMEL

The parameters 1, f, τ, χ, and M, as well as the angle δ, are all functions of the rotational speed of the trommel. Table 6-3 lists these calculated variables for the MSW trommel analyzed by Alter et al. Figure 6-16 shows these parameters, based on the percentage of critical speed.

APPENDIX G

Sampling and Analysis of Refuse and Refuse Products

In order to determine the properties of refuse, its products and residues, it is necessary to collect samples and analyze them. Since refuse and its derivatives are highly variable, it is necessary to take many samples in order to obtain average values that are representative of the stream or quantity under consideration.

The properties of MSW are of interest in planning facilities in order to estimate the nature of the material to be fed to the plant, the operations needed for processing, the properties of the processed streams, and the products and residues.

Due to the large size of most of the components of MSW, sampling to determine its properties requires large samples of material, as well as many of those samples to represent the continuous variability of the refuse. The size of the samples is greatly reduced by size-reducing the MSW through methods such as shredding. Analysis of the shredded product can be done by the methods appropriate for RDF and its products.

The two basic reasons for sampling processed MSW are: (1) to let the producer know how the process and equipment are functioning so the desired product can be assured and (2) to reach agreement between buyer and seller as to the actual qualities of the product. In order to achieve these objectives, standardized methods of testing and analysis, such as the ASTM standards, which have been developed for this purpose, must be agreed upon.

When measurements of a product are made, buyer and seller must agree on a method of determining how accurately the measurements have been made and how accurately they describe the product. The only true measurement is to test the entire lot since sampling gives only an indication and is subject to error. From a practical point of view a compromise must be reached as to an acceptable level of error in measurements. In a highly variable material such as MSW and its derivatives, a ± 10% estimate of properties is often a reasonable expectation for a given set of measurements. If the product is generated month after month, the monthly averages can be accumulated into an annual average that may be more accurate than those obtained over shorter periods. Errors tend to average out unless there is a consistent bias or error in one direction. Testing methods must be carefully set up to avoid bias.

Properties of variable materials are measured by mean or average values and by the deviation from the mean. By using a standard method of analysis of deviation of individual analyses from the mean of a group of analyses, a definable property of the material is obtained that can be reproduced by others and, hence, agreed upon. The standard deviation, S, is defined as the square root of the sum of the deviations, squared, and divided by the number of analyses, or:

$$S = \sqrt{\frac{\Sigma(x_i - X)^2}{(n - 1)}} \tag{G-1}$$

For calculations, it is simpler to use this alternate expression:

$$S = \sqrt{\frac{[\Sigma x^2 - (\Sigma x)^2/n]}{(n - 1)}} \tag{G-2}$$

Because the square of the standard deviation (variance) has additive properties it is often the more convenient term to deal with:

$$S^2 = \frac{\Sigma(x_i - X)}{n} = \frac{[\Sigma x_i^2 - (\Sigma x_i)^2]}{n} \tag{G-3}$$

The standard error, e, is defined in terms of the standard deviation. It takes into account the fact that the more samples used, the lower the error, and that the number of samples influences the probability that the sample will not be representative:

$$e = \frac{t\,S}{\sqrt{n}} \tag{G-4}$$

where t = Student's t is itself a function of n (see Table G-1), and of the confidence level.

A confidence level of 95% means that 95 out of 100 samples will have values within the range of error defined by this equation, with the value of t set by the number of samples minus one (n–1), representing the degrees of freedom.

Precision and accuracy are defined in terms of the measured value rather than the standard deviation. Precision is strictly applied to a set of samples, and accuracy to the whole population for cases where this distinction has meaning; otherwise, the terms are often used interchangeably.

$$\text{Accuracy} = A = \frac{e}{X} \quad \text{in fractional terms, and}$$

$$= 100 \times \frac{e}{X} \quad \text{expressed as a percentage.}$$

Table G-1. Student's t at Various N, and Confidence Limits

n	$n-1$	t_{90}	t_{95}
2	1	6.31	12.71
3	2	2.92	4.30
4	3	2.35	3.18
8	7	1.90	2.36
10	9	1.83	2.62
21	20	1.72	2.09
31	30	1.70	2.04

When comparing the variances from tests performed with different numbers of samples, it is necessary to use the above expressions, which make appropriate corrections for the number of samples.

FACTORS CONTRIBUTING TO ACCURACY AND TO ERRORS

When sampling streams or large amounts of refuse and its products, a number of errors must be taken into account. It is useful to determine the portion of total error or variance that is contributed by each of these factors, so that major errors can be attended to and perhaps reduced. It is also valuable to know the relative size of these errors so efforts are not devoted to reducing errors that are already small. A practical objective is to maintain an acceptably high level of precision or accuracy while minimizing the sampling effort and the cost of laboratory analysis.

These are major factors contributing to error:

- Method of collecting sample: Is it representative?
- Weight of sample: How small without serious error?
- Number of pieces in sample: How few is too few?
- Method of handling and shipping samples: Is moisture lost, or do components become segregated?
- Method of division of sample into smaller samples: Are all subsamples equally representative?
- Method of mixing samples: Is mixing thorough?
- Analysis: What errors can occur due to procedures?

The effects of these factors can be measured by the following variances in accordance with procedures developed for coal and published in ASTM Standards (such as D-2234 and D-2013):

1. Variance of division (or reduction) and analysis,
2. Random variance (affected by sample increment weight),
3. System variance (affected by the refuse itself and the processing and handling equipment).

DETERMINING THE RELATIONSHIP BETWEEN THE NUMBER AND WEIGHT OF SAMPLES OR INCREMENTS AND THE PRECISION OF THE ANALYSIS

The precision represented by averages of property values obtained by analysis of samples of RDF collected during a sampling period can be determined by calculations. These calculations require the determination of appropriate variances obtained by carrying out sampling procedures in the producing or consuming facilities.

The number and size of samples required to achieve a given precision in the estimate of measured average properties depends upon the total variance, consisting of the random and system variances, and the variance of division and analysis. These variances reflect the properties of the RDF and the method of processing and sampling.

USE OF PUBLISHED INFORMATION FOR ESTIMATES

The variances needed for these determinations may be estimated from published information on similar RDF systems and products when no better data are available. The weight and number of sample increments to be taken each day during a test period of twenty days may be estimated from Table G-2 for ash content and Table G-3 for HHV-1 for several types of RDF, for a precision in the estimate of the mean value of ±10% of ash content and as-sampled higher heating value, with a confidence level of 95%.

DETERMINING THE VARIANCE OF REDUCTION OR DIVISION AND ANALYSIS

The procedure for estimating the variance of division and analysis of materials requires that four subsamples be taken from a number of sets of five gross samples taken on successive days or weeks, and analyzed in order to obtain the variance:

$$S_{da}^2 = \frac{\Sigma(X)^2 - \Sigma(X^2)/4}{3} \quad \text{(G-5)}$$

A factor, F, is calculated to determine whether a large enough number of samples has been obtained. If the prescribed test fails, additional sets of data must be obtained. A detailed procedure is given in ASTM D-2013.

DETERMINING VARIANCE AS A FUNCTION OF NUMBER AND WEIGHT OF SAMPLES

The weight of the sample or sample increment has an effect on the number of samples needed to attain a given precision, or the precision that will result from a given number of increments and gross samples. In order to determine the effect of sample weight it

Table G-2. Number of Increments of RDF Samples to Be Taken in 20 Gross Samples over a 20-Day Period, to Achieve an Accuracy of ±10% in Ash Content at 95%

Type of RDF	*Top Size (in.)*	*Wt. (g)*	*Mean Ash (%)*	*Var. (S_o^2)*	*Var. (S_T^2)*	*Number of Increments (N_i)*	*Total Number of Increments ($N_g \times N_i$)*
A*	1.0	200	10.0	4.4	0.28	0.80	16
M	1.0	160	32.7	48.2	2.95	0.81	16
M	1.0	300	30.2	35.8	2.49	0.67	14
M	1.0	750	29.0	29.6	2.90	0.58	12
M	1.0	2,700	30.3	35.6	2.51	0.76	14
W	2.0	2,000	29.6	38.6	2.48	0.77	16
B	6.0	2,000	36.3	62.7	3.64	0.86	17

*Double-screened, double shredded RDF.

Table G-3. Accuracy in the Estimate of As-Received Heating Value, Determined by Taking Two Increments in 20 Gross Samples (95%)

Source of RDF	*Top Size (in.)*	*Incr. Wt. (g)*	*Mean HHV-1 Btu/lb*	*Variance (S_o^2)*	*Accuracy (%)*
Eco-Fuel	0.02	200	7,868	155,879	± 0.8
Ames, IA	1.0	200	6,230	422,410	± 2.9
St. Louis, MO	1.0	2,000	4,950	1,600,325	± 3.7
Baltimore, MD	6.0	2,000	4,475	1,283,000	± 6.9

is necessary to carry out a sampling program wherein samples of substantially different weights are collected and analyzed. Calculations are carried out on this data to determine the weight of sample or increment that has relatively significant effect on the overall precision of sampling.

The procedure set forth in D-2234 calls for systematically collecting two sets of thirty samples A and B over a period of time (such as one week or a series of weeks) using the sample collection procedure normally used, or intended to be used.

Samples A represent the smallest increment that can be obtained by the sampling device without incurring bias. The collection device should have an opening at least three times the top particle size and enough capacity to take a representative sweep of the falling stream. Alternately, the increment may be collected from a stopped belt, including a sample over three times the top particle size in length, and the full belt width.

Samples B, representing large samples, must be at least twenty times the weight of samples A, collected in large containers from a free-falling stream; or, if from a stopped belt, collected adjacent to the A samples.

The samples are weighed and reduced to two sets of analysis samples by the procedures anticipated for normal use. The means, the standard deviations, and the variances, S_A^2 and S_B^2, are calculated and the random and system variances calculated using expressions proposed by Bertholt and Visman, as follows:

Random variance:

$$S_r^2 = \frac{W_A W_B (S_A^2 - S_B^2)}{(W_B - W_A)} \quad \text{(G-6)}$$

System variance:

$$S_s^2 = S_B^2 - \frac{S_r^2}{W_B} \quad \text{(G-7)}$$

The total variance for any weight, W, is:

$$S_t^2 = S_s^2 + \frac{S_r^2}{W} \quad \text{(G-8)}$$

where W_A = the A sample average weight,
W_B = the B sample average weight,
W = any selected sample average weight.

DETERMINING THE OVERALL VARIANCE OF INCREMENTS OF FIXED WEIGHT

A procedure has been developed that calculates estimates of the overall variance (including the variance of division and analysis) for sample increments obtained by a mechanical sampler or other procedure resulting in increments of approximately fixed weight (see Table G-4).

Table G-4. Variance Ratio Limit and F Factor for Overall Variance

Increment per Set	*Variance Ratio Limit*	*F Factor*
10	3.18	1.92
20	2.17	1.53
30	1.86	1.40
40	1.70	1.33

In this procedure, one sample set consisting of two or more samples taken each day on five successive days is followed by a second set of samples taken on five successive days in the next week, using the same procedure to obtain all samples.

The variances of the two sets A and B are used to calculate the overall variance for increments as follows:

$$S_o^2 = F \times \frac{S_A^2 + S_B^2}{2} \tag{G-9}$$

where S_o^2 = the overall variance for single increments of fixed weight (including division and analysis).

F = 1.92 for 10 samples per set,
F = 1.53 for 20 samples per set,
F = 1.40 for 30 samples per set.

VARIANCE OF A SINGLE GROSS SAMPLE

The variance of a single gross sample includes the random and system variances and the variance of reduction or division and analysis. When the alternate overall variance for a fixed sample weight is used, it includes the variance of division and analysis.

The variance of a single gross sample can be reduced by taking more than one increment, N_i, per gross sample, and the variance of division and analysis can be reduced by taking more than one analysis sample, P. The resulting expressions for the variance of a single gross sample are:

$$S_g^2 = \frac{S_s^2 + S_r^2/W}{N_i} + \frac{S_{da}^2}{P} \tag{G-10}$$

or, alternately:

$$S_g^2 = \frac{S_o^2 - S_{da}^2}{N_i} + \frac{S_{da}^2}{P} \tag{G-11}$$

By multiplying S_g^2 by the number of gross samples in a test period, N_g, we obtain the overall variance of the test program, S_T^2, upon which the overall accuracy is based:

$$S_T^2 = \frac{S_s^2 + S_r^2/W}{N_i N_g} + \frac{S_{da}^2}{N_g P} \tag{G-12}$$

or, alternately:

$$S_T{}^2 = \frac{S_o{}^2 - S_{da}{}^2}{N_i\, N_g} + \frac{S_{da}{}^2}{N_g P} \qquad \text{(G-13)}$$

CALCULATION OF PRECISION OR SAMPLING ACCURACY FROM VARIANCE

Precision or sampling accuracy, A, is a function of a probability and the standard deviation of the test period or sampling period, as follows:

$$A = \pm F \times S_T \qquad \text{(G-14)}$$

where A = sampling accuracy or precision, expressed as percentage of the constituent being tested or of heating value,
S_t = the standard deviation of the test or sampling period,
F = 1.95 for a confidence level of 95%,
F = 1.65 for a confidence level of 90%.

This can be rearranged to obtain an expression for the variance of a test or sampling period during which N_g samples are collected, having a variance of $S_T{}^2$ at a confidence level of 95%, as follows:

$$S_T{}^2 = 0.263\ A^2 \qquad \text{(G-15)}$$

This can be rearranged to solve for accuracy, or precision, as follows:

$$A = 1.95 \sqrt{S_T{}^2} \qquad \text{(G-16)}$$

The variance of a gross sample, $S_g{}^2$ is defined as the test period variance, $S_t{}^2$ multiplied by the number of samples taken during this test period and can be related to the accuracy as follows:

$$S_g{}^2 = S_T{}^2 \times N_g = 0.263 \times N_g \times A^2 \qquad \text{(G-17)}$$

where $S_g{}^2$ = variance of a single gross sample,
N_g = number of gross samples during test period,
$S_T{}^2$ = test period variance for N_g.

PRECISION, OR ACCURACY, IN TERMS OF WEIGHT AND NUMBER OF SAMPLES

The precision, or accuracy, can be calculated from S_o^2 for overall variance (Eq. G-9), or S_T^2 for total variance (Eq. G-8) by substituting these variances in Eq. G-13.

Accuracy in terms of overall variance can be determined as follows:

$$A = 1.95 \times \sqrt{\frac{S_o^2 - S_{da}^2}{N_i} + \frac{S_{da}^2}{N_g p}} \qquad \text{(G-18)}$$

where p = number of analysis samples per gross sample.

Accuracy in terms of random and system variances can be determined as follows:

$$A = 1.95 \sqrt{\frac{S_s^2 + S_r^2/W}{N_g N_i} + \frac{S_{da}^2}{N_g p}} \qquad \text{(G-19)}$$

where S_s^2 = system variance,
S_r^2 = random variance,
W = weight of sample increment,
S_{da}^2 = variance of division and analysis,
N_i = number of increments per gross sample,
N_g = number of gross samples,
p = number of analysis samples per gross sample.

NUMBER OF INCREMENTS IN A SINGLE GROSS SAMPLE

The number of increments in a single gross sample can be calculated from overall variance by transposing Eq. G-10:

$$N_i = \frac{S_o^2 - S_{da}^2}{0.263\, A^2 S_T^2 - S_{da}^2/P} \qquad \text{(G-20)}$$

The number of increments in a single gross sample can be calculated from random and system variances by transposing Eq. G-12:

$$N_i = \frac{S_s^2 + S_r^2/W}{0.263\ N_g A^2 - S_{da}^2/P} \qquad \text{(G-21)}$$

These equations show clearly the effect of the various components of variance on the number of increments needed to attain the desired or practical level of accuracy.

PLOTTING VARIANCES

It is helpful to plot the variances obtained by this method as an alternate means of determining the random and system variances. This method helps to visualize the data and the significance of the random and system variances, as well as to determine whether the small sample was small enough to introduce a significantly different variance to permit determination of the two variances. If variances are not different enough, the small sample is too large, and the procedure must be repeated with a significantly smaller weight increment.

Figure G-1 shows variance versus the weight of the sample increments on which are plotted the variances S_A^2 and S_B^2 versus the average weights of these samples. The coordinates are logarithmic in order to condense the graph. On the assumption that the large sample contains only system variance and the small sample contains only random variance, a diagonal line is drawn through point 0, which represents the random variance for unit weight, and a horizontal line through B. Adding the values of variance represented by these lines gives the total variance, S_t^2, and passes through points A and B.

If the weight of sample A is too large, the difference between the two variances may be too small to give an accurate determination. If so, the test must be repeated with a small enough sample weight to result in a significant error.

VARIANCES AND NUMBER OF SAMPLE INCREMENTS OF TYPICAL RDF

Table G-2 lists the ash content and variances of some typical RDF types along with the number of sample increments needed to achieve an accuracy of ±10% of ash content, with a confidence level of 95% of the samples. In spite of the wide range in ash content, the total number of increments required to achieve the same accuracy does not show a significant variation. One increment collected in 20 gross samples would produce an accuracy of ±10% in all cases.

Samples M were collected at different weights to demonstrate the significant effect of weight on variance. The size of the material has a considerable influence on overall variance, roughly proportional to the square root of particle size, as indicated by the 6-inch material as compared with 1-inch material.

Table G-3 shows the accuracy that would be achieved in a sampling program wherein two increments were taken in 20 gross samples. The wide range of RDF types

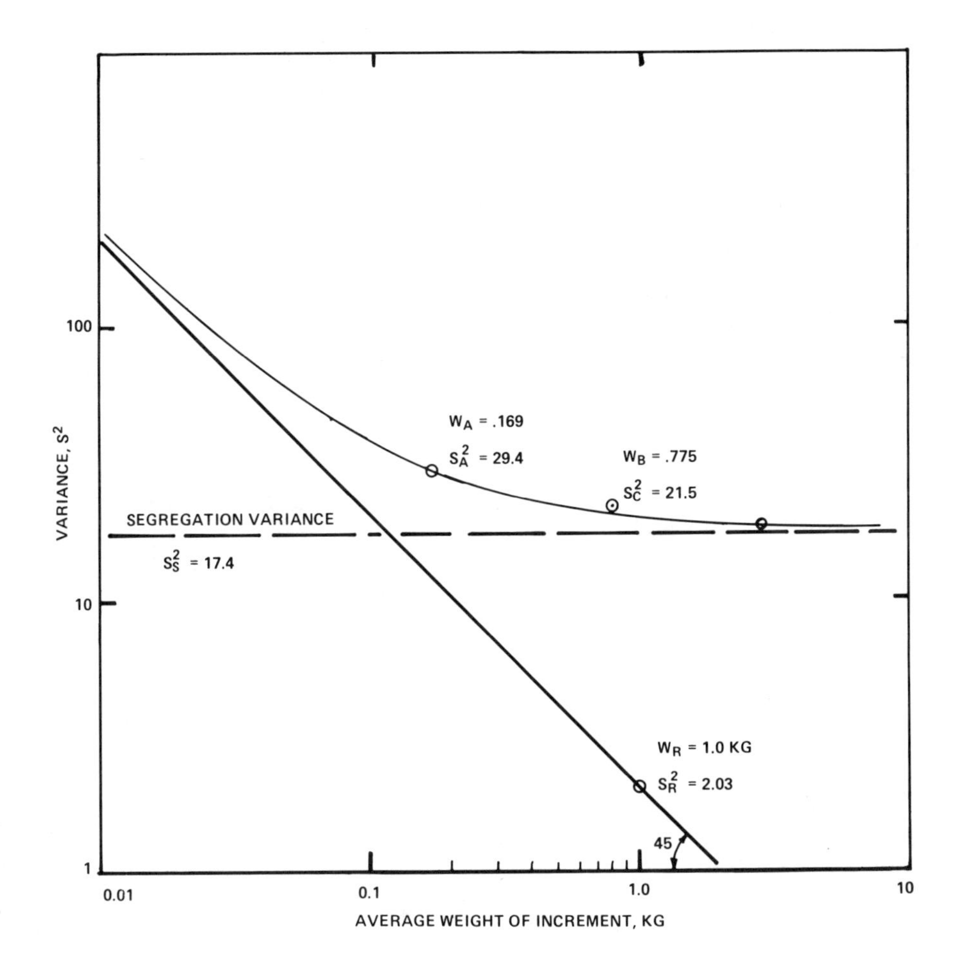

Figure G–1. Graphical representation of system, random, and total variances obtained at weights W_A and W_B of RDF–3 samples.

from powdered Eco-Fuel to 6-inch single-shredded material shows a range of 0.8% to 6.9% in heating value.

Table G–5 shows the results of several determinations of the variance of division and analysis for various RDF types. The last two values present the variances between a number of different laboratories, whereas the first three were determined by a single laboratory. The values range from 5% to 20% of the overall variance and, hence are significant. These values can be reduced by improved procedures for division and reduction of the sample or by making more than one analysis on each laboratory.

Table G–5. Variance of Division and Analysis for Various RDF Types

RDF Source	*Variance of Division and Analysis for Ash*
Ames, IA	1.06
Monroe County, NY	1.50
WEPCo., Milwaukee, WI	2.00
NCRR, Washington, DC (RR)	2.50
Americology, Milwaukee, WI	3.20

INDEX

Abrasive materials, 74
Acceleration, 171, 311
Accepts, 106
Accuracy, 324
  in estimate, 327
  sampling, 330
AENCO rotary classifier, 189
Aerodynamic properties, 249
Air classifier, zig-zag, 160
Air classifiers, 173
Air velocity, through shredder, 127
Americology, 211
Ames disc screens, 246
Ames, 248
Ames, Iowa, 27, 37
Amplification factor, 204
Analysis, proximate, ultimate, elemental, size, 30, 301
Angle
  of inclination of conveyors, 72, 76, 82
  of repose of materials, 73, 76, 249
  of separation, 255
  of slide of materials, 77
  of surcharge of belt conveyors, 77
As-disposed waste, 4
As-generated waste, 4
ASR, air/solids ratio, 213
As-received heating value, 14
Ash, 2, 7
  element concentrations, 303
  inherent, 291
  retention, 219
  removed, 219
  true, 291
Aspirator, 181
ASTM
  E-38 on Resource Recovery, 107, 159, 277, 301
  screen series, 239
Atlas storage and feeding bins, 62
Austin and Luckie, 112
Attrition, 87, 96
Autogenous grinding, 96
Available thrus, 264
Available lights, 217
Available heat in MSW, 12

Bag-breaker, 92, 270
Ball mill, 95, 139
Bed depth, 321
Boeing, 175
Bond work index, 112
Bomb calorimeter, 11
Breakage, 103
  bag and glass, 260
Breaker bars, in shredders, 104
Bridgeport, CT, 39
Bureau of Mines, 27, 178, 275
Brittle materials, 103
Bulk density, 44, 70
Burden, 152

Capacity
  of conveyors, 69, 77, 84
  of loaders, 47
  of screen, 226
  volumetric, 245, 252
Cardboard particle size, 105
Cascade
  of material, 250
  mill, 95, 108, 126, 139
Cataracting, 251
CEA, 175, 177, 182, 276
CEMA (Conveying Equipment Manufacturers' Association) Handbook, 47
Chemung County, NY, 39, 210

College Park, MD, 277
Cellular moisture, 8
Characterization of MSW, 5, 26
Characteristic particle size, 102, 106
Characteristics of MSW, 5
Chemically bound water, 8
Chemical composition, 8, 17
Chippers, 88
Choking velocity, 184
Classifier, horizontal, 179
Coefficient of drag, 311
Coefficient of variation, 21, 30
Column velocity, 180, 201, 207
Combustible fraction, 7, 8, 9
Combustibles, 109
  true, 291
Commercial waste, 2, 110
Communication, 86
Compactors, stationary, 48
Compaction, 44, 46
Components of MSW, 7
Composition
  generic, 289
  of MSW, 5
Computer programs, 288
Concentrator, ferrous, 153
Concentration of component, 22
Confidence, 3
  limits, 25
  interval, 33
Configuration table, 290
Conservation
  of momentum, 183
  of energy, 183, 309
Constant flow phase, 234
Constituents of MSW, 7
Containers, stationary, hauled, 43, 48
Contamination, of combustibles, 282
Contaminants, 2, 158, 206
Cost of collection, 47
Crane, 57, 60
Crew size, 50
Critical speed, 251, 254, 270
Critical column velocities, 218
Cumulative capacity distribution, 55
Curtains, at transfer points, 74
Cut, 200
Cutters, 88

Dade County, FL, 161
Daily capacity, 55
Degree of variability, 98
Density, free-fall, bulk, true, compressed, 17
Densification, 88
Depletion phase, 234
Difficult grains, 227, 241
Disc screens, 137, 233, 243
Discharge rate, of bins, 63
Displacement
  of rotary shear shredders, 136
  of particles in rotary drum, 315, 317
Distance traveled, 188, 191, 311
Distribution, of lifting velocities, 198
Distribution slope, n, 106, 308
Double-shredded MSW, 101
Double-rotor flail mill, 133
Drag force, 171, 184, 310
Drop tests, 208
Dry heating value, 14
DuLong formula for heating value, 15
Duke University, 181
Dust collection, 73

East Bridgewater, MA, 39, 117
  flail mill, 133, 135
  trommel, 267
Eco-Fuel, lifting velocity, 196
Eco-Fuel powdered RDF, 36, 39, 95, 108, 112, 121
Effectiveness, 145
  of collection, 46
Efficiency
  energy, 145
  mechanical, 145
  recovery, 226
  removal, 227
  of screen, 226
  separation, 148
  thermal, 144
Elemental composition, 8, 303
Empirical estimates, 242
Energy required for shredding, 109
Energy recovery, 217, 221
Energy consumption, 117
EPA, 107
Equilibrium moisture, 8

Error, 325
Essex County, 26
Estimating solid waste quantity, 51
ETEF (Experimental Test and Evaluation Facility), 107, 113
Extraneous ash, 218

Faraday's law, 149
Feed capacity, 249
Feed enclosures (for conveyors), 81
Feed rate, 211, 320
Ferromagnetic materials, 150
Field intensity, 149
Field tests, of air classifiers, 216
Fillage, 263
First-in, first-out, 58
Flail mill, 92, 94, 104, 135, 139
Flat-deck screens, 243
Flow analysis, 287
Flow characteristics, 45, 64
Flowsheet of pilot plant, 283
Franklin, OH, 161, 276
Free fall, 184, 311
Front-loader, 58, 60, 61
Froth flotation, 167
Fuel efficiency, 16

General Electric, 189
Generic composition, 289
Gondard mill, 126
Goutal's equation for heating value, 14
Granulators, 88
Grate openings, 105
Grate size, 126
Grateless shredders, 123
Grinding and attrition, 108
Gritty fines, 108
Gruendler hammermill, 132, 137

Hanford, U.K., 157
Half size, 241
Hammermills, 92, 104
Hammer wear, 98, 127, 129
Hard-facing of hammers, 127
Hardness, 127, 129
Hazards of fire and explosion, 73
Heating value, 2
Heavy media separator, 166
Heil shredder, 177
Hempstead, NY, 161
Height of burden of conveyors, 84
Higher heating value, 11
Histograms, 52, 54, 99, 107, 258, 305
Holdup, 122
Hole size, 264
Hollander, 30
Holmes, 112
Horizontal
  air stream, 187
  shaft hammermills, 92, 126, 139
Hydrasposal wet process, 161
Hydradenser, 162
Hydraulic rams for feeding, 61
Hydraulic drives, 75, 88

Idlers, idler angle, 75
Idling power, 133
Impact, 87
Impact on conveyors, 85
Impingement, 171
Increments, 331
Indecisive particles, 200
Industrial waste, 2, 110
Inertial, 171
Inherent ash, 282
Inspecting and picking, 60
Interaction, of particles, 209
Interference, 172
Irregular particles, 193, 195

Jamming, 87, 267
Jig, 166

Kinetic energy, 171
Koppers storage and feeding bins, 64

Last-in, first-out bins, 62
Lifting velocities, 192
Lint, dirt and glass, 109
Linty materials, 96
Live-bottom conveyors, 61
Loaders, front, side, and rear, 47, 49
Log-normal distribution, 21, 308
Lower heating value, 11

Magnet, 149
drum, 155
electropermanent, 151
permanent, 150
Maintenance, 98
Mass balance
air classifier, 283
primary trommel, 280
of process, 294
Mass-burning incinerators, 60
Material burden on conveyors, 72
Material velocity, 214
Matrix calculation, 280
Matrix analysis, 285
McNabb, 208
Mean, 30
Mechanical conveyors, 68
Mechanical friction, 113
Mechanical efficiency, 130
Miller-Hofft storage and feeding bins, 64
Moisture content, 2, 8, 112, 124, 211
Moisture and ash-free heating value, 14
Momentum, 218
Monroe County, NY, 35, 38
Mixtures of materials, 106, 107
Motor friction and windage, 130
Motor amperage control, 137
Moving auger bins, 63
Multistage classifier, 202

NCRR (National Center for Resource Recovery), 76, 114
Newton's laws, 183
No-load power, 133
Nodes, 288
Number of tries, 256

Occidental Research, 182
Open area ratio, 256
Operation and maintenance, 86
Optimizing, 293
Optimum efficiency, 269
Optimum speed, 253
Overs and unders, 225
Overs capacity, 249
Oversize bulky waste (OBW), 3
Oversize and undersize, 225, 241

Packer truck, 47
Paper, 105
Particle
acceleration, 186, 312
size distribution (PSD), 98, 239, 266
size distributions of MSW and RDF, 110
Path of particles, 251, 318
Pelletizing, 93
Performance, 143
of air classifiers, 215
wet process, 168
curves, 131
Picking station, 60, 153
Powdered RDF (Eco-Fuel), 108
Power consumption, 98, 113
Power recovery, 296
Precision, 22
Preventive maintenance, 74
Primary shredder, 138
Primary shredded MSW, 106
Primary (MSW) conveyors, 75
Probabilities, 172
Probability
distributions, 138
phase, 234
of passing, 255, 258
Properties of MSW, 2, 7
Protocol, for sampling survey, 25
PSD, particle size distribution, 266
Purity, 146

Quality, 146

Rader
air classifier, 175
disc screen, 244
Raytheon air classifier, 175, 178
Recovery One, 153, 156, 159
Rags, 104
and wires, 63
Range of error, 34
Rate of arrival, 57
Rated capacity, 139
Raw household refuse, 99
Receiving conveyor, 60
Recovery of organics, 214
Recovery, power, 296

Regression analysis, 125, 211
Rejects, 106
Relative velocity, 171
Relative standard error, 31, 33
Removal rate, 200
Research in size reduction, 97
Residence time, 254
Retention time, 255
Retrieving from storage, 62
Reynold's number, 310
Rollback, 72
Rosin-Rammler (RR), 102, 103, 239, 307
Rotary
  grinding mills, 95
  screens, 247
  shear shredders, 91, 136, 139
Rotational speed, 260
Rotor speed, 128
Round robin, 107
Rupture
  breakage, 111
  energy, 111, 120

S-curve, 107
Salt Lake City, UT, 206
Saturn shredder, 136
Savage, 116, 127
Sample size, 21, 326
Sampling, 1, 19, 144, 323
Sand and grit, 108
Scalping screen, 230
Screens, air-assisted, 174
Screen, double-deck, 160
Screen inclination, 268
Screening capacity, 262
Screening time, 237
Screw conveyors, 66
Seasonal variation, 35
Secondary shredder, 138
Secondary shredded MSW, 106
Separator, eddy current, 160
Separation efficiency, 202
Separating velocities, 180
Separator
  heavy media, 166
  rising current, 163
Separation curve, 200
Sharpness, 205
Shear shredder, Saturn, 158
Shear, 87
Shredded textiles, 104
Shredder performance, 122
Sieves, 113
Single-shredded MSW, 100
Size distribution, 19
Skirt boards (for conveyors), 75
Slope, of trommel, 249
Sorter, optical, 169
Sorting, 27
Species of refuse, 289
Specific energy, 115, 124, 131, 220
Specific capacity, 115, 118, 131, 228
Specifications, feed, 289
Speed of rotation, 249
Speed, critical, 251
Speed, optimum for trommel, 253
Speed-controlled conveyors, 62
Spillage from conveyors, 78, 80
Split table, 290
Splits, 283
SPM screening tests, 271
St. Louis, MO, 27, 37
Stratton and Alter, 112
Stages, multiple, 170
Standard deviation, 21, 30
Static pressure, 212
Storage and feeding bins, 64-66
Storage, 45, 57
Student's t, 324
Surface moisture, 8
Surge, 132

Tensile force to rupture, 109
Terminal velocity, 172, 186, 193
Testing, 1
Time of fall, 197
Tipping floor, 57, 59
Trace elements, 1
Trajectories, 72, 184, 192, 318
Transfer points of conveyors, 71
Transitions between conveyors, 72
Transverse conveyor, 63
Trezek, 97, 120
Triple/S Dynamics, 177
Trommel screens, 231
Trommel, primary, 230

True density, 44
Trunnion, 248
Tumbling action, 252
Turbulence, 252

Ultimate stress in rupturing, 110
Underflow, 155
Unders capacity, 249
Undersize material, 235, 242
Union Electric, St. Louis, 29
University of California, 181
University of Eindhoven, 204
Unit operations, 142
Utilization factor, of containers, 48

Variability, 2, 30
Variance, random, system, division, and
  analysis, 325
  overall, 328
Velocity, choking, 184
Velocities, column, 180
  separating, 180
Vertical air stream, 188
Vertical motion, 171
Vertical-shaft hammermills, 92, 126, 139
Vesilind, 98
Vibrolutriator, 175
Vibrating conveyors, 81–84

Warren Springs, 279
Wayne County Municipal Incinerator, 23
Weinberger, 98, 133
Wemco-Remer jig, 166
Worrel, 209
Weigh scale, 62
Wet process, 161
Work done in rupturing, 110
Work index, 113, 115, 118, 119

Yard waste, 24
Yield, 225, 281, 292, 295

Zig-zag classifier, 203